I0032773

Francis Campin

Constructional Iron and Steel Work

Francis Campin

Constructional Iron and Steel Work

ISBN/EAN: 9783743435643

Printed in Europe, USA, Canada, Australia, Japan

Cover: Foto ©berggeist007 / pixelio.de

More available books at **www.hansebooks.com**

CONSTRUCTIONAL
IRON AND STEEL WORK

AS APPLIED TO

PUBLIC, PRIVATE, AND DOMESTIC BUILDINGS

A PRACTICAL TREATISE FOR ARCHITECTS,
STUDENTS, AND BUILDERS.

BY

FRANCIS CAMPIN, C.E.

PAST PRESIDENT OF THE CIVIL AND MECHANICAL ENGINEERS' SOCIETY
AUTHOR OF "MATERIALS AND CONSTRUCTION," "MECHANICAL ENGINEERING"
ETC. ETC.

With Illustrations

Capio Lumen

LONDON
CROSBY LOCKWOOD AND SON
7, STATIONERS' HALL COURT, LUDGATE HILL
1896

[All rights reserved]

PREFACE.

THE object of the present work is mainly to place before the Architectural Student, in a concise and practical form, correct methods of designing the Constructional Iron and Steel Work now so generally used in all classes of buildings; and it is believed that the volume will be found acceptable, not only by Students, but by many Architects in actual practice, and by persons engaged in Building. For any one who is concerned in this branch of Engineering, it is impossible to overlook the fact that it has not hitherto received the attention from Architects which it deserves; its importance is paramount, for if the constructional part of a fabric fails all must become a ruin; and it is therefore incumbent upon those preparing to practise as Architects, to make themselves acquainted with the theoretical laws of stress and strength, and with the practical conditions upon which depends the safety of the buildings they are commissioned to design.

I have endeavoured throughout to explain everything in

the easiest manner, and where mathematical investigation is indispensable to present it in the simplest form consistent with reaching the issues in view.

The examples taken for illustration are such as occur in actual practice, and they are worked out in the same way as they would be in the Architect's or Constructional Engineer's office.

In order to follow properly those passages which deal with English and foreign rolled girders and joists, it will be necessary for the reader to furnish himself with tables of weights and strengths such as are issued by the different manufacturers and their agents, for without these, or sheets of sections, the sizes required for any particular purpose cannot be selected. These tables are to be found in most Architects' offices, as the manufacturers are naturally anxious to give all information that may lead to business, so that in this respect the articled pupil is not likely to find any difficulty.

FRANCIS CAMPIN.

LONDON, 1896.

CONTENTS.

—◆—

LIST OF ILLUSTRATIONS.

CONSTRUCTIONAL IRON AND STEEL WORK.

INTRODUCTION.

MANY of the difficulties by which architects and builders were formerly beset in the designing and execution of large structures intended to carry heavy floor loads, have been swept away by the adoption of iron and steel as materials for the constructional parts of such fabrics. The thick walls, which formerly occupied much of the floor room of warehouses and railway depôts, have given place to a skeleton of columns and girders which support everything within the external walls, and allow a freedom of movement and extent of ventilation previously unattainable.

The use of such materials as iron and steel demands increased accuracy in the preparation of plans, and a more rigid adherence to them in the execution of the work than when brickwork and timber alone are relied upon ; in the latter case, if a slight alteration occurs in the brickwork the timber joists are easily sawn shorter, or may be put in stock and longer ones supplied without serious loss, but the cutting of an iron or steel girder is a different matter and adds a

B

very noticeable percentage to the cost of the work, especially if it has to be done on the site of the building.

If a large building is to be erected into the structural parts of which iron, or steel, or both, enter as the leading material, the most economical course to pursue is to order the girders from the rolling mills, where they can be cut off to the required lengths without extra cost; but this cannot be done at short notice, and therefore the plans of the work should be absolutely settled in every detail before active operations are commenced; the tentative way of building—designing as you go on—is totally inappropriate in any structure in which iron or steel is to be used, nor is there any need for it if competent men are employed upon the plans, for it is to be presumed that works of importance are not projected without their requirements and uses being known to the projectors, and this being the case, the rest is a mechanical adaptation of the structural elements to the positions in which they must be placed to support the superstructure. Frequently complicated combinations of girders must occur, but there will be no difficulty in designing them, if the general arrangement has been properly considered, and the dimensions are absolutely fixed.

It is not only in the preparation of the floor plans that this careful prevision is to be exercised, but also in the vertical section of the building. As will be shown in a subsequent chapter, the depth of an iron or steel girder should bear a certain proportion to its span, below which it is injudicious to let it fall, and from an economical point of view the greater the load the deeper should the girder be for a given span. Then, again, if other circumstances are the same the deflection of the girder will vary inversely as its depth; and this matter of deflection causes considerable anxiety in the case of long spans in any building which has finished and ornamental ceilings, as any material vibration is apt to crack them. The thicknesses of the floors must therefore evidently be determined with due regard to the spans of the girders required to run through them, though of course it is not necessary that main girders should be kept within that thickness, as they may project below and, if the nature of the building demands it, be encased in ornamental work, and so divide the ceiling into a series of bays.

In a general way buildings of the kind here dealt with may be divided into two classes : those in which the floors, and perhaps some of the walls, are carried on girders and iron or steel joists, supported upon brick or stone walls or piers ; and those in which the superincumbent load comes upon stanchions or columns running down to the ground and there resting on their own foundations. In the latter case the work should be so arranged that the loads come entirely upon the stanchions in every case, and not, where a stanchion happens to be enclosed in a wall or pier, partly on the stanchion and partly on the wall ; when occurring in such positions the stanchions may be enclosed in the brickwork, but should be free, not built into it, then any settlement of the walls will be independent, and they also will not be affected by stresses upon the main elements of the structural iron work. This I know is contrary to general practice, but it is so obviously the correct course to pursue that further argument is unnecessary, especially as in many such matters convenience of execution often overrides constructional accuracy.

There is one point I would here strongly impress upon the architectural student, and that is, that in matters purely constructional he should not be biassed by appearances ; he cannot tell the strength of a column, or girder, or a connection, by looking at it, and a little paint will cover a serious defect. The full force of this observation will be seen when the subject of joints and connections is reached. In those buildings of which the main structural elements only are of iron or steel the difficulty about dimensions will not be so apparent, as the general plan is not likely to be departed from after the foundations are put in ; it is in the internal details that trouble may arise.

If wood joists are used, these can be cut and trimmed about to suit altered spans and altered arrangements of staircases and lift openings, without serious increase of expense ; but where concrete floors are used, carried upon rolled iron or steel joists, the case is very different, for the floor becomes a net-work of metal joists, each one of which should be sent to the site of such length that it will pass into its place with sufficient bearing on the main girder, but not too long to be placed without cutting.

Men who have been employed all their lives upon brickwork are not in the habit of thinking much of an inch either way, but an inch out, in some floors carried on metal joists, when these run between girders—not over them—is a very serious matter. Without going into details here, I wish to impress on the mind of the student the gravity of matters apparently insignificant, but in practice calling for very close observation.

Assume that the plan of a floor intended to be made of metal joists filled in with concrete, and carried by main girders into which the joists run, being there supported on the bottom flanges, or on angle irons riveted to the webs, to have been delivered to the general contractor for a large building, and a copy of that plan sent to the constructional engineers ; the latter will order the joists to the dimensions shown on such plan, and if the bearing is, say, three inches from the web of the main girder, will allow one inch play in the length of the joists, that is half an inch at each end, sufficient to allow their being swung into place without jamming against the webs of the girders ; this would leave the joists two and a half inch bearing at each end. Now if one main girder is one inch out of position, the joists will be jammed on one side, and left with short bearing on the other side of such girder, and if the displacement is more than one inch, some of the joists must be cut.

The arrangement of constructional iron and steel work cannot be thoroughly dealt with, unless the other materials in conjunction with which it is used receive due consideration, and the nature of such materials will determine how far iron and steel shall enter into the construction of any particular building ; for instance, if a large building is to be erected in a district where granite is available at a moderate cost, iron stanchions which would otherwise be necessary to take the load off the external walls would not be required, though they might be indispensable if brick were the material to be used.

The distribution of the floor loads upon the walls must be very carefully watched, for it is a widely different thing to spread the weight of a heavily loaded floor along a wall by using joists at a moderate distance apart, say two feet, to concentrating the load from a considerable floor area upon main girders, placed at distances of ten feet along the wall. In the latter case the wall may be strengthened by piers upon which the ends of the main girders rest ; but then another danger is incurred, which is that there being a considerable load upon the piers, and but little on the walls, the subsidence of the former may lead to cracks and dislocations. To obviate this a proper foundation is to be built under each pier, one which shall so spread the load that the settlement of the piers will not exceed that of the walls.

How far it may be safe, or even economical, to concentrate loads will depend upon the nature of the subsoil upon which the structure rests, and this must also be considered in preparing the preliminary designs. The architect who would evolve a perfect structure must, from the inception of the design, regard the exigencies of every trade which will be employed in the execution of the work.

CHAPTER I.

MATERIALS.

THE materials with which we shall have principally to deal are, cast iron and wrought iron and steel. Cast iron is especially suitable for columns and stanchions on account of its rigidity and high resistance to compressive stress, but the wrought metal is more suitable for tensile and bending stresses, as not only is its resistance greater under such stresses, but from its malleable character it is not liable to snap or break without warning.

CAST IRON.—The qualities of cast iron are very numerous and are generally designated from the district in which the ore is obtained, but they are all comprised under two classes, "hot-blast" and "cold-blast" iron, according to whether a heated or a cold blast of air is used in working the furnaces. The "pig" iron is usually classified by numbers, that containing the most free carbon being No. 1, and other numbers up to seven are used to indicate less proportions of carbon or graphite. No. 1 pig shows a coarse, granular fracture, with large dark grey scales, and when melted it is very fluid. Castings made from this are very soft, and it is chiefly used for mixing with other qualities of metal. A mixture of numbers 1, 3 and 4, makes a very good casting, it is close-grained and hard. The higher numbers are not used in the foundry, but are made for manufacture into wrought iron. Scotch iron is considered the best for foundry purposes; it is strong and uniform in quality, and will mix well with other brands. Cleveland iron is harder than Scotch iron, and not so strong. It is much used in the Cleveland district, a mixture of Nos. 1 and 3 being suitable for large work. Lincolnshire iron is very similar to Cleveland. Cumberland "hematite" iron is generally used for steel-making, but it is also employed to improve other irons for foundry use ; it is very tough and strong, but does not run well when melted by itself ; a little mixed with good Scotch irons makes a strong mixture. The best-known cold-blast iron is the Blenavon, it shows great strength and fineness of grain, and is used to close the grain and increase the strength of other irons.

All cast iron is improved by remelting up to ten to twelve times, but after this, remelting will diminish its strength. For thick plain castings, wrought-steel scrap—shearings and punchings of boiler plates—may be added to the cast iron up to about 10 per cent.

The variations in strength of different samples of cast iron have a very wide range ; its tensile resistance varies from 4·85 to 14 tons per sectional square inch, and its crushing resistance from 22·5 to 58 tons.

Averages of strength are of course utterly useless for practical purposes where each casting must be strong enough for the duty assigned to it ; therefore the quality of iron to be used must be decided upon before the work is designed, and in large work sample bars should be cast at the same times and from the same charges as the columns or other elements of the structure, and subsequently tested to ascertain if the quality of the material is up to the standard fixed upon.

For our purposes the cast iron should not break under tension at less than 7 tons per sectional square inch, and a short block should not crush with less than 38 tons per square inch. In transverse strength a bar 2 inches deep by 1 inch wide, supported at the ends with a clear span of 3 feet, should not break under a less central load than 24 cwt. Cast iron is incapable of permanent extension, and therefore when its limit of elasticity is reached, it breaks suddenly in tension, or under transverse stress, though under compression it may give some warning by cracking.

Wrought iron and steel will stretch or become otherwise crippled before actually breaking asunder, and this constitutes their great advantage under transverse and tensile stresses, because when they are used the fear of a sudden catastrophe does not exist.

It is not only the great resistance of cast iron which makes it so suitable for columns, but also the fact that it can be cast into any required shape, to receive girder ends which it is required to support, and lugs may be cast in any position for the attachment of the bars and struts. Connections can thus be easily made, which with wrought iron or steel would be impracticable. The resistance of cast iron to shearing stress—such, for instance, as that which would tend to shear or split a bracket off the side of a column or stanchion, is about 12 tons per sectional square inch for metal of the quality indicated above.

In designing castings the thicknesses should be kept as nearly uniform as possible, and any variations from one thickness to another should be gradual, as sudden changes are apt to lead to cracks occurring through inequality of contraction in cooling.

WROUGHT IRON in the form of plates; bars, flat, angle, tee and channel ; and rolled joists and girders, is largely used in constructional engineering. For comparatively short spans the rolled joists are more economical than girders built up with plates and angle irons, as the labour of punching or drilling and riveting is all saved, but solid girders cannot at present be rolled deep enough to be used with very long spans.   20 inches is about the maximum depth for a rolled girder, and this should not be used for a span much over 30 feet.   Iron girders and joists are not rolled in England or Scotland, and when used are obtained from Belgium or Germany.

Wrought iron is made in this country by the process of puddling, from cast iron.   This process consists briefly in working molten cast iron about in contact with air until the carbon and other impurities are burnt out of it, when it gradually assumes a pasty form which can be made into balls and "blooms" to be worked under the hammer. Hammering squeezes out the slag and consolidates the iron. The tendency at present seems, however, to be towards substituting hydraulic pressure for hammering, whereby a great saving in the cost of production is expected to be effected. The product in either case is very nearly pure iron.   The strengths of the irons of different districts vary very widely, and it is therefore necessary to specify what resistance is required.   The puddled bars have too low a tenacity for use as they are and have to be cut up and piled together, reheated, and again rolled into bars, which constitute what is known as "merchant bar" iron.   This, however, is still of low tenacity and shows inequality of strength, and is therefore not used for important structural work, but only for such ironmongery as gratings, ladders, and small bearing bars.   Best bar is made by re-heating "faggots" of merchant bar and again rolling it into bars, by which its strength is raised to a standard sufficient for constructional purposes. Its tensile resistance per sectional square inch varies with different makers from 21 to 26 tons.   By again rolling bars from faggots of selected best bars, a yet stronger iron is obtained and of a quality so malleable and ductile that it will bend double cold without breaking, and elongate about 25 per cent. before fracture, with a reduction of area at the point of fracture of about 50 per cent.; its tensile strength is 27 to 28 tons per sectional square inch. Very little of this quality of iron finds its way into constructional iron work for buildings.

Rivet iron must be soft and of superior quality to close up without loss of strength; the best quality is made by rolling from "piles" of ordinary rivet iron, which is generally made from scrap. It has a tensile strength of about 24 tons and a shearing strength of 20 tons or a little more, but it should not be taken as higher than this in ordinary work.

Yorkshire plate iron runs to a tensile strength of about 22 tons with the grain and 20 tons across the grain, the elongations before fracture being 16·7 and 11·2 per cent. Annealing slightly reduces the strength but increases the elongation. The elastic limits, after which permanent set occurs, are 12 tons in tension and from 11 to 13 tons in compression.

Staffordshire plates are harder than Yorkshire, and have an ultimate strength of about 22·5 tons per sectional square inch in the direction of the grain and about 19 tons across it; the elongations 5 per cent. and 4·35 per cent.

Cleveland iron has only about 20 to 21 tons strength with and 17 tons across the grain, and this iron is very largely used in constructional work, where the strength across the grain is not a matter of much importance, but this iron has only a slight elongation before fracture. Scotch iron is between Staffordshire and Cleveland in its qualities.

Iron of higher qualities than these can be obtained by specifying it, and is often obtained without so doing. I have tested many samples of Staffordshire girder iron which stood 25 tons when 22 only was specified, and Yorkshire plates under 21 ton specification which did not break under 26 tons to the square inch.

Iron suitable for girder work should, when broken, show a grey fibrous fracture ; it should have no crystalline appearance, and no black scales imbedded in its substance.

It is highly important that wrought-iron plates should be of uniform strength throughout the section, otherwise one part will yield more than another, and the plate will have a tendency to curve or buckle sideways under longitudinal stress.   In order to test the uniformity of the material a strip should be cut out on one side of the centre line of the plate, planed to exact width to give a known sectional area, and each edge carefully chalked.   The bar is then placed in the testing machine and tension applied to it, then if one side is yielding more than the other, the chalk will crack away more rapidly on that side than upon the other, and thus indicate the inequality.

All iron that has to be forged or bent should be specified to be not less than 22 tons per square inch tensile resistance with the grain and 19 tons across ; it should also have an elongation not less than 15 per cent. with the grain and 8 per cent. across it.

The resistance of the foreign joist iron is 19 to 20 tons per sectional square inch.

The crushing resistance of average quality wrought iron is about 18 tons per sectional square inch, but in the forms in which this material is used to resist compression, it would fail by crippling, and therefore transverse stress also comes into action.   This will be dealt with subsequently.

STEEL is used in the same forms as wrought iron in constructional engineering.   In some special cases where elements are required which cannot be built up conveniently in wrought metal, and cast iron is objected to, the castings may be made in steel; but it is very seldom used in this way in buildings, it is therefore with wrought steel that we are especially concerned.

Formerly steel was made from the best kinds of wrought iron by a process called "cementation." In applying this method small bars of wrought iron are bedded in powdered charcoal and the whole kept at a high temperature for a considerable time, during which the iron absorbs a portion of the carbon, necessarily in an irregular way. The product, which is far from being of uniform texture, is known as "blister" steel. This is broken up and the pieces sorted according to the fracture shown, after which they are packed in closed crucibles, melted and cast into ingots; these are known as "cast" steel. The ingots are rolled into bars which are almost exclusively used for the manufacture of cutting tools, and is therefore known as tool steel. It is very hard and has a tensile resistance from 50 to 65 tons per sectional square inch, with an elongation of about 5 per cent. only. The milder kinds, used for drifts and other percussion tools, have a tensile strength of 44 to 60 tons per square inch, with an average elongation of 13 per cent.

The steel used for building purposes is almost exclusively made by direct processes, thus avoiding the cost of the wrought iron. The two kinds now in use are the "Bessemer" and the "Siemens-Martin" steels.

BESSEMER STEEL.—In the Bessemer process there are two distinct operations. Molten cast iron is first converted into pure iron, and this is then turned into steel by the addition of a definite amount of carbon. Cast iron free from phosphorus, and comparatively free from sulphur, is melted and poured into a vessel termed a "converter." A strong blast of air is forced through the molten metal to consume the carbon, and this at the same time raises the temperature of the mass. When the carbon is all eliminated, a small quantity of Spiegeleisen—an iron containing a known proportion of carbon and manganese—is added, and this changes the charge from iron to steel; it is then run into ingot moulds, and submitted afterwards to hammering and rolling into the forms required.

SIEMENS-MARTIN STEEL is made in the hearth of a reverberatory furnace heated by gas to an intense violet heat, by the mixing of certain qualities of wrought and cast iron, or by mixing cast iron with certain kinds of ore, the latter method being that most generally employed for the production of steel on a large scale and at a moderate cost.

The strengths of both the above steels may be varied by altering the proportions of the materials combined; for structural work it is usual to specify an ultimate strength in tension of from 28 to 32 tons per sectional square inch : an upper limit is assigned as well as a lower, because it is found that the lower ultimate strength is accompanied by a higher elastic limit, and plates of this material which have an elongation of 20 per cent. before fracture possess an ultimate tensile strength of 30 tons only per sectional square inch.

Rolled steel girders and joists are made in England from both descriptions of steel and can be turned out of a quality up to the higher limit if specified, but this strength is not to be relied upon in buying this material out of stock, or on an order given without requiring a special guarantee.

In some experiments recently made upon some German rolled steel joists taken at random out of a parcel ordered without guarantee of strength the ultimate tensile resistance varied from 24 to 26 tons per sectional square inch, and the ratio of elastic limit to ultimate strength was found to be about 60 per cent. One sample showed a tensile strength of 36 tons per sectional square inch, but this must be regarded as exceptional.

The crippling resistance of mild steel may be taken as equal to that of good wrought iron, and will be dealt with when treating of columns and struts of the latter material.

Each section of rolled girder, both English and foreign, can be supplied at slightly varying weights, but the most economical to use to resist transverse stress is in each case the minimum weight, as the heavier sections are obtained by spacing the mill rolls farther apart, whereby the additional weight is principally in the web, where it is of much less use than that in the flanges, as will appear when we come to consider the stresses upon girders and their moments of resistance.

CHAPTER II.

COLUMNS AND STANCHIONS.

THE ultimate resistance of materials to crushing stress does not alone give the measure of the strength of a column, stanchion, or strut; as soon as the length becomes a multiple of the diameter, the crushing force may become complicated by bending stress. This bending stress arises from the imperfections of section; we can imagine that if a solid column, say, were perfectly homogeneous throughout, the superposed load would only bring a crushing stress upon it, but this cannot be obtained in practice, one part of the column will be sure to yield more than another, and thus a deviation from a straight line will be brought about.

We cannot infer from theory where, or to what extent, this deviation will occur, and it is therefore only possible to obtain an empirical formula based upon experiments. The experiments carried out by Hodgkinson have supplied the data, which have been used as the basis for several formulæ, more or less intricate. The most convenient form to which any of these has been reduced is, I consider, that of Professor Gordon, which is as follows: for cast iron—

Let $BW =$ the breaking weight of the column per square inch of horizontal section in tons; $r =$ the ratio of the length of the column to its diameter;

Then for a cast-iron column, well bedded and with the

load bearing uniformly upon its cap—for either solid or hollow columns—

$$BW = \frac{36}{1 + \frac{r^2}{400}}$$

For example, let there be a cast-iron column required 18 feet high, to carry a load of 80 tons, and let its diameter be 11 inches outside, then reducing both measurements to the same name (inches in this case)—

$$r = \frac{18 \times 12}{11} = 19.63 \, ; \, r^2 = \{19.63\}^2 = 385.34$$

and the breaking weight required is,

$$BW = \frac{36}{1 + \frac{r^2}{400}} = \frac{36}{1 + \frac{385.34}{400}} = 18.34 \text{ tons.}$$

If the column is required to carry a steady load with no violent vibrations, such as would occur if it supported a floor on which machinery works, a factor of safety 6 may be used for obtaining the working resistance, which will then be $18.34 \div 6 = 3.05$ tons per sectional square inch; we may drop the decimal and call it 3 tons—then the sectional area of column necessary will be $80 \div 3 = 26.6$ square inches. The area of a circle 11 inches diameter, corresponding to the outside of the column, is 95.03 square inches; as 26.6 square inches of sectional area of metal are wanted, a circle corresponding to the internal diameter of the column will have an area of $95.03 - 26.6 = 68.43$ square inches. The nearest convenient diameter is $9\frac{1}{4}$ inches, with an area of 67.2 square inches, which will give a slight excess of strength. The thickness of metal in the column will then be $\frac{1}{2}(11 - 9\frac{1}{4}) = \frac{7}{8}$ inch.

C

It is a very common practice to allow some excess of thickness in columns to allow for defective workmanship, such, for instance, as inequality of thickness through rising of the core which forms the interior; this being much lighter than the molten iron will float up if not made sufficiently stiff to hold its position, or packed down in the mould by wrought-iron studs. It is not easy to determine such variations of thickness without either drilling holes in the column, or going to the expense of testing its deflection when put under cross-stress. The latter course would determine the question, as if the metal is of equal thickness all round, the column will show the same deflection under a given load, no matter on what side it is laid, but if there are irregularities the deflection will vary as the column is turned; for if the thin side is at the bottom the deflection will be greater than when that part is on the top.

The difficulty about equality of thickness is avoided by casting the columns in an upright position, but it is only in some districts that foundries are to be found in which the work can be thus done. In every way it is better to cast the work erect, as then the gases which form in the mould have a better chance of escaping to the refuse "head" which is cast on for this purpose. If the column cannot be cast vertically, the mould should be inclined as much as possible; for in a column cast in an absolutely horizontal position, the metal on the upper side is almost certain to have air holes in it.

Every care should be taken to thoroughly dry the column cores in the stove before they are placed in the moulds, in order to prevent the formation of gas from the moisture contained in them.

Columns have been used with round ends to allow of alterations of length in the superstructure carried by them

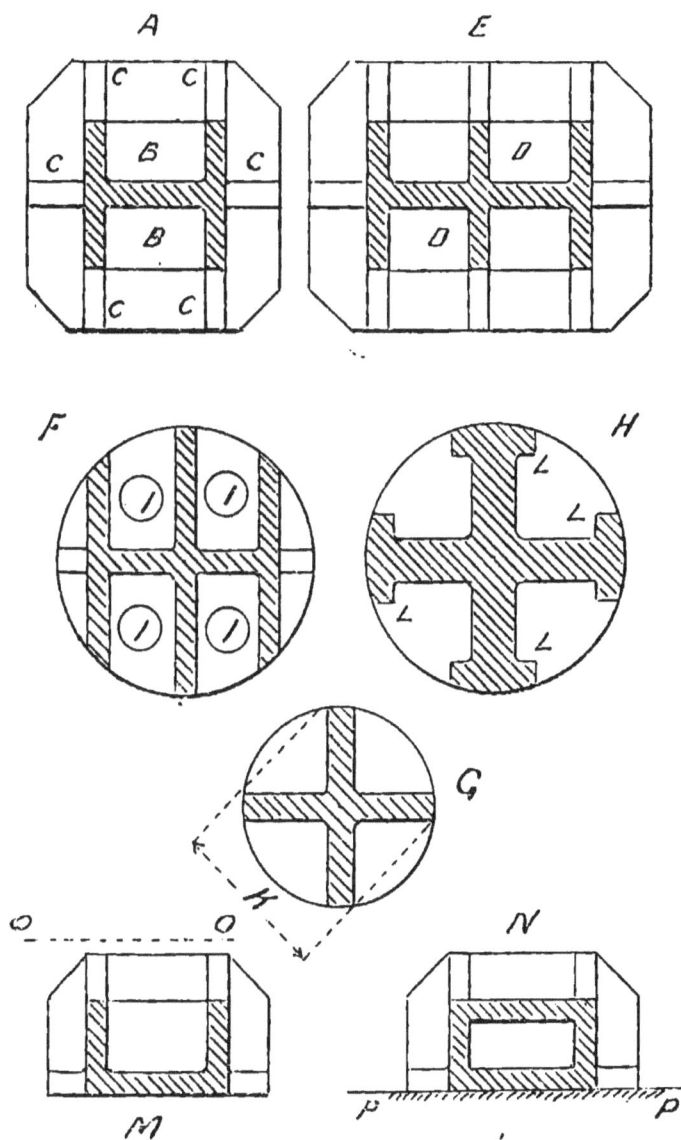

Fig. 1.

when changes of temperature occur, but when this is done, the divisor 400 in the above formula must be replaced by 100, as the tendency to bend is much greater with the round end than when the base of the column is flat.

In many positions the circular form of the column is inconvenient, and a stanchion is therefore used instead. Horizontal sections of the shafts of several forms of stanchions are shown by Fig. 1, all being of cast iron. *A* is a horizontal section of the commonest form of stanchion in use, the etched part shows the continuous section of the shaft; *B B* are cross webs, or stiffeners, to brace the flanges, and these are usually placed about three feet apart throughout the height of the stanchion. The effect of these stiffeners is to tie the flanges in such a way that the stanchion is practically as strong as a column of the same sectional area, and the formula given for columns will therefore apply to stanchions thus made. The least width of the stanchion is treated as the diameter in working the formula. Let a stanchion be required of the same height as the column in the previous example, to carry the same load, and let it be 11 inches by 9 inches, the two flanges being each 9 inches wide. Then the value of r will be $(18 \times 12) \div 9 = 24$, of which the square is 576; the breaking stress of the stanchion, per sectional square inch, will be

$$BW = \frac{36}{1 + \dfrac{576}{400}} = 14 \cdot 75 \text{ tons.}$$

Using the same factor of safety as before, the working resistance of the stanchion will be $14 \cdot 75 \div 6 = 2 \cdot 46$ (nearly). The sectional area of stanchion required will therefore be $80 \div 2 \cdot 46 = 32 \cdot 52$ square inches. The sectional area of the stanchion is equal to the sum of the width of the flanges, plus (the width parallel to the web less twice the thickness of the metal) all multiplied by the thickness of metal; or,

Let $b =$ breadth of flange, $w =$ width of section, and $t =$ thickness of metal, all in inches; and $a =$ sectional area in square inches; then

$$a = \{2 . b + (w - 2t)\} . t$$
$$= \{2 \times 9 + (11 - 2t)\} t$$
$$= 29 t - 2 t^2.$$

This is a quadratic equation thus solved—

$$t^2 - \frac{29 t}{2} = -\frac{a}{2} = -\frac{32 \cdot 52}{2} = -16 \cdot 26$$

$$t^2 - \frac{29 t}{2} + 52 \cdot 56 = 52 \cdot 56 - 16 \cdot 26 = 36 \cdot 3$$

therefore

$$t - 7 \cdot 25 = \sqrt{36 \cdot 3} = \pm 6 \cdot 025$$

and

$$t = 7 \cdot 25 \pm 6 \cdot 025 = 13 \cdot 275, \text{ or } 1 \cdot 225 \text{ inches thick.}$$

The latter is evidently the result required.

This, the academical mode of obtaining the result, is too lengthy for ordinary office use, we should therefore proceed thus: we require 32·52 square inches; adding twice the flange breadth to the width, we get $2 \times 9 + 11 = 29$ inches, and dividing the area by this, $32 \cdot 52 \div 29 = 1 \cdot 12$ inches, for the thickness. This would reduce the effective width by 2·24 inches; and so the section length would become $29 - 2 \cdot 24 = 26 \cdot 76$ inches, and the thickness $32 \cdot 52 \div 26 \cdot 76 = 1 \cdot 21$ inches thick. This very nearly approaches the thickness obtained by the more tedious method, and as in practice the metal would be made $1\frac{1}{4}$ inches thick, the latter is covered by it.

The size of the base of the stanchion will be determined

by the resisting properties of the foundation beneath, but if we assume that it is to be bedded directly on a York stone template of sufficient thickness to spread the load over the pier, or footings, below, the area of stanchion base may be determined on the basis of 7 tons per superficial foot. The area required will then be $80 \div 7 = 11\cdot43$ square feet nearly.

A base of 3 feet 6 inches square would give an area of $12\cdot25$ square feet, and if we cut off the corners in the way shown in the figure, 6 inches from the corners of the square in each direction, half a square foot must be deducted, leaving $12\cdot25 - 0\cdot5 = 11\cdot75$ square feet of effective bearing.

The load upon the shaft of the stanchion must be spread over the base by means of brackets c, for it will not be safe to rely upon the cross-breaking resistance of the base to distribute the load.

Where heavier loads have to be carried, the section shown at E is convenient, the parts $D D$ forming a third flange; but the narrowest dimension must be used in determining the ratio of height to width. In determining the form of stanchion to be used, there is not only the load upon it to be considered, but also the character of the building of which it is intended to form a part. In a storage warehouse a naked stanchion may not be objectionable, but in public buildings, such as town-halls, schools, and assembly rooms, these, where used, must be clothed with some encasing material, which shall in itself be ornamental, or otherwise serve for the attachment of adventitious decoration. In the first case, the surrounding material will be marble, granite, or some imitation in artificial stone; and a circular or square column will then be all that is required, its encasement being prepared in segments to fit around it.

If, however, stanchions are to be used as the skeletons of columns, they must be made of such sections as will hold securely the concrete upon which the finish is to be carried. At *F* is shown the section of a stanchion similar in design to *E*, but with its flanges narrowed, and its *web* extended in both directions to fit a circumscribed circle, and the horizontal stiffeners will fill this circle. In order to give a "key" for the cement, or fine concrete, with which the stanchion is filled up, it should be surrounded with wire netting tightly laced together, and the stiffeners may be made with holes in them, as shown at *I, I, I, I*, to prevent the cement from being cut into short lengths by their interposition; the cement should reach to about an inch outside the circle of the stiffeners. Care should be taken that the cement or fine concrete used is solid throughout; but this is a matter with which every builder is acquainted, and the mode of execution need not be described here.

At *G* is shown the horizontal section of a stanchion, the worst, from a constructional point of view, that could be imagined. Its adoption is no doubt due to the fact that it may conveniently be buried in brickwork, a course the evils of which I have already pointed out. The minimum diameter of this section is shown by *K*, and as soon as bending stress sets in, the four "wings" of the stanchion are exposed to twisting stresses which do not come upon any other type of stanchion. At *H* a similar section is shown, but the ends of the webs are widened out to form little flanges *L*, to act as "keys" to the encasing concrete. This, besides possessing all the disadvantages of section *G*, would be very expensive to make, as the flanges on two webs would have to be cored in the mould; the pattern, if made of the section shown, could not be withdrawn from the sand.

The channel section, shown at *M*, is much used to carry the ends of main girders when they enter a party-wall, and it is usually embedded in the wall, of which the front is shown by the line *O O*. A better mode of construction is to use a rectangular pilaster *N*, placed against the surface *P P* of the wall.

If the constructional iron work in buildings is regarded by the architects as a subordinate matter, a thing to be left to others to arrange, generally to those who have the supplying of it, that part of the structure will be of a jobbing character; it will be fitted in, more or less, to suit the requirements of the general design and the alterations that occur during its execution.

The arrangement of the columns, or stanchions, in the basement of a high building requires very close attention, especially if the floors of the structure will be heavily loaded. One common-sense rule may be set down, from which no deviation should ever be permitted; it is this: do not design a girder to be carried partly on a wall and partly on a stanchion imbedded in the wall; when a girder end has a stanchion to support it, let the whole load be taken upon the stanchion, and keep the brickwork, or masonry, as the case may be, a little *below* the top surface of the cap of the column. This will prevent the possibility of the wall being cracked while the girder is finding its bearing upon the cap of the stanchion.

For stanchions of the section shown at *G*, a factor of safety of ten at least should be used, and for the section *M* a factor eight.

We must now turn our attention to the caps of cast-iron columns and stanchions, in regard to the provision necessary for receiving and supporting the ends of main girders. In this respect they may at once be divided into two classes : those that carry only girders on their caps, and those which also support columns carrying upper floors.

In large warehouses, factories, mills, and breweries, the columns or stanchions will be continued tier upon tier from the basement to the roof, and at each floor will be required to pick up the main girders by which it is carried.

In these cases the columns, or stanchions, must be directly connected together ; on no account should the connection be made through a girder, that is, by bolting the girder to a column below, and then bolting the base of another column to the top flange of such girder. There cannot be the lateral rigidity of connection which is obtained by the direct bolting of column to column, neither is there the same certainty of absolute vertical alignment of the columns as when the joint is made by their machine-faced ends.

The carrying of a column upon a girder at any point between its supports should never be permitted, and I do not think that any architect who gives a thought to his iron work when designing his building, would allow the necessity for such a course to arise. A column is an element in constructional work whose value rests in its vertical stability, its compression is inappreciable and is not affected by passing loads in such a way as to cause vibration of the floor supported by it. A girder under transverse stress must deflect to some degree, and under a varying or moving load must vibrate ; the vibrations may not be perceptible unless there is rapid change of load, as in dancing, but they occur, and if a column or stanchion is carried by a girder which also carries a floor load, the deflections of the girder will lower and lift the column, and tip it out of the perpendicular as a load passes from one side to the other of the bearing girder. The racking stress upon the connecting bolts must be very severe and cause them to become slack, whereby the stability of the column will be still more imperilled.

Whether columns or stanchions are used will not affect the fact that the "stiltings" must be made of the form suitable to receive the girder ends, and also have a horizontal sectional area equal to that of the stanchion immediately above it, for the stilting is relieved of the load brought upon the stanchion by the main girders which merely run into it. A simple and convenient form of stilting is shown in Fig. 2. *A* is an elevation, and *B* a horizontal section on the line *c c*; *e e* shows the cap of the lower stanchion or column, upon this is cast a square stilting *d d*, which is hollow, and therefore particularly suitable for a column, as the core used in casting can be withdrawn through it.

Rolled girders *m m* are shown with their ends resting upon the cap *e e*, and secured to the square stilting by angle irons *i*, riveted to their webs and held to the stilting by the bolts *n n*. The angle irons *i* are termed "cleats" by the dealers in merchant iron, and therefore it will be convenient to use that name in reference to their application to rolled girders.

In this case it is obvious that no vertical stress comes upon either the rivets or the bolts, as the load is taken directly upon the cap of the stanchion. On the upper part of the stilting *d* is cast a flanged cap *f f*, to receive the base *g g*, of the stanchion *h*, which forms one of the next tier. With a stilting of this form four girder ends may be conveniently carried; the other two are shown by the dotted lines at *k k*. When four main girders thus come against one stilting, the bolts *n n* must of course be arranged to clear each other.

In almost all constructional iron-work in buildings— unless they are structures of considerable magnitude—we

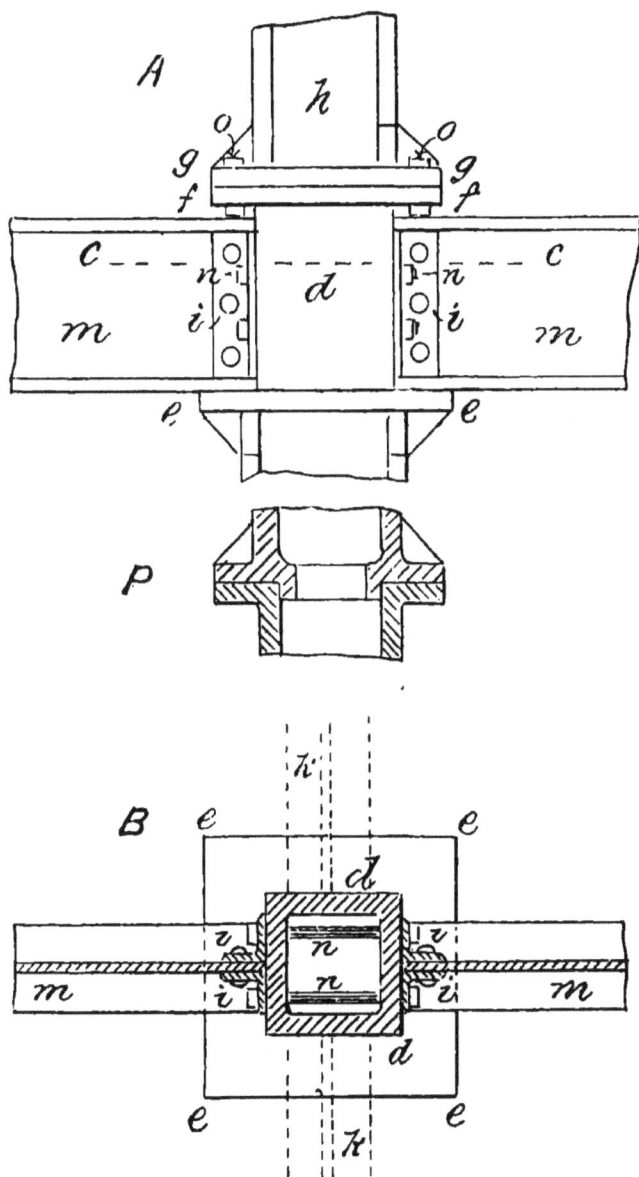

Fig. 2.

arc cramped for room in the matter of bolts, and therefore this should not be neglected in the preparation of the general design.

All faces of contact of the stanchions or columns must be faced, so as to meet truly, and, for the best class of work, spigot and faucet joints, as shown at P, are used to prevent the possibility of a stanchion or column slipping upon the one beneath it; if this is omitted, there are only the bolts o o, to retain the upper stanchion or column in alignment with the lower.

If the spigot and faucet joint is not used, the stanchions or columns should be connected together by turned bolts passing through holes drilled for their reception; if the holes are cast, with allowance to receive bolts without rhymering out, the bolts will be a loose fit, and the stability of the joint a matter of chance.

In some large buildings—breweries—the columns have merely been fitted together with a spigot and faucet joint, no connecting bolts being used, but I do not consider this safe, even when the main floor girders are carried on the bases of the columns in the upper tier, instead of the caps of the lower tier of columns.

If the main girders are designed to run over two or three or more bays, the square stilting is inadmissible and the cap must be made with lateral stiltings as shown in Fig. 3. A and B are side elevations of the upper part of the stanchions; C is a horizontal section through the stilting on the line c c, and D a horizontal section through the shaft on the line d d.

It will be seen that the girder e runs clear across the cap of the stanchion, which is made with vertical jaws f f, with flanges g g at the top to receive the base of another stanchion l above, which being bolted down at the four

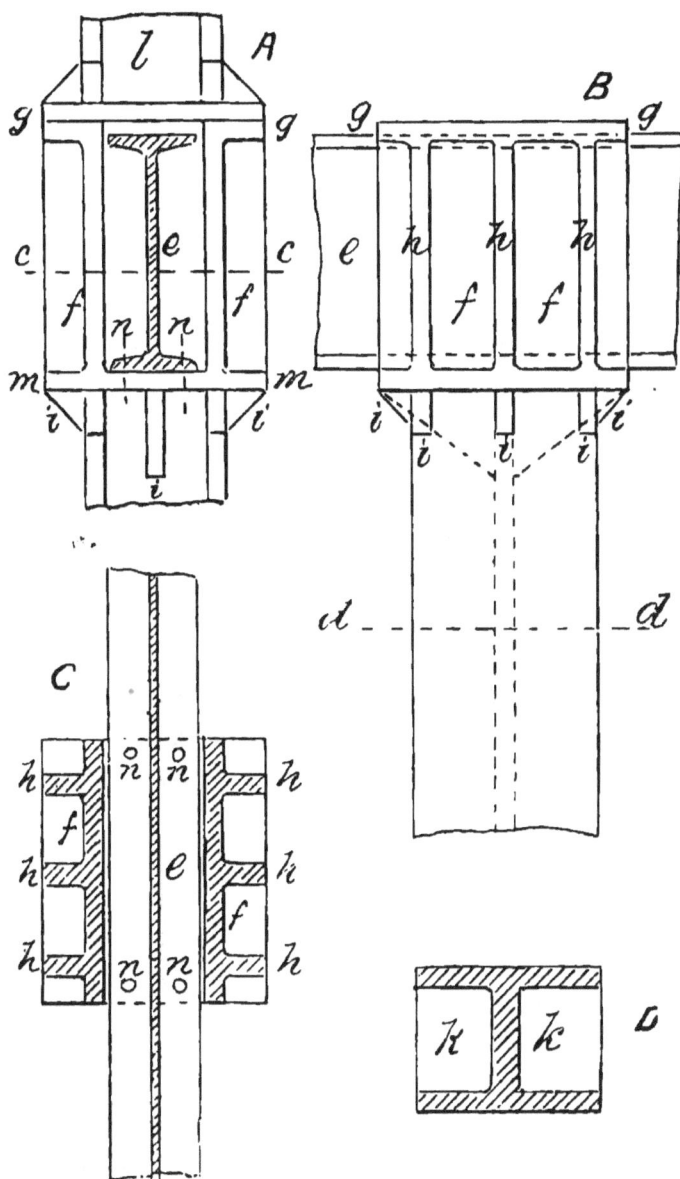

Fig. 3.

corners forms a tie to the two jaws and saves them from the danger of being broken by a side blow. The jaws are also further stiffened by vertical webs *h, h, h.* The bearing plate or cap proper *m m* on which the girder *e* rests, is strengthened by brackets *i, i, i,* &c. Horizontal stiffeners *k k* also occur about three feet apart in the height of the stanchions to strengthen the edges of the side flanges of the shaft.

If the girder is light it may be bolted to the stanchion by bolts *n n,* though this is not necessary when the girder is continuous. It is optional also to bed the girder upon a layer of felt to equalise the bearing, but this is certainly advisable if the plate *m m* is rough at all, and the girder flange not quite flat. The girder should not when bolted be screwed down very tightly, as then it is likely to strip the thread of the bolt when its full load comes upon it and causes deflection.

When the work assumes very large proportions it will not be convenient to cast the jaws on the columns or stanchions, and they must then be made in separate pieces, as shown in Fig. 4, and bolted on to the column or stanchion cap.

In this figure *A* and *B* are side elevations, and *C* is a horizontal section through one of the stiltings *d.* The stiltings are stiffened by ribs *g, g, g,* and are planed to true surfaces top and bottom, care being taken that these surfaces are parallel to each other, the cap *e* of the column and the base *f* of the stanchion or column above are both planed true, or faced up in a lathe, generally the latter, which affords great facilities for insuring the parallelism of the caps and bases of each column, as they can both be faced simultaneously. Holes for the connecting bolts *i, i, i, i,* are to be drilled in the bottom flanges of the stiltings, and in

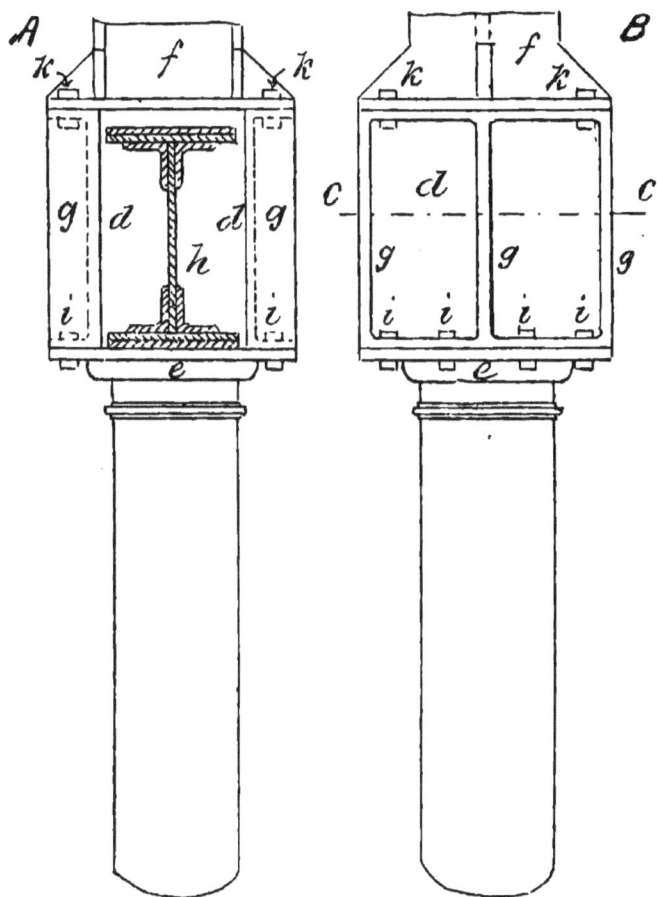

Fig. 4.

the cap e, their centres being first marked from the same template, so that the holes shall coincide exactly, and these holes are to be fitted with bolts accurately turned, so that the work shall be solidly joined up without the possibility of a shake.

The upper stanchion f should also be bolted down with turned bolts k fitting in drilled holes. A built-up girder h is shown in this case resting upon the column cap. The thickness of metal in the stiltings must be such as to give a horizontal sectional area equal to that of the shaft of the stanchion or column carried by it. The centre of the bolt-holes should not be nearer any edge of the casting than two diameters of the bolts ; thus the centre of a ¾-inch diameter bolt should not be nearer the edge of the casting than 1½ inches.

Fig. 5 shows another form of stanchion designed to support the ends of three girders which are discontinuous. A is a side elevation, B a vertical section of the line b b, and C a horizontal section in the line c c. Two simple girders i i in elevation, and e e in horizontal section, have their ends resting in pockets into which they are lowered ; the central web d of the stanchion is continued up between the ends of three girders, so that the vertical tie to the flanges of the stanchion is unbroken. In this case the horizontal area of the section C must be equal to the sectional area of the shaft of the upper stanchion h.

The third girder f is supported upon a bracket g, and a fourth, if required, might be carried by a similar bracket on the other side of the stanchion. All the girders may be fastened down by bolts h.

The stress brought upon the bracket g may be of a bending or of a shearing character, according to the way in which the girder f takes its bearing ; if it happens to rest

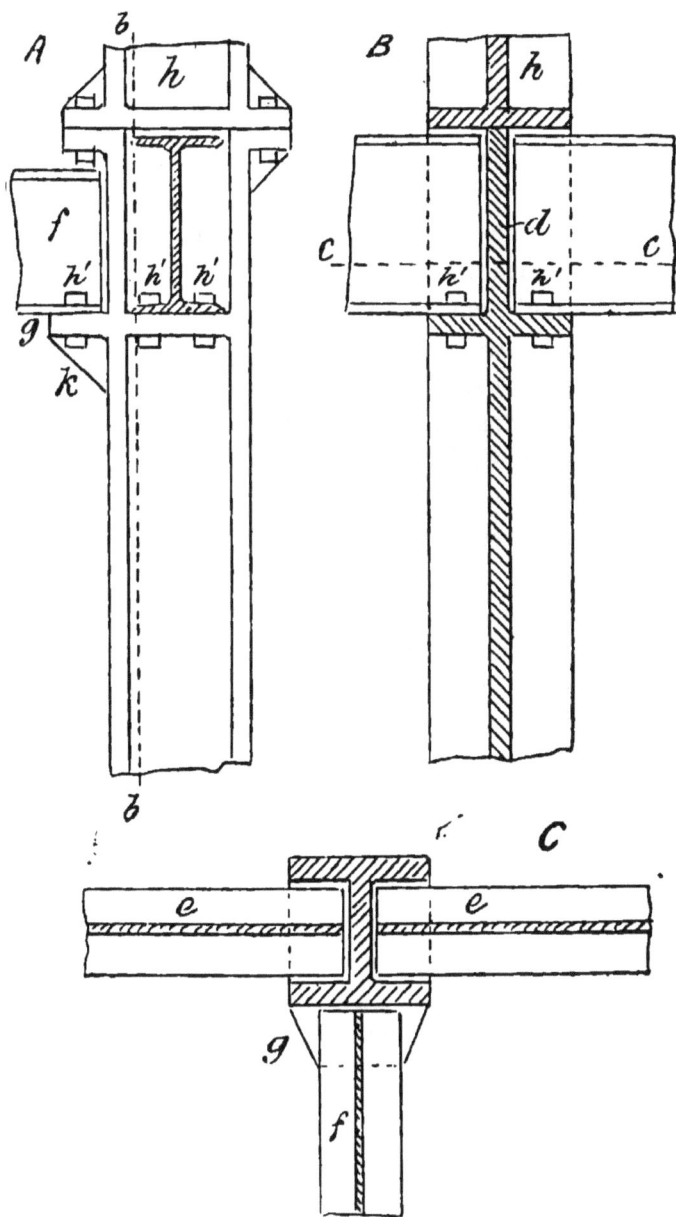

Fig. 5.

upon the edge of the bracket, as it will when a load comes upon it, the stress will be bending, assuming that the bracket is cast flat, and for this reason it is necessary to use stiffening brackets k. The stress upon these will be of a somewhat mixed character, compressive on the outer edge, but either tensile or shearing where it joins the flange of the stanchion. Shearing stress in cast iron reaches 12 tons per sectional square inch before rupture, but working stress in a bracket should never be taken at more than 1½ tons per sectional square inch, we shall then have an equal resistance when the stress changes from shearing to tension.

Cast-iron columns or stanchions made with brackets should be most carefully examined before erecting them, as cracks are apt to occur in the re-entrant angles of this part of the work. All such castings should be allowed ample time to cool in the foundry, and should a bracket crack off it should on no account be permitted to be "burnt on." This process consists in remoulding the bracket and placing the column, heated to a very high temperature at the part to which it is desired to attach the bracket, so as to close the mould. The molten metal poured in to form the bracket may, or may not, further raise the temperature of the column surface to melt on to and cool down with it, but in any case it is a very risky thing and may very likely cause a crack in the shaft of the column itself. If the job will not afford a new column, it is best to make a new bracket and bolt it on.

In Fig. 6 is shown a column upon which the ends of four girders are supported. It has cast upon it a hollow square stilting d to carry a column above, and on this stilting are cast four "lugs," c, c, c, c, to which the webs of the girders are bolted, the load being taken upon the cap of the column, which, however, does not afford facilities for bolting the

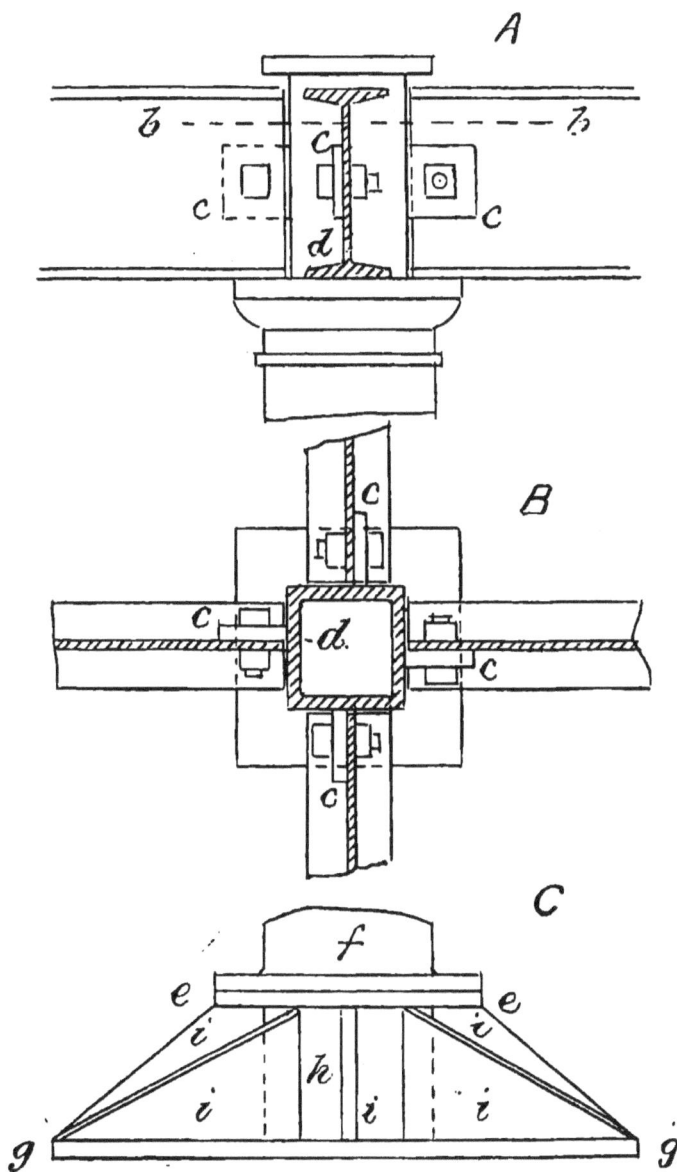

Fig. 6.

girders down to it.   The bolt holes in the lugs and girder webs must be made with ample clearance in this case, so that the bolting up of the girders may not interfere with their taking a complete bearing upon the column cap.

When, on account of the low bearing capacity of the foundation, it is necessary to spread the base of a column or stanchion to a width disproportionate to that of its shaft, it is advisable to cast a separate base of the form shown at C. $e\,e$ is a flanged top to receive the base f of the column or stanchion above it; $g\,g$ the wide base, h a central column, and $i\,i$, &c., brackets to distribute the load on $e\,e$ equally over the base $g\,g$.

CHAPTER III.

GIRDERS—WITH SOLID WEBS.

WE have now to turn our attention to the characters of the different stresses that are brought, directly or indirectly, upon the main supporting girders used in buildings. Gir ders are subject to loads which cause them to deflect, and thereby set up tensile and compressional stresses, and also shearing stresses at certain points depending upon the construction of the girder. It is obvious that when a load is placed upon a girder some deflection must occur—even if it is not noticeable—because if the form of the girder were not altered it would not be pressing upward, and if it were not pressing upward it would not resist a load pressing downward and that load would descend. As a matter of fact the load descends until the resistance of the girder to further deflection is equal to the superincumbent load. Opposing forces can only be at rest when their intensities are equal and their directions opposite. An ounce weight placed upon a spring letter weigher will cause its pan to descend until the spring presses upwards with the force of an ounce, when it will come to rest if the weight has been gently lowered on it ; if the weight is dropped upon the pan it will descend below the position corresponding to one ounce pressure and will vibrate up and down for a while, but when it comes to rest the spring will be pressing up with a force of one ounce. It is the same with everything in nature ; wherever rest—

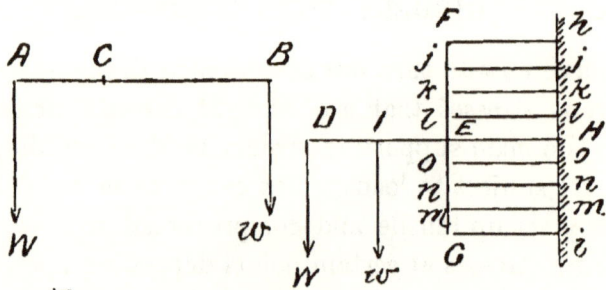

Fig. 7.

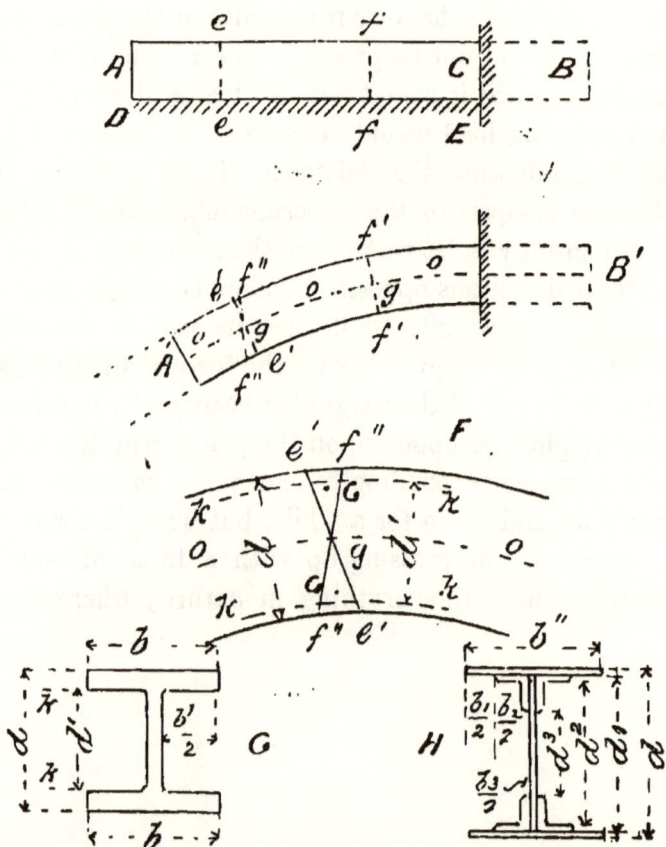

Fig. 8.

or equilibrium—is established there must be at least two forces acting, and these must be opposite and equal. If a number of forces are acting towards one point, then if a straight line be drawn through the point, and the forces on each side of the point concentrated up this line, the effect in the direction of the line on one side of the point from all the forces on that side must be equal to the effect along the line, but in the opposite direction, of the forces on the other side of the point to produce equilibrium. Similarly, if $A B$, Fig. 7, is a lever fulcrummed at C and the lever is at rest with the weights W and w suspended from the ends, it is evident that the effort of W to turn the arm $A C$ about the bearing C, must be just equal to the effort of the weight w, to pull the end B down around the bearing C. If $A B$ is horizontal the turning effort of W about C is $W \times A C$, and that of w is $w \times B C$. If W is equal to 19 lbs., and $A C =$ 6 inches; then $W \times A C =$ 19 lbs. $\times$ 6 inches $=$ 114 inch-lbs. If $C B =$ 14 inches, then 114 inch-lbs. $= w$ (lbs.) $\times$ 14 inches; and $w =$ 114 inch-lbs. $\div$ 14 inches $=$ 8·14 lbs. The bearing of this argument upon the resistance of girders may not appear evident to the student at this stage, but, if he will follow me through it, he will soon see the advantage of proceeding in so elementary a way.

Now let us suppose we have a lever shaped in another way, let it be supported on an axle or fulcrum at E, and have a horizontal arm $D E$, at the end D of which is suspended a weight W. The lever has also two vertical arms, $E F$ projecting upward from the fulcrum, E and $E G$ projecting downward from it. To the right of the vertical arms of the lever $E F$, $E G$, is shown the face of a solid wall H. If the ends F and G are connected to the wall H by bars $F h$ and $G i$, it is evident that the weight W acting about the point E will put a pull upon the bar $F h$ and a thrust upon the bar $G i$, and if the whole is at rest, the tension upon $F h$ multiplied by $F E$, plus the compression upon $G i$, multiplied by $G E$, must be equal to the weight W multiplied by the distance $D E$. Stresses and weights are of course taken in terms of the same name, as, for instance, both in pounds or both in tons, and the lengths of the arms of the levers all in inches or all in feet.

Now there is no reason why the force exercised by the weight W—that is $W \times D E$, should not be resisted by more than two bars or strips of material connecting the vertical arms with the wall H. We may have on the upper limb strips $j j$, $k k$, and $l l$; these will all be in tension, and on the lower limb strips $m m$, $n n$, and $o o$, which will all be in compression. By the way, the amount of the force multiplied by the distance from the point (E) about which it acts, is called the "moment" of the force.

In the case illustrated the moment of the weight W about E will be equal to the sum of all the "moments of resistance" of the strips connecting the vertical arms with the wall. The moments of resistance of these strips will not all be of the same value for two reasons: first, the distances of them from the point E, and therefore their leverages, vary, and as E is approached they are less stretched or compressed—a strip from E to H would be neither stretched nor compressed—and therefore, if they are all of one material, exert less resistance, so that an equal-sized strip midway between F and E would only give a moment of resistance one quarter of that due to the strip $F h$, because it is only stretched half as much, and has only half the leverage ; the resistance of any elastic material—while its limit of elasticity is not exceeded—being in direct ratio to its extension or compression. The effort of the weight W about the point E is known as the "moment of stress." There might be another load w at I, then there would be two "moments of stress," and in all cases the total "moments of resistance" must be equal to the total "moments of stress," otherwise equilibrium cannot occur.

As there is no limit to the number of strips, or layers, Fh, jj, &c., except their being so numerous as to be in contact, we may consider that they are so, and in fact form a solid beam, then we have to find out the moment of resistance of such a beam.

When we have determined a formula for the moment of resistance of any beam, it is easy to equate it with any formula of stress, and so ascertain the necessary dimensions of the beam to resist such stress.

The section of a beam, or girder, being given, its moment of resistance in, say inch-tons or foot-tons, may easily be calculated, and does not vary according to the moment of stress, it is constant for a constant section. I shall therefore consider the moments of resistance of various sections before dealing with the moments of stress accruing from different dispositions and intensities of load.

In Fig. 8 $A B$ represents a cantilever beam fixed in a wall which is sufficiently stable to resist the overturning effort of any load that will be put upon the cantilever. For this beam to be entirely free of stress it must be supported by some surface such as $D E$; if this surface is withdrawn, the beam will have to support its own weight, and will therefore deflect and become curved, as shown by $A'B'$. While the beam is supported, in a horizontal position, let two vertical lines, $e\,e$ and $f\,f$, be drawn upon it at any convenient distance apart. When the support is removed and the beam deflects, these lines will no longer be either vertical or parallel to each other, their directions will be shown by the lines $e'\,e'$ and $f'\,f'$. The upper surface of the cantilever will be extended, while its lower surface is compressed, and there will be a central surface *o, o, o,* where there is no stress; this is called the neutral surface, and its position in the cross section of the beam is the neutral axis of the section.

The line *e' e'* intersects the neutral surface *o, o, o,* in the point *g*; through this point draw the straight line *f" f"*, parallel to *f' f'*, then *e' f"* represents the extension of the upper layer of fibre in the length *g g*, and *f" e'* on the lower surface shows the compression in the same length, and the triangles *e' f" g*, above and below the neutral surface, represent respectively the extensions and compressions of the layers and fibres which constitute the substance of the cantilever. These triangles are shown to a larger scale at *F*, Fig. 8, the same letters being used.

The maximum resistance of the material is represented by the length *e' f"*, and all layers between that outer line and *g* must offer less resistance, because their extension is less; when the ultimate resistance at the outer surface is overcome the inferior layers of fibres will be torn asunder one after another, and the beam destroyed. From this it is evident that the area of the triangle *e' f" g* will represent the sum of the resistances of the layers of fibres on one side of the neutral surface for a unit of width; these resistances may be considered as concentrated at the centre of gravity *G* of the triangle, and therefore the distance of this point from the neutral surface will be the distance at which the direct resistances act about the neutral axis of the section. Let all the dimensions be taken in inches, and let s = the maximum resistance per sectional square inch of the material; d = depth of beam; b = breadth of same.

The distance from the upper or lower surfaces of the beam to the neutral surface $= \frac{d}{2}$, and this represents the height of the triangle $e'\, f''\, g$, of which the centre of gravity is one-third of the height from the base, and therefore two-thirds of the height from the point g, so the distance at which the resistance acts about the neutral axis is—

$$= \frac{d}{2} \times \frac{2}{3} = \frac{d}{3}$$

The area of the triangle of resistance is, $\dfrac{s}{2} \times \dfrac{d}{2} = \dfrac{s\,d}{4}$;

and, as there are two triangles of resistance, the total moment of resistance of the section will be—

$$= 2 \left\{ \frac{d}{3} \times \frac{s\,d}{4} \right\} \times b = \frac{s \times b \times d^2}{6}$$

For example, let it be required to calculate the moment of resistance to transverse stress of an iron beam, of which the direct resistance is 20 tons per sectional square inch, its size being three inches deep and two inches wide ; then the breaking weight will be—

$$= \frac{s \times b \times d^2}{6} = \frac{20 \times 2 \times (3)^2}{6} = 60 \text{ inch-tons.}$$

For practical purposes we do not calculate the breaking moment, but the safe working moment of stress, which will of course be kept well within the elastic limit of the material; thus, for iron of the quality assumed above we should use a factor of safety 4, and the working moment of resistance of the section would be 15-inch tons. This calculation is more true for the working than for the ultimate stress of the material, for in the former the limit of elasticity is not exceeded, and the modulus of elasticity remains the same for tension as for compression.

This formula refers to a solid rectangular beam, but others suitable to different sections may be derived from it.

In Fig. 8, G shows a transverse section of a rolled girder; it is obvious that the strength of this girder will be less than that of a solid beam of the same depth and width, by the strength of the parts $\left\{ d' \times \dfrac{b'}{2} \right\}$ omitted on each side of the vertical web. The direct resistance of the inner layers, k, k, shown dotted in the view F, will evidently be less than that of the outside layers of the whole section in the ratio of d_1 to d, because their extension will be less in this ratio; so the formula for the strength of the omitted part will be, calling M the " moment of resistance—

$$M = \frac{b_1 d_1^2}{b} \times s \times \frac{d_1}{d} = \frac{s \cdot b_1 \cdot d_1^3}{6 \cdot d}$$

Deducting this from the strength of the whole section, the remaining moment of resistance will be—

$$M = \frac{s \cdot b \cdot d^2}{6} - \frac{s \cdot b_1 \cdot d_1^2}{6 \cdot d} = \frac{s}{6\,d} \cdot \left\{ b\,d^3 - b_1 d_1^3 \right\}$$

Let it be required to determine the safe working moment of strength of a rolled girder 12 inches deep by 6 inches wide, of which the flanges have a thickness of one inch each, and the web a thickness of ¾-inch; then the different values are $b = 6$; $d = 12$; $b_1 = 5\cdot25$; and $d_1 = 10$. Assume that the girder is rolled from steel having an ultimate tensile strength of 28 tons per sectional square inch, then taking 4 as the factor of safety the value of s for working stress will be 7 tons per sectional square inch, and the moment of resistance of the section in inch-tons will be,

$$M = \frac{s}{b\,d} \cdot \left\{ b \cdot d^3 - b_1 \cdot d_1^3 \right\} = \frac{7}{6 \times 12} \left\{ 6 \times \overline{12}^3 - 5\cdot25 \times \overline{10}^3 \right\}$$

$$= 497\cdot58 \text{ inch-tons.}$$

For a girder built up with plates and angle irons, to form a section such as that shown at H, the formula is merely an extension of the above, as from its form the parts to be deducted necessitate the use of more dimensions; for this particular section it would be—

$$M = \frac{s}{b\,d} \cdot \left\{ b\,d^3 - b_1\,d_1{}^3 - b_2\,d_2{}^3 - b^3\,d_3{}^3 \right\}$$

In this case, however, there is loss in the tension flange by the rivet holes, and this loss should be deducted from the width; if, for instance, there are two rows of $\frac{3}{4}$-inch rivets, and the plate is 12 inches wide over all, the effective width should be taken as $10\frac{1}{2}$ inches.

We will now look into the "moments of stress" caused by loads upon girders, and the course to be pursued in adapting the strengths of the latter to the loads which they are intended to carry.

In Fig. 9 A B shows a side elevation of a girder carrying a concentrated load W, and for the present the weight of the girder itself is neglected. We want to find the moment of stress about any point c, which is distant x from the point of support B. The load is distant y from the point of support A; l equals the clear span of the girder. The dimensions must all be taken in terms of the same name, such as all in inches, or all in feet, but comparing the moments with those of resistance care is to be taken that the two moments are stated in similar accordance. If the moments of resistance have been determined in inch-tons, and those of stress in foot-tons the latter must be multiplied by twelve before being equated with the former.

The load W is supported by the upward resistance of the piers at A and B, which must obviously press upwards with the same force that the load presses down, otherwise it would sink. As a matter of fact the piers do, on the first application of the load, suffer compression until the elastic resistance is equal to the superincumbent weight; but if the load is not in the centre of the span it will not be equally divided between the supports. The upward reactions of the piers we will call R_1 and R_2.

Regarding $A B$ as a lever and A the fulcrum; the moment of the weight about the point A will be $W \cdot y$; the moment of resistance of the pier B will be $B \cdot l$ about A, and these two must be equal, or, $Wy = R_2 l$, whence $R_2 = \dfrac{W \cdot y}{l}$. Similarly the moment of W about the point B is $= W \{l - y\} = R_1 l$, whence $R_1 = \dfrac{W \{l - y\}}{l} = W - \dfrac{Wy}{l}$. Now it is evident that the two reactions together must equal the load, or—

$$R_1 + R_2 = W - \frac{W \cdot y}{l} + \frac{W \cdot y}{l} = W.$$

The moment of stress about the point c will be equal to the reaction R_2 multiplied by its horizontal distance x from the support B, and—

$$M = R_2 x = \frac{W y}{l} \times x.$$

If the moment about the point C is calculated from the reaction at A, that reaction acts upward; but there is another moment about the same point on that side, the moment of W acting at the distance z and in a downward direction, therefore the difference of the two moments will be the resultant moment at c, and,

$$M = R_1 \times \{l - x\} - W \times z.$$

But $z = l - x - y$; and—

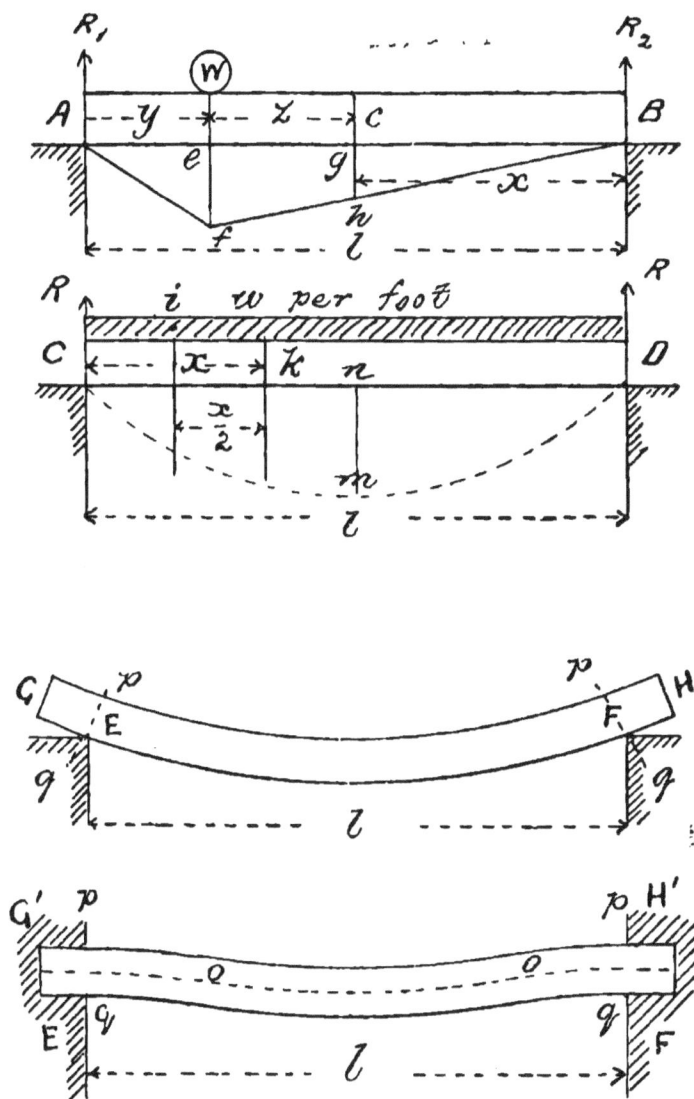

Fig. 9.

$$M = \left\{ W - \frac{Wy}{l} \right\} \cdot \left\{ l - x \right\} - W \times \left\{ l - x - y \right\},$$

$$= Wl - Wx - Wy + \frac{W \cdot y \cdot x}{l} - Wl + Wx + Wy = \frac{Wyx}{l}$$

which is the same as that obtained in the previous calculation. The maximum moment of stress is immediately beneath the load, and is—

$$M = \frac{W \cdot y}{l} \left\{ l - y \right\} = Wy - \frac{Wy^2}{l}$$

If below the straight line $A\,B$ the vertical line $e\,f$ is made to scale, to equal the moment directly under the load, and the point f is joined by straight lines to the points of support A and B; then vertical lines drawn from $A\,B$ to $A\,f$ or $f\,B$ will give, to the same scale, the moment of stress at other points, as, for instance, $g\,h$ under the point c.

As an example, let the girder be 16 feet span, and required to carry one end of another girder in the position indicated by W, and let the load thus brought upon it be 7 tons, also let $y = 6$ feet, then the maximum moment of stress will be—

$$M = Wy - \frac{Wy^2}{l} = 7 \times 6 - \frac{7 \times 6^2}{16} = 29 \cdot 25 \text{ foot-tons}$$
$$= 351 \text{ inch-tons.}$$

When rolled girders are used we must choose some commercial section, and therefore it would be a waste of time to calculate a special section for each case which occurs. The usual and most convenient practice is to have a table of the strengths of standard sections at hand and select the most suitable for the purpose. If a very large parcel of girders of one section is required, it may pay to have rolls specially cut for them, but it would be a very large building that would warrant this expenditure.

The manufacturers of rolled joists, and also the merchants, publish books of tables showing transverse strength of the sections they supply, and these judiciously used will serve all practical requirements. About the method of applying them I shall have more to say, but must first deal with the stresses produced by a uniformly distributed load upon a girder freely supported at each end, and upon one with the ends absolutely fixed.

In Fig. 9, $C\,D$ represents a girder freely supported at each end, and loaded with an equally distributed weight, which, together with the weight of the girder itself, is equal to w tons per lineal foot of span. It is required to find a formula for the moment of stress at any point distant x feet from one of the points of support.

The load being uniformly distributed will be equally borne by the two piers, and the reaction R of either of them will therefore be equal to half the load, or—

$$R = \frac{w \times l}{2}$$

This reacting force will act upwards about the point k, and its moment will be—·

$$= R \times x = \frac{w \,.\, l \,.\, x}{2}$$

There is also another force acting downwards, that is the load between C and k; its weight is $w \times x$. The whole of this weight is to be regarded as acting at its centre of gravity i, which, as the form of the load is symmetrical, will be in the centre of its length; therefore, the distance from k at which it acts is $\frac{x}{2}$ and its moment about R—

$$\cdot \quad = w \times x \times \frac{x}{2} = \frac{w \,.\, x^2}{2}$$

E

The differences between these two will be the resultant moment, and is given by the formula—

$$M = \frac{w \cdot x^2}{2} - \frac{w \cdot l \cdot x}{2} = \frac{w \cdot x}{2}\left\{x - l\right\}$$

For example, let the span of the girder be 32 feet; the load $w = 0.6$ ton per foot; and $x = 12$ feet, then—

$$M = \frac{w \cdot x}{2}\left\{x - l\right\} = \frac{0 \cdot 6 \times 12}{2}\left\{12 - 32\right\} = -72 \text{ foot-tons.}$$

I must now explain the meaning of the *minus* sign in the result. The moments might have been so placed as to give the plus or positive sign; but it is usual to consider the downward moments as positive, and the upward as negative, and it is here to be remarked that the negative sign indicates compression in the top and tension in the bottom flange, the top being concave and the bottom convex, when deflected, and the positive sign before the resultant moment signifies the reverse.

The maximum moment of stress accruing from this load will occur at the centre of the span, where $x = \dfrac{l}{2}$, and

$$M = \frac{w \cdot x}{2}\left\{x - l\right\} = \frac{w \cdot}{2} \times \frac{l}{2}\left\{\frac{l}{2} - l\right\}$$

$$= \frac{w \cdot l^2}{8} - \frac{w \cdot l^2}{4} = -\frac{w \cdot l^2}{8}$$

If $W = w\,l =$ the total load, this expression becomes

$$M = \frac{W \cdot l}{8}. -$$

If a number of consecutive values of x are taken, and the moments calculated for each are plotted vertically downwards to scale from the straight line $C\,D$, a curve may be drawn through the ends of the ordinates so plotted; it is shown by the line $C m D$, and is parabolic in form, as will be recognised by the student of conic sections from the construction of the formula. If, then, the central ordinate $n m$ is calculated, and a parabolic arc is drawn through the three points $C m D$, the moment at any point can be scaled off.

Let us turn now to the consideration of a girder with its ends absolutely fixed. A girder $G H$, carried upon supports $E F$, will, under an equally distributed load, be deflected as shown, and if the ends are not fixed down, they will rise as shown, and the condition of the girder between the points E and F will be that of one freely supported. The top surface will be shortened by compression to the length $p p$, and the lower surface extended to $q q$; from E to G, and F to H, there will be no stress upon the girder.

If the ends $E G$ and $H F$ be now loaded with sufficient weight to bring them down to a horizontal bearing on the surfaces of the two piers E and F, as shown by $G' H'$, the sections $p q$ will become vertical, and the length of the layers of fibres in the girder restored to their length when unloaded, except for the deflection, which makes so slight a difference that for our purpose it may be neglected. The result is exactly the same, whether the girder is first loaded and then the ends pulled down, or the ends first fixed before the loading of the girder. In order to bring the sections $p q$ to, or to keep them vertical, it is evident that a moment of stress equal to the *average* moment of stress on the free girder must be established over the points of support E and F. As the curve of stress $C m D$ is parabolic, its area is found by multiplying its base $C D$ by two-thirds of its height $m n$; this area may be regarded as the sum of the moments of stress on the girder, and therefore the average moment of stress will be two-thirds of the maximum moment of stress, that is—

$$\frac{Wl}{8} \times \frac{2}{3} = \frac{Wl}{12}$$

and this moment will act over each point of support, and will be the maximum moment upon the girder, as that at the centre of span will then be—

$$= \frac{Wl}{8} - \frac{Wl}{12} = \frac{Wl}{24}$$

One result of this form of construction is that the stresses on the top half are not all of the same kind, neither are those on the bottom half; the holding of the ends of the girder will put a pull upon the upper part, and resist extension in the lower, so that over the points of support the top will be in tension and the bottom in compression, while at the centre of the span the top is in compression and the bottom in tension; there must therefore be a point on each side of the centre of the span, where there will be *no bending stress.*

Now the moment of stress at any point on the fixed girder will be equal to that on the same girder freely supported less the moment over the pier $\left(\text{which is } \frac{Wl}{12} = \frac{w \cdot l^2}{12}\right)$ or,

$$M = \frac{w\,l^2}{12} + \frac{w\,x^2}{2} - \frac{w\,l\,x}{2}$$

To find the value of x when the bending stress is nothing, we must make $M = 0$; then

$$M = 0 = \frac{w\,l^2}{12} + \frac{w\,x^2}{2} - \frac{w\,l\,x}{2}$$

therefore

$$\frac{w\,l^2}{12} + \frac{w\,x^2}{2} = \frac{w\,l\,x}{2};$$

dividing both sides of this equation by $\frac{w}{2}$ we get

$$\frac{l^2}{6} + x^2 = l\,x$$

which transposed gives us quadratic equation

$$x^2 - lx = -\frac{l^2}{6}$$

thus solved ;

$$x^2 - lx + \frac{l^2}{4} = \frac{l^2}{4} - \frac{l^2}{6} = \frac{l^2}{12}$$

$$x - \frac{l}{2} = \pm \sqrt{\frac{l^2}{12}} \,;$$

$$x = \frac{l}{2} \pm \sqrt{\frac{l^2}{12}} = \frac{l}{2} \pm \frac{l}{3\cdot464}$$

$$= 0\cdot5\,l \pm 0.288\,l = 0\cdot788\,l \text{ or } 0.212\,l.$$

It is obvious that the two points thus formed should be equidistant from the centre of the span, therefore the two added together should make the whole span, thus,

$$0\cdot788\,l \times 0\cdot212\,l = l.$$

If a girder thus fixed and loaded were cut through at the points thus indicated and hooked together, so that the central part shall not fall out, no change in the deflections curve will take place ; therefore, if such a girder is too long to be made in one piece, these points of *contrary-flexure*, *o o*, in the figure, are those which should be chosen for the joints. If the girder can be made in two lengths the joint should be made in the centre of the span, where the moment of stress is half that over the points of support. These remarks apply to rolled girders which are necessarily of uniform section throughout; in built-up girders in which more than one plate is used in each flange the case becomes modified, but of this I shall treat subsequently.

I have dealt with this question by supposing the ends of the girder to be *weighted* down for a special purpose. An inexperienced student might perhaps think that any iron or steel joint built into a wall at its ends has those ends fixed, but if the weight of the wall above such joist is insufficient to cause a resisting moment of $\frac{Wl}{12}$, the end of the girder will not be fixed—neither will the wall, for it will be lifted when the girder gets its load.

I will take an example to show how much may be expected from an ordinary wall when the girder ends are built in.

In Fig. 10 let AA and BB represent parallel walls of a warehouse having a clear internal width of 24 feet, and its length is supposed to be divided into bays by iron or steel girders placed 10 feet apart from centre to centre, with their ends built into the side walls. E shows a part elevation of the wall taken as 14 feet high, and it is assumed that a roof girder C' comes directly over each floor girder. H is a vertical section of one wall showing the end of the girder C as built in. The floor load is assumed to be 4 cwt. per superficial foot of floor area, including the dead weight of the floor itself. The floor joists, d, d, &c., of either wood or metal, run from girder to girder, so each girder has to support the weight of half a bay on either side, equal one whole bay. The floor area for each girder will be $24 \times 10 = 240$ square feet, and therefore the total load upon the girder will be 240 square feet $\times$ 4 cwt. $= 960$ cwt. $= 48$ tons. The load per lineal foot of girder $= w =$ $48 \div 24 = 2$ tons, and if the girders ends are firmly fixed

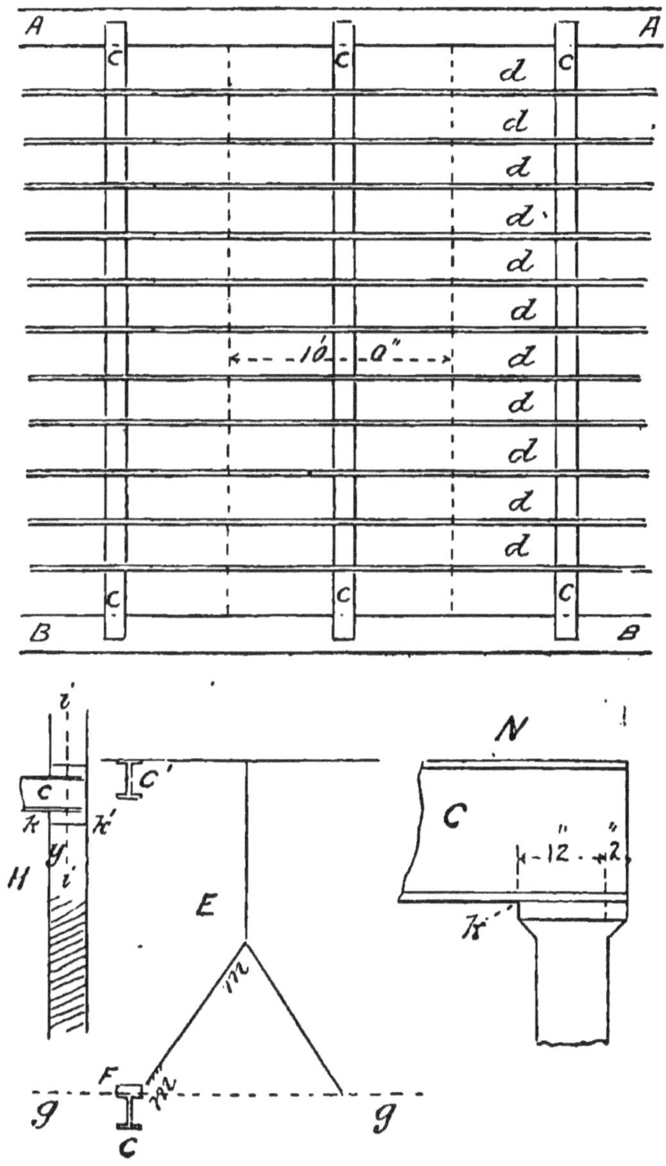

Fig. 10.

and held in the walls, the moment at each point of support will be,

$$M = \frac{w \cdot l^2}{12} = \frac{2 \times (24)^2}{12} = 96 \text{ foot-tons.}$$

A solid bed $k\,k'$ in the section is built in the wall to receive the girder so that the inner edge k shall not crumble under the pressure of the girder when it deflects, throwing the weight thereby on to this edge. Hard stone is generally used for this, but iron would be better. To prevent the lifting of the girder end there is the weight of the wall above it—with any load it may carry at a higher level—acting in the direction of the centre of gravity of the wall, shown by the line i—i, which, as the wall is taken of the same thickness throughout, will be in the centre of its thickness, and the wall being 18 inches thick the leverage of its weight and load about the point k will be 9 inches, $= 0.75$ foot. Upon the end of the girder C is shown a stone, or plate, F, to give the brickwork above a fair bearing on it.

In case of rupture it is not to be supposed that the brickwork will rise in a rectangular mass, parting from that below along the horizontal line g—g; the mortar is not strong enough to hold the bricks together in such a solid mass. The most that can be expected is a mass of wall spreading as it rises from the stone F, the angles of the bounding lines depending upon the bond of the brickwork. If the break of joint in each course is one quarter of a brick or $2\frac{1}{4}$ inches, the spread of the line $m\,m$ will be $2\frac{1}{4}$ inches for every 3 inches rise. As the line $m\,m$ starts at the bottom 6 inches from the centre of girder C, it will meet a similar line drawn from the next girder, when it has spread laterally to the extent of 4 feet 6 inches; bringing this to inches and dividing by the break, the number of courses is found thus,

$$\frac{54 \text{ ins.}}{2\cdot25 \text{ ins.}} = 24 \text{ courses of bricks} = 6 \text{ feet.}$$

the remaining 8 feet (vertical) of the wall load will be 10 feet wide. The lower part of the wall load has an average width of (1 ft. × 10 ft.) ÷ 2 = 5·5 ft.

The face area of the triangular part of the wall will be 5·5 × 6 = 33 square feet, and that of the part above 10 × 8 = 80 square feet, making a total of 113 square feet; and as the wall is 18 inches = 1·5 feet thick, the cubic contents will be 113 × 1·5 = 169·5 cubic feet. Good brickwork weighs 1 cwt. per cubic foot, therefore the wall load will be 169·5 cwt. = 8·475 tons. To this is to be added the roof load brought upon the wall through the girder *C''*. There will be the same area of roof surface as of floor, but the roof load will not exceed ¾ of 1 cwt. per superficial foot, and of this one half will be taken on each wall, making this load

$$= \frac{240 \times 0\cdot75}{2} = 90 \text{ cwts.} = 4\cdot5 \text{ tons.}$$

Adding this to the wall load the total acting along the line *i—i* is, 8·475 × 4·5 = 12·975 tons, which has a leverage of 0·75 feet about the point *k*, and therefore acts with a maximum moment,

$$M = 12\cdot975 \times 0\cdot75 = 9\cdot73 \text{ foot-tons,}$$

towards resisting the girder moment of 96 foot-tons.

From this it is evident that in such cases the girder ends are not fixed, and they must be regarded and calculated as girders with freely supported ends. This it is very important to keep in mind, as a free girder only supports two-thirds the load it would carry with its ends fixed.

Neither can a girder be considered as fixed because its ends are bolted down to stanchions, the stanchion tops will be drawn inwards when the girder deflects, or if they are so fixed they cannot, the connecting bolts will break. Suppose the girder C has its ends bolted down to a stanchion cap 14 inches wide as shown at N. If the bolts are 2 inches in from the back it will give them a leverage of 1 foot, so a total tensile resistance of 96 tons would be called for.   Taking the bolts to be of metal, having a working tensile strength of 7 tons per sectional square inch, a total bolt area of $96 \div 7 = 13\cdot71$ square inches will be required. The strength of a screw bolt must of course be taken at the bottom of the thread—its weakest section. The sectional area at the bottom of the thread of a bolt $\frac{3}{4}$ inch in diameter, is $0\cdot3$ square inches, so the number of bolts required would be $13\cdot71 \div 0\cdot3 = 46$, as there cannot be a part of a bolt; and if there were eleven steel bolts only, they would all be torn asunder. It is scarcely necessary to say that no such number of bolts could be put in.   It might also be shown that a greatly stronger stanchion than ordinarily needed would be required to resist the bending moment which would come upon it were its base absolutely fixed, but I have said enough about fixed ends without occupying more space.

In a building sufficiently long to require the use of continuous girders, the advantage of fixed ends can be used to a certain degree.   In Fig. 11 A B represents part of a continuous girder supported upon walls, or piers, C, D, E; and B G represents the end span of the girder.   All the spans except the end ones should be equal, and the ends should be about four-fifths of the intermediate span, as its

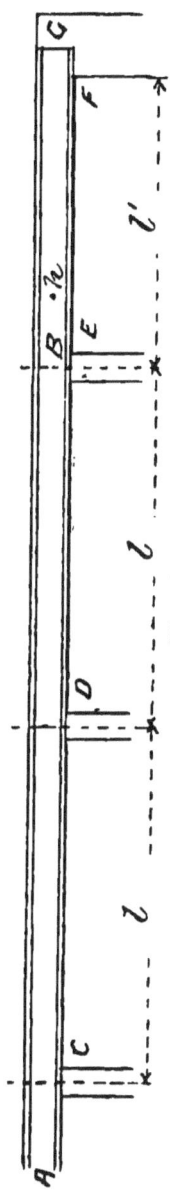

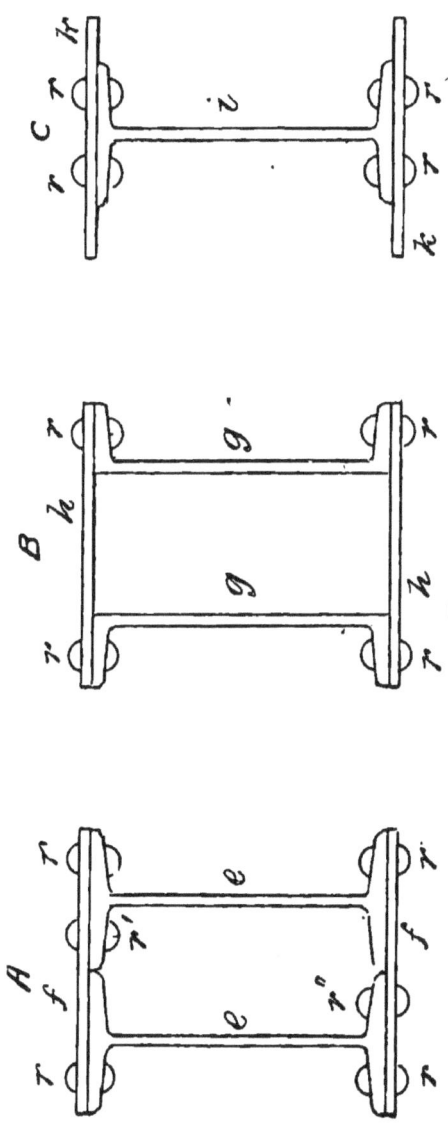

Fig. 11.

Fig. 12.

terminal end will be freely supported, and should therefore take a bearing at its point of *contrary flexure.*

Now the principle of continuity will only obtain in regard to permanent or dead load, and therefore in determining the moments of stress, two distinct loads must be dealt with—the permanent load and the working or live load.

For the dead load I have already shown that the moments of stress are,

Over the point of support,

$$M = \frac{w \cdot l^2}{12}$$

and at the centre of the span,

$$M = \frac{w \cdot l^2}{24}$$

This latter moment of stress will be the maximum, for dead load, in the end span between the point of contrary-flexure h and the point of support F.

The worst case that can occur for any span is that in which that one span is fully loaded, and no other spans have any except their dead loads. If $w' =$ the working load per lineal foot, the maximum central moment of stress is, for free ends,

$$M = \frac{w' l^2}{8}$$

As soon as this comes into action the moments over the points on each side of the span are disturbed, and this full stress does not come upon the girder; but as this is a very complicated problem to work out, which in the end makes very little difference in the maximum moment, it is best to

determine the section of the rolled girder to suit the maximum moment, which will be either,

$$M = \frac{(w + w') \, l^2}{12} \; ; \text{ or } M = \frac{w \cdot l^2}{24} + \frac{w' \, l^2}{8}$$

Having considered the transverse stresses on rolled
girders I now return to the use of the published tables of
strengths issued by the manufacturers and their agents.

In some cases the strengths given are prefaced by the
formulæ used and the tensile resistance of the metal adopted
as a basis. Where these data are given there is no trouble
about determining any section required, for if we know
that the tabular numbers refer to steel of 32 tons per
square inch ultimate tensile resistance, and if we only intend
to specify for 28 ton steel, we lower the tabular number by
one-eighth, and so on for other specified strengths. Where
the data are not given, but merely working strengths, said
to be one-third of the breaking weight, we must find what
strength of metal was accepted as the basis of calculation,
by calculating the strength from the sheets of full-sized
sections of rolled girders issued by the manufacturers.

I have in this way checked several published tables, and
in some cases have found that a much lower tensile strength
has been assumed than that which is shown by tests made
in England, of bars taken promiscuously from a parcel from
abroad.

Another point to be noticed is that different agents pub-
blish tables which do not give the same resistances for
similar sections, although they probably all come from the
same mills; this, however, cannot be proved, and therefore
we can only safeguard ourselves by specifying the strength
we require and rejecting all parcels of which samples, when
tested here, fail to meet the requirements prescribed.

In connection with the use of these tables it is desirable to treat of the so-called "compound" girders, which are made by using rolled girders with plates riveted on to the flanges to give additional strength.

In Fig. 12 three sections of compound girders are shown; that at *A* is built up with two rolled joists *e e*, and a top and bottom plate *f f*. The strength of this section will not be equal to that derived from theory on the assumption that the whole section is solid or continuous, for—setting aside the loss by rivet holes—the distribution of longitudinal stress by the rivets *r*, *r*, *r'*, *r'*, is very irregular, and if the most that can be done is, the partial riveting of contiguous flanges to the plates by rivets *r'* and *r''*, the arrangement still remains a shaky one, and the loss of strength amounts to from ten to twenty per cent. The section *B* which is made up of two channel irons or steels, *g g*, and two plates *h h*, is a better section, as the rivets *r* get a fair hold of the channel irons, and after deducting the rivet holes this form does not probably lose more than ten per cent. in strength as compared with a solid section. The same may be said of section *C*, which consists of a rolled girder *i*, with top and bottom plates riveted on as shown, for in this case both sides of each flange of the rolled girder are riveted to the plates. Of course more than one plate may be riveted upon each flange if required.

In Fig. 13 *A B* shows a side elevation of a girder built up with plates and angle irons, and *C*, *D*, and *E* show transverse sections of girders of this type. These are used for longer spans than those for which solid rolled girders can be used, as there is a limit to the depth of the latter, while that of the built girder can be made anything that is desirable.

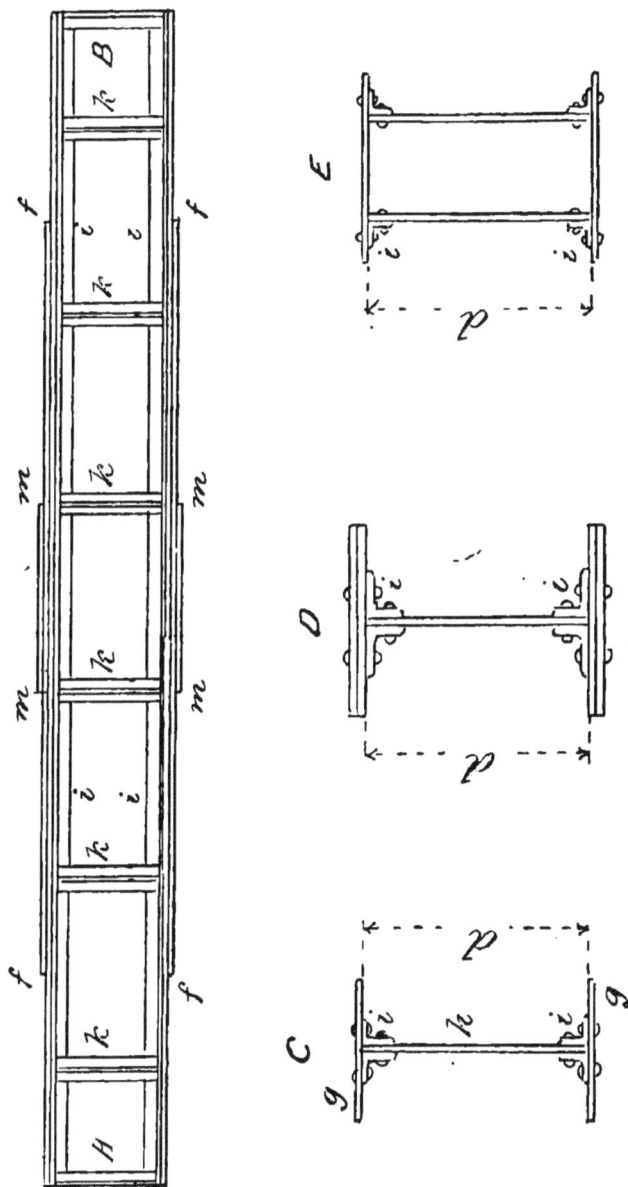

Fig. 13.

The common practice is to regard the web h as carrying the load vertically, by virtue of its resistance to shearing stress, the flanges taking the horizontal stress and keeping the web rigidly stretched so that it cannot fail by twisting. If the flanges of the girder are of a thickness which is inconsiderable in comparison with its depth, the horizontal stress may be regarded as uniform throughout their thickness, so that the section multiplied by the direct resistance and by the distance from the centre of gravity of the flange from the neutral axis of the girder will give the moment of resistance. In order to be on the safe side the distance of the plate from the neutral axis is usually measured from the nearest surface, instead of from the centre of gravity of its section, so that for purposes of calculation the depths of the several sections shown would, in each case, be represented by d.

Now leaving out for the present the value of the connecting angle iron i, we will see what difference exists between the two methods of calculating the moments of resistance of the flanges for an average case. Let the girder be 2 feet 6 inches deep and have each flange formed by a plate 12 inches wide and $\frac{1}{2}$ inch thick. If the girder is iron we take the outside working stress as 4 tons per sectional square inch; this will be right for the compression flange, and will make allowance for loss by rivet holes in the bottom flange, where the working tensile resistance for nett sectional area is five tons.

The moment of resistance of the two flanges, worked out by the formula for solid beams is,

$$M = \frac{s}{b\,d}\left\{b\,d^3 - b\,d_1^3\right\} = \frac{4}{6 \times 30}\left\{12 \times (30(^3 - 12 \times (29)^3\right\}$$
$$= 696 \cdot 26 \text{ inch-tons.}$$

Calculating this moment by the second method we have,

$$M = s \cdot b \cdot \left\{ d - d_1 \right\} \times \frac{d_1}{2} = 4 \times 12 \left\{ 30 - 29 \right\} \times \frac{29}{2}$$

$$= 696 \text{ inch-tons ;}$$

a difference which is practically unnoticeable.

Some engineers ignore the angle-irons in calculating the sectional areas of the flanges, but I think this waste and always take in the horizontal members of the angle-irons as parts of the flange.

As an example I will assume that a wrought-iron girder is required 40 feet span to carry a load of 60 tons equally distributed along its length. It is in the nature of builder's work that there should often be restrictions as to the allowable depth and width of girders, but sometimes there is freedom in this matter, and then the most economical section should be adopted. The effective resistance of the flanges will obviously increase as their distance from the neutral axis increases, and their sections may therefore be reduced in the same ratio. When the flanges consist of more than one plate each it is not necessary to continue the central sectional area throughout the length of the girders, the plates may be reduced in number as the stress diminishes in either direction towards the points of support. I have found that the actual weight of plate girders thus designed amounts to an average of 12 per cent. over their theoretical weight; this includes rivet-heads and cover, plates on the flange, and loss because the plates are not tapered down to fit the curve of stress exactly. The stress upon the web being vertical shearing stress is not affected by the depth of the web, but is constant, therefore the web sectional area should be constant; but this is impracticable, because there is a practical limit to web thickness below which it is unadvisable to go.   The webs of deep girders being merely their plates, require securing from buckling by rivetting on the iron stiffeners, k, k, &c., Fig. 13, and these form parts of the web and should be included in its weight.   As the load is equally distributed one-half is carried on each support and the shearing stress on the web at the centre of the span is nothing.   From this point it increases in each direction directly as the distance from the centre, so that at any point distant x from a support it becomes $= w \left\{ \dfrac{l}{2} - x \right\}$; when the pier is reached $x = 0$ and the vertical stress $= \dfrac{w \cdot l}{2}$.

As the laws under which the web stress occurs vary from those relating to the flange stresses, it follows that for each girder, there must be a ratio of depth to span, corresponding to the lightest weight of girder to carry a given load, and a formula for this ratio must be found.   Let $l =$ the span in feet; $d =$ the depth of girder in feet between flanges; $w =$ load per lineal foot in tons; $s =$ direct mean resistance of iron in tons per square inch. $A =$ sectional area of both flanges together in square inches ; $a' =$ vertical sectional area of web (including allowance for stiffeners, &c.) in square inches ; $t =$ average thickness corresponding to this sectional area in inches; $A =$ total sectional area of web and flanges.

I have already shown that the moment of stress at any point, under an equally distributed load, is

$$M = \frac{w \cdot x^2}{2} - \frac{w \cdot l x}{2}$$

The moment of resistance in a plate girder will be equal to

the area of the flanges multiplied by their mean resistance
and by half the depth of the girder, therefore as the moment
of resistance is equal to the moment of stress,

$$M = \frac{w \cdot x^2}{2} - \frac{w \cdot l \cdot x}{2} = a \times s \times \frac{d}{2} = \frac{a \cdot s \cdot d}{2} \; ;$$

therefore,

$$a = \frac{w \cdot x^2}{s \cdot d} - \frac{w \cdot l \cdot x}{s \cdot d}$$

The weights of homogeneous bodies vary as their sectional areas, so the flange weight—adding in the 12 per

cent. waste—will vary as $1\cdot 12 \left\{ \dfrac{w\,x^2}{s \cdot d} - \dfrac{w \cdot l \cdot x}{s \cdot d} \right\}$; and the

weight of the web will vary as its sectional area (in square
inches the same as the flanges are taken), or $a' = 12 \cdot d \cdot t$.
The whole weight will vary as A, and

$$A = 1\cdot 12 \left\{ \frac{w\,x^2}{s \cdot d} - \frac{w\,l\,x}{s \cdot d} \right\} + 12 \cdot d \cdot t.$$

Here we find the positive quantity $1\cdot 12 \dfrac{w\,x^2}{s \cdot d} + 12 \cdot d \cdot t$,

and the negative quantity $- 1\cdot 12 \dfrac{w\,l\,x}{s\,d}$, varying with d, and

it is evident that if a point is reached when the value of A
is a minimum, and about to change from the descending to
the ascending value, if the increment of d is indefinitely
small the two values will be equal. Let f be an indefinitely
small increment in the value of d, and $A^1 =$ the corresponding value of total area, then

$$A^1 = \frac{1\cdot 12 \cdot w \cdot x^2}{s\,(d+f)} + 12\,t\,(d+f) - \frac{1\cdot 12\,w\,l\,x}{s\,(d+f)}$$

Deducting this from the value of A, the remainder should,
when the value is a minimum, $= 0$, therefore

$$\frac{1\cdot12 \cdot w\, x^2}{s \cdot d} - \frac{1\cdot12\, w\, x^2}{s\,(d+f)} - \frac{1\cdot12\, w\, l\, x}{s \cdot d} + \frac{1\cdot12\, w\, l \cdot x}{s\,(d+f)} + 12\, d\, t$$
$$- 12\, t\,(d+f) = 0 ;$$

therefore,

$$12\, t f = \frac{1\cdot12 \cdot w \cdot x^2}{s} \left\{ \frac{1}{d} - \frac{1}{d+f} \right\} - \frac{1\cdot12\, w\, l\, x}{s} \left\{ \frac{1}{d} - \frac{1}{d+f} \right\}$$

$$= \frac{1\cdot12\, w\, x^2}{s} \times \frac{f}{d^2 + d f} - \frac{w \cdot l \cdot x}{s} \times \frac{f}{d^2 + d f}$$

But as f is taken as indefinitely small in regard to d, the term $d f$ as compared with d^2 may be neglected, hence—dividing both sides of the equation by f we get—

$$12\, t = \frac{1\cdot12\, w\, x^2}{s \cdot d^2} - \frac{1\cdot12\, w\, l\, x}{s \cdot d^2} ;$$

multiplying both sides by d,

$$12\, t d = \frac{1\cdot12\, w\, x^2}{s \cdot d} - \frac{1\cdot12\, w\, l\, x}{s \cdot d}$$

Which shows that the vertical area of the web should be equal to the sum of the areas of the flanges, plus 12 per cent.

The value of d for any point is found, for as—

$$12\, t = \frac{1\cdot12\, w\, x^2}{s \cdot d} - \frac{1\cdot12\, w\, l\, x}{s \cdot d} ;$$

therefore

$$t = \frac{0\cdot093\, w\, x^2}{s \cdot d^2} - \frac{0\cdot093\, w\, l\, x}{s \cdot d^2} ;$$

whence

$$d^2 = \frac{0\cdot093\, w\, x}{s \cdot t} \left\{ x - l \right\}$$

$$d = 0\cdot305 \sqrt{\frac{w\, x}{s \cdot t} \left\{ x\, l \right\}}$$

To get rid of the negative sign this may—without altering the arithmetical value—be written—

$$d = 0·305 \sqrt{\frac{w\,x}{s\,.\,t}\left\{l - x\right\}}$$

This formula, worked out for different points in the span, will give the girder the form of a semi-ellipse, but for our purpose the girder flanges will be required to be parallel, so the depth at the centre, where $x = \dfrac{l}{2}$, is the only one required, and the expression becomes—

$$d = 0·305 \sqrt{\frac{w\,x}{s\,.\,t}\left\{l - x\right\}} = 0·305 \sqrt{\frac{w\,l^2}{2\,.\,s\,.\,t} - \frac{w\,l^2}{4\,.\,st}}$$

$$= 0·305 \sqrt{\frac{w\,.\,l^2}{4\,.\,s\,.\,t}}$$

In the example taken there is a load of 60 tons upon 40 feet, that is, 1·5 tons per lineal foot. Taking into consideration the loss of rivet-holes in the bottom, tension, flange, I shall take the direct working resistance of the flanges all round as 4 tons per sectional square inch for both tension and compression, measured by the gross sectional area.

The shearing stress upon the web at the points of support is half the total load—20 tons, and this, taking the working resistance to shearing as 4 tons, will require 5 square inches of vertical sectional area. At one quarter of an inch thickness the web would give this sectional area if it were only 20 inches in height, but a less thickness may not be used, and this would require to be stiffened on both sides by tee-irons, which may be 6 inches × 3 inches × ⅜-inch thick, spaced 4 feet apart. The sectional area of the two tee-irons will be,

$$2 \left\{6 + (3 - 0·375)\right\} 0·375 = 6·47 \text{ square inches.}$$

This averaged over 4 feet ($= 48$ inches) will give the average thickness of

$$\frac{6\cdot47}{48} = 0\cdot135 \text{ inch thick.}$$

This added to 0·25, the thickness of the web-plate, makes 0·387 inch, and with 5 per cent. for rivet-heads the effective thickness totals up to 0·4 inch. The most economical depth will be,

$$d = 0\cdot305 \sqrt{\frac{w\cdot l^2}{4\cdot s\cdot t}} = 0\cdot305 \sqrt{\frac{1\cdot5 \times (40)^2}{4 \times 4 \times 0\cdot4}}$$
$$= 5\cdot906 \text{ feet.}$$

Practically 6 feet. In some places this might be admissible, as, for instance, to carry the wall of a factory across a stream, so I shall detail the girder under this supposition; and also on the supposition, for a second example, that the depth is limited to 3 feet between the flanges.

Let $S =$ the total resistance of the top flange, and $T =$ the total resistance of the bottom flange, then the moment of resistance of the whole section will be,

$$M = (S + T) \times \frac{d}{2} = \frac{w\cdot l^2}{8}$$

The flanges will be so proportioned as to give equal direct resistances, therefore the moment of resistance of one flange will be $S \times \frac{d}{2}$, and this must be made equal to half the moment of stress, or

$$\frac{S\cdot d}{2} = \frac{w\cdot l^2}{8} \times \frac{1}{2}; \text{ and } S = \frac{w\cdot l^2}{8\cdot d}$$

The stress on either flange will therefore be,

$$S = \frac{w\cdot l^2}{8\cdot d} = \frac{1\cdot5 \times (40)^2}{8 \times 6} = 50 \text{ tons,}$$

The top flange is in compression, so the gross sectional area, at a working stress of 4 tons per sectional square inch, will be, $50 \div 4 = 12.5$ square inches. If the angle-irons connecting the web to the flanges are 3 in. $\times$ 3 in. $\times \frac{1}{2}$ in., the horizontal limbs of the two angle-irons under the upper flange will give a sectional area of 3 square inches, leaving $12.5 - 3 = 9.5$ square inches of sectional area to be made up in the flange-plates. Two plates 10 in. $\times \frac{1}{2}$ in. would be used, thus giving a slight margin of strength.

The strength of the bottom flange is equal to its nett sectional area multiplied by 5 tons per square inch, so the total nett area will be $50 \div 5 = 10$ square inches. Deducting the rivet-holes from the horizontal limbs of the angle-irons there is left as nett area, the rivets being $\frac{3}{4}$ inch diameter,

$$\tfrac{1}{2} \{3 + 3 - (2 \times 0.75)\} = 2.25 \text{ square inches.}$$

The area remaining to be made up by plates will be $10 - 2.25 = 7.75$ square inches. If two 10 in. $\times \frac{1}{2}$ in. plates are used, the same as in the top flange, the nett sectional area is

$$\tfrac{1}{2} \{10 - (2 \times 0.75)\} \cdot 2 = 8.5 \text{ square inches,}$$

which gives also a slight margin of strength. We should not make any difference between the two flanges, and it will generally be found that the bottom and top flanges in moderate-sized plate girders come out nearly equal, the loss by rivet-holes in the bottom flange nearly balancing the extra resistance per square inch and tension over compression.

It will not be necessary to carry the outside plates the whole length of the girder, as the flange-stresses diminish

from the centre towards each end, and to save the trouble
of calculating the whole curve of stress, it will be convenient to have a formula to show where the plates may
be discontinued.   We know that when we stop off the outside plate the sectional area of the flange will be reduced
to $13 - 10 \times \frac{1}{2} = 8$ square inches.   As there is a formula
for calculating the required sectional area at any point, it
can be transposed to give an expression for any known
flange area, such as at ff, Fig. 13.   Let $a =$ sectional area
of one flange.

The equation for the sectional area of the flange at any
point distant x from one point of support is,

$$a = \frac{w\,x^2}{2\,.\,s\,.\,d} - \frac{w\,.\,l\,.\,x}{2\,.\,s\,.\,d}$$

w, l, x, and d are the quantities required for solving this,
but now the formula wanted is to solve the equation in
regard to x when a, w, l, and d are given, from the above

$$x^2 - l\,x = \frac{2\,.\,s\,.\,a\,.\,d}{w}$$

As l can never be less than x, it follows that the equivalent
of $x^2 - l\,x$ must have a negative sign, or

$$x^2 - l\,x = -\frac{2\,.\,s\,.\,a\,.\,d}{w}$$

To solve this equation the square on the left-hand side
must be completed by adding $\frac{l^2}{4}$ to it, and the same quantity
must of course be added to the other side of the equation ;
then

$$x^2 - l\,x + \frac{l^2}{4} = \frac{l^2}{4} - \frac{2\,.\,s\,.\,a\,.\,d}{w}$$

Taking now the square roots of both sides,

$$x - \frac{l}{2} = \pm \sqrt{\frac{l^2}{4} - \frac{2 . s . a . d}{w}}$$

$$x = \frac{l}{2} \pm \sqrt{\frac{l^2}{4} - \frac{2 . s . a . d}{w}}$$

The square root can be either plus or minus, for $a \times a = a^2$, and so does $-a \times -a$. So the quantity under the radical sign can be added to or deducted from $\frac{l}{2}$ = half the span. This shows there are two points equidistant from the centre where the outer plate can be stopped, the distance being,

$$= \sqrt{\frac{l^2}{4} - \frac{2 s . a . d}{w}}$$

Referring to the top flange, the area a is reduced, at the termination of the outer plate, to 8 square inches, therefore the distance the outer plate extends on each side of the centre of the span is,

$$= \sqrt{\frac{40^2}{4} - \frac{2 \times 4 \times 8 \times 6}{1\cdot5}} = 12 \text{ feet},$$

and the whole length of the plate will be 24 feet, and it will terminate therefore at 8 feet from each pier; the accuracy of this may be checked by calculating the required area at those points—

$$a = \frac{w . x^2}{2 s d} - \frac{w l x}{2 s d .} = \frac{1\cdot5 \times 8^2}{2 \times 4 \times 6} - \frac{1\cdot5 \times 40\cdot8}{2 \times 4 \times 6}$$
$$= 8 \text{ sq. inches.}$$

The minus sign here has only the significance that the top flange is in compression and the bottom in tension.

The outer plates may be obtained in one length, but

the inner plates cannot—at least, in iron—so a cover-plate will be required at the centre. The total sectional area of the rivets on each side of the joint should be equal to the sectional area of the plate connected by them to the cover-plate, as the resistance to shearing is the same as that to compression.   Rivets $\frac{3}{4}$ inch in diameter will be used, and the sectional area of each is 0·44 square inch, therefore the number required on each side of the joint will be 5 square inches $\div$ 0·44 = 12 rivets, as we cannot have part of a rivet.   The rivets will be in two rows and 4 inches pitch, so the cover-plate reaches 6 × 4 = 24 inches on each side of the joint and 4 feet long in all.   Each angle-iron has a sectional area of 2·75 square inches—deducting the root—and will therefore want 2·75 $\div$ 0·44 = 7 rivets on each side of the joint.   As the angle-irons are riveted zigzag, 4 rivets may be put in one limb and three in the other.   Angle joint covers are rolled with round, instead of square, corners, so as to fit properly in the angle-irons they join.   If the edges are kept flush with those of the main angle-irons, they must be made thicker.   A round-back angle-iron 2¼ in. × 2¼ in. × $\frac{3}{8}$ in. thick, will fit in a 3 in. × 3 in. × ½ in. angle-iron, and will give nearly the same sectional area, the one being 2·75 and the other 2·734 square inches.

I will now take the second example, in which it is assumed that the depth of the girders must not exceed 3 feet. The flange area at the centre of the span will be,

$$a = \frac{w \cdot l^2}{8 \cdot s \cdot d} = \frac{1 \cdot 5 \times 40^2}{8 \times 4 \times 3} = 25 \text{ square inches.}$$

The sectional area being twice that required in the previous example, more flange-plates will be required, and therefore larger angle-irons to connect them to the web.   Let 4 in. × 4 in. × ½ in. angle-irons be used for this purpose, then the angle-iron area available in the top flange will be 4 square inches, which will leave $25 - 4 = 21$ square inches to be made up in plates. If the flange is made 12 inches wide the total thickness required at the centre of the span will be $1\frac{3}{4}$ inches, and this may be made up with a 12 in. $\times$ $\frac{1}{2}$ in. plate next the angle-irons and web, and a 12 in. $\times$ $\frac{1}{2}$ in. plate next to that, and a 12 in. $\times$ $\frac{3}{4}$ in. plate outside.

The nett area required at the centre of the bottom flange will be;

$$a = \frac{w \cdot l^2}{8 \cdot s \cdot d} = \frac{1 \cdot 5 \times 40^2}{8 \times 4 \times 3} = 20 \text{ square inches.}$$

If the same plates are used for the bottom flange as for the top, and angle-irons also of the same size, the total nett area of the flange will be, deducting rivet-holes,

$$(12 - 1 \cdot 5) \, 1 \cdot 75 + \tfrac{1}{2} \{4 + 4 - (2 \times 0 \cdot 75)\}$$
$$= 21 \cdot 625 \text{ square inches,}$$

which allows a slight margin over the area actually required.

When the outer plate is stopped off the remaining sectional area will be $21 \cdot 625 - 9 = 12 \cdot 625$ square inches, and therefore the length of the outer plates on each side of the centre of the span should be

$$= \sqrt{\frac{l^2}{4} - \frac{2 \cdot s \cdot a \cdot d}{w}} = \sqrt{\frac{40^2}{4} - \frac{2 \times 4 \times 12 \cdot 625 \times 3}{1 \cdot 5}}$$
$$= 14 \cdot 071 \text{ feet.}$$

The total length would be $28 \cdot 142$ feet; but to fit the rivet-pitch the length will be fixed at 28 feet 4 inches.

When the second tier of plates is stopped the remaining flange area will be $12 \cdot 625 - 6 = 6 \cdot 625$ square inches, therefore the length of the second tier each side of the centre will be,

$$= \sqrt{\frac{l^2}{4} - \frac{2 \cdot s \cdot a \cdot d}{w}} = \sqrt{\frac{40^2}{4} - \frac{2 \times 5 \times 6 \cdot 625 \times 3}{1 \cdot 5}}$$
$$= 17 \cdot 146 \text{ feet};$$

so the total length would be made 34 feet 4 inches. All the tiers of plates will have joints in them and require cover-plates.

Having now before us an example of a plate girder, in which each flange comprises several plates, a convenient opportunity occurs for dealing with the disposition of the rivets. Fig. 14 shows—to a distorted scale—a diagram elevation of the girder now under consideration; its total length is 43 feet, 18 inches being allowed at each end for bearing. The distortion of the plate thickness allows their divisions to be clearly shown. The rivets are all to be $\frac{3}{4}$ inch in diameter, and 4 inches general pitch. I say general pitch, because we may come upon some connections where it must be varied; but it must be evident that keeping to a uniform pitch facilitates the execution of the work and renders its appearance when complete more symmetrical than it would be, were this point neglected.

The outer and second tiers of plates both suit for odd numbers of pitches, 85 in the outer, and 103 in the second tier, it will therefore be necessary to have a rivet in the centre of the span and set out both ways from it. If the number of pitches were even there would be a rivet on each side of the centre, and two inches from it. Both flanges will be made alike, and the corresponding tiers of plates in each are indicated by the letters a, b, c. For ordinary iron girder-plates it is not economical to run the plates over 20 feet long, as the cost per ton increases with the length,

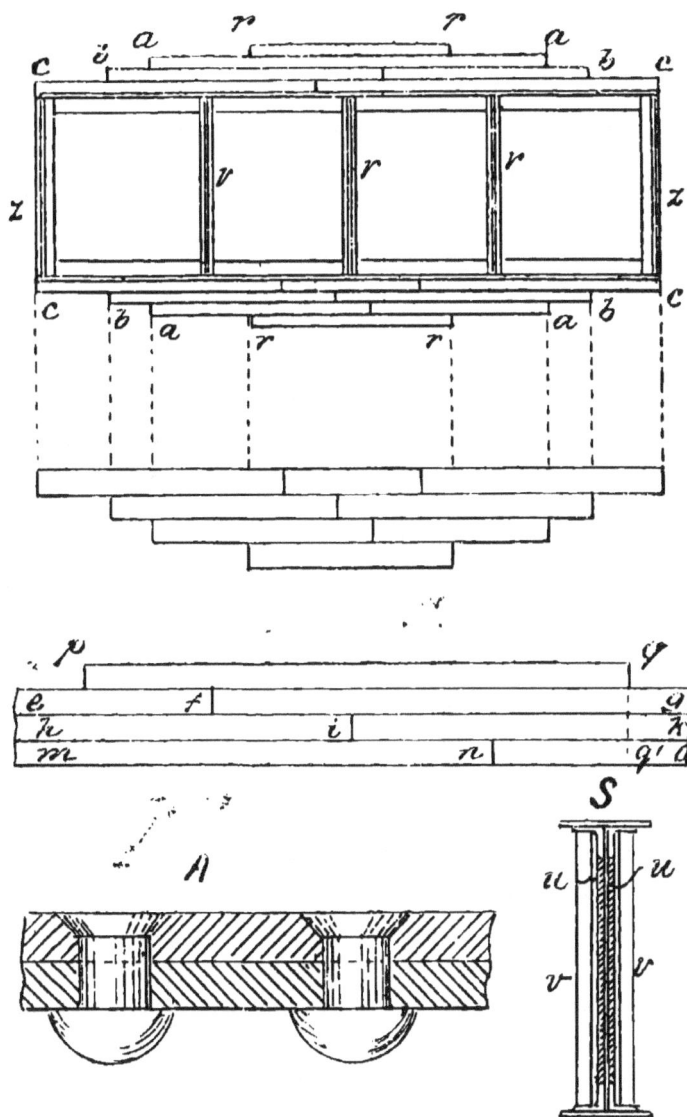

Fig. 14.

and it is better to keep them shorter than this where convenient. Economy in cover-plates has also to be considered,
and by bringing several joints under one cover-plate there
is a saving. A cover-plate pq is shown in the side elevation, covering the three joints f, i, and n, in the plates eq,
hk, and mo. Assuming these plates to be all of the same
thickness, and that pf is the lap required on each side of a
joint to receive the necessary number of rivets, it is evident
that, if all the joints had separate cover-plates, the total
length of these would be equal to twice the number of laps
that there are joints—in this instance $pf \times 6$; but the joints
in the plates can be brought together so that $ef = fi = in$
$= nq'$; q' being vertically below the point q, then the length
of cover-plate is only $ef \times 4$, saving two laps, and the
greater number of joints that can be thus brought together
the greater is the saving in cover-plates. The theory of
the duty of the continued cover-plate is this :—The stress
upon the plate ef is picked up by the cover-plate and trans-
mitted through the rivets to no; that upon hi passes to
fg; and the stress on mn to ik. Similarly, if a joint in a
plate can be brought to the normal termination of the plate
above, this latter may be made longer to form a cover-plate
to the joint beneath.

This matter being thoroughly understood, we are prepared to arrange the joints in the flanges. The inside plate
tier being 43 feet long must be in three lengths; the others
may be in two lengths each. The cover-plate for the outside tier must of course be $\frac{3}{4}$ inch thick, and one lap for
that tier will require rivets equal in total sectional area to
that of the plate, $12 \times \frac{3}{4} = 9$ square inches; and $9 \div 0.44$
$= 21$ rivets, but as there are two rows of rivets the length
of lap will correspond to 11 pitches or 44 inches $= 3$ ft.
8 ins. The lap for the other tiers will be $6 \div 0.44 = 14$

rivets = 7 pitches = 28 ins. = 2 ft. 4 ins. Now if we can place this cover-plate in the centre of the span the extension of its ends will cover also the joints of the inner tier, for its total length is 9 ft. 8 ins. and 43 ft. −9 ft. 8 ins. = 33 ft. 4 ins., of which half = 16 ft. 8 ins. would be at each end, and the 9 ft. 8 ins. in the centre of the inner tier of plates, to pick up the joint in which the cover-plate would require to be lengthened to 14 ft. 4 ins.

Placing this cover in the centre of the span would throw the joints in the outer and second tiers of plates 1 ft. 2 ins. on each side, giving in the outer tier a maximum length of 14 ft. 2 in. + 1 ft. 2 in. = 15 ft. 4 in., and in the second tier 17 ft. 2 in. + 1 ft. 2 in. = 18 ft. 4 in.—none of which would be inconvenient lengths.

The strength of joints such as these does not depend alone upon the shearing resistances of the rivets but is certainly helped by the friction between the plates, which are drawn closely together by them. This frictional resistance has by some writers been estimated as greater than the shearing resistance of the rivets, but their modes of calculating have not always been practical. When the rivets have cooled they cannot press the plates together with a force exceeding their limit of elasticity, that is, the point at which permanent elongation occurs; this is taken at eight tons, and assuming all this to act, and the co-efficient of friction of iron upon iron to be 0·25, the frictional resistance of the plates to moving would be $8 \times 0\cdot25 = 2$ tons per sectional square inch of rivet area, so without using a factor of safety this shows half the amount of the working shearing resistance. If half of this can be relied upon, the strength of the riveted joint will be 25 per cent. more than that calculated from shearing resistance, and this is really wanted, as the bearing surfaces in the rivet-holes are commonly deficient in area. In single shear a $\frac{3}{4}$ rivet has a shearing sectional area of 0·44 square inch $= 0·44 \times 4$ tons $= 1·76$ tons working resistance to shearing. This rivet in a half-inch plate has a bearing $= 0·75 \times 0·5 = 0·375$, which in compression shows a working resistance of $0·375 \times 4$ tons $= 1·5$ tons only.

In regard to the matter of rivet bearing, it is obvious that if it passes through plates of different thicknesses, its diameter cannot be proportioned to all; but if it passes through, say an angle-iron and one plate, each of the same thickness, the ratio of plate thickness to the diameter of the rivet may easily be determined. The resisting surface upon which any rivet bears is its diameter multiplied by the thickness of the plate through which it passes, for this represents the area at right angles to the line of thrust. The shearing and compressive working resistances are equal, each being 4 tons per sectional square inch, therefore the shearing area should be equal to the bearing area. If $D =$ diameter of rivet, and $t =$ the thickness of the plate, then the shearing area is equal to $D^2 \times 0·7854$, and the bearing area $= t \times D$. Equating these quantities,

$$D^2 \times 0·7854 = t \times D; \; t = 0·7814 \cdot D; \; D = 1·273 \cdot t$$

For the class of work with which we are dealing, the use of $\frac{3}{4}$-inch rivets is almost universal, and to give it a bearing equal to its shearing resistance, the plate thickness should not be less than

$$t = 0·7854 \cdot \times 0·75 = 0·589 \text{ inch,}$$

which is nearly $\frac{5}{8}$ inch.

In the top or compression flange long cover plates are not required theoretically, as, if the plates butt truly against each other over the whole of their contiguous end surfaces, there can be no stress upon the rivets, and the only object of the cover is then to keep the ends of the plates in juxtaposition. In practice, however, this accurate meeting of the ends cannot be relied upon, for although such joints no doubt often are made, we have no means of proving the contact, and therefore, for safety, the compression joints are dealt with in the same way as those under tension.

The rivets required for the web-plate joints will also require calculating, and if the web is of insufficient thickness to afford adequate bearing area to one row of rivets on each side of the joint, the cover-plates must be widened to take more rivets. This joint must be calculated, not only as regards the number of rivets necessary to carry the vertical stress, but also in reference to the web thickness.

The shearing stress upon the web at the points of support is half the total load, or 30 tons, and this will require in the end web plates a vertical sectional area of $30 \div \frac{1}{2} = 7\cdot5$ square inches. The height of the web is 36 inches, therefore its thickness must not be less than $7\cdot5 \div 36 = 0\cdot2$ inches, less than $\frac{1}{4}$ inch. Plates $\frac{1}{4}$ inch thick, with stiffeners every 4 feet over the clear span, would be used, and the whole length made up of four plates, one on each side of the centre of the span 10 feet long, and two end plates 11 feet 6 inches long each. The vertical stress on the web at any point is equal to the distance of such point (y) from the centre multiplied by the load per lineal foot of span (w). At the centre there is no stress upon the web, but at the joint 10 feet nearer the pier it is,

$$= w \times y = 1\cdot5 \times 10 = 15 \text{ tons.}$$

The bearing surface required in the rivet-holes to sustain

this will be $15 \div 4 = 3.75$ square inches. At 4-inches pitch there will be nine rivets in the depth of the web, and the total bearing area corresponding to this number will be

$$= 9 \times 0.75 \times 0.25 = 1.687 \text{ square inches.}$$

which is quite inadequate to the stress; more rivets must, therefore, be used, although the nine rivets in one row— being in double shear as a cover is put on each side of the web—give more than twice the shearing resistance necessary. The number of rivets required will be,

$$9 \times \frac{3.75}{1.687} = 20 \text{ rivets}$$

on each side of the joint. This means more than two rows, and therefore for the vertical joints the pitch of the rivets will be made 3 inches, to give twelve rivets in one row. The rivets may be zigzagged, as shown on the elevation in Fig. 14; and the cover-plates u will be made $\frac{1}{2}$ inch thick, so that the stiffening irons v, of which the ends are riveted through the angle-irons and web, can be flush, and so save joggling their ends. On each side of the joint twelve rivets go through the 6 in. $\times$ 3 in. $\times$ $\frac{3}{8}$ in. tee-irons and the cover-plates and web, and eleven rivets through the cover-plates and web only. This makes a very rigid stiffening, and the other tee-irons will be packed similarly, but with only six-inch wide bars between the main angle-irons. There will not be any stiffeners at the centre web joint, and there is no vertical stress there except under a changing load, so there a $\frac{1}{4}$-inch cover-plate on each side of the web with a row of nine rivets on each side of the joint will be ample. A four-inch rivet pitch will also be sufficient for the tee-iron stiffeners that do not come over web joints. The ends of the girder will be strengthened by end plates

x, z, of the same width as the flanges and the same height as the web, and connected to the web by 4 in. × 4 in. × $\frac{1}{2}$ in. angle-irons, meeting the main angle-irons in a mitre-joint.

Directly over the front edges of the supporting piers will be riveted 4 ins. × 4 ins. × $\frac{1}{2}$ in. angle-irons on each side of the web, but these will pass on to the main angle-irons, and be packed up in the same way as the tee-iron stiffeners.

The stiffening of the web is very necessary over the piers, as there the vertical force is at its maximum, and becomes essentially a crippling or buckling force on the web, which being there on a solid bed cannot yield at its lower edge.

When the stresses on the flanges require very large sectional areas, broad plates are used to avoid having to make the flanges of an excessive number of plates, and broad plates are also sometimes used where the girder is required to carry a thick wall. In such cases the single web is not sufficient, as the flange-plates and the angle-irons would have to do heavy duty as cantilevers, and therefore a double webbed or box-girder of the section shown at F, Fig. 13, should be used.

The work imposed upon the rivets connecting the angle-irons with the web, and those connecting them with the flanges, must be considered in order to decide upon a minimum rivet-pitch. It is obvious that the increase of stress from one rivet to the next must be carried from the web to the angle-irons and from the angle-irons to the flange through the rivets connecting them together. The angle-irons are connected to the web by rivets in double shear, and the angle-irons to the flanges by twice the number of rivets each in single shear, so that for each increment of stress the pitch required will be the same in both connections.

Starting from one point of support the flange stress increases in a gradual decreasing ratio to the centre of the span, therefore the closest pitch is required near the supports. If $x =$ the distance to one web rivet, from the pier, and x_1 that of the next rivet from the same point, the shearing stress to be borne by the further rivet is equal to the difference of stress between the two points indicated. At the pier face the stress on the flange $= 0$; four inches $(= \frac{1}{3}$ ft.) from it the flange stress is,

$$S = \frac{w \cdot x^2}{2 \cdot d} - \frac{w \cdot l \cdot x}{2 \cdot d} = \frac{w}{2 \, d} \left\{ x^2 - l \, x \right\}$$

$$= \frac{1 \cdot 5}{2 \times 3} \left\{ (\tfrac{1}{3})^2 - 40 \times \tfrac{1}{3} \right\} = 3 \cdot 305 \text{ tons.}$$

The sectional area of a $\frac{3}{4}$ rivet is $0 \cdot 44$ inches, it is double-shear, and therefore its working strength is $0 \cdot 44 \times 2 \times 4$ tons $= 3 \cdot 52$ tons. The strength being sufficient with a four-inch pitch at this part of the girder, a longer pitch might be used towards the centre of the girder. At the centre of the span the increment in four inches will be,

$$S - S' = \frac{w}{2 \, d} \left\{ (x^2 - l \, x) - (x_1^2 - l \, x_1) \right\}$$

$$= \frac{1 \cdot 5}{2 \times 3} \left\{ (\overline{20}^2 - 40 \times 20) - (\overline{20\tfrac{1}{3}}^2 - 40 \times 20\tfrac{1}{3}) \right\}$$

$$= 0 \cdot 02\dot{7} \text{ tons.}$$

We must, however, have the constituent parts of the girder held securely together, and for those built up of plates and angle-irons where $\frac{3}{4}$-inch rivets are used, I should never make the pitch longer than four inches. Where the number of flange-plates is such as to require rivets of greater diameter to draw them together, the pitch may be increased—for $\frac{7}{8}$-inch rivets to 5-inch pitch, and for inch rivets to 6-inch pitch.

I know that in builders' iron-work these pitches are commonly exceeded, presumably to save money, but such practice is altogether bad, and should not be tolerated.

When a wall is carried directly upon a built-up, or on a compound girder, a flush top is required—there should be no rivet-heads projecting above the level of the top flange plates, and then the rivets are countersunk in the top plate, as shown at A, Fig. 14. The heads of rivets so made are of course inferior in strength to the "cup" heads, but as there is no axial stress upon them in this case, this is not of any great importance; where, however, this course has to be followed it is better to use one thick flange-plate than two thin ones, for the countersink should not be carried more than one-third down into the thickness of the plate.

In order to carry out the designs of the architect it is frequently necessary to use inclined, bent, and curved girders, and with these I will now deal.

In Fig. 15, $A B$ represents an inclined girder with its ends cranked into a horizontal position, and supported on rollers which rest upon horizontal beds, on piers C and D. If a load, W, is suspended from the girder it will not cause it to move laterally, because the only available resistances are the vertical reactions of the piers. If the student has any doubt about this matter, he should make a model—it is a very simple affair—and so set his mind at rest as to the fact.

The reasoning is very clear—all weight acts, in the first instance, vertically, the re-acting stress equilibrating the weights acts at right angles to the plane of the supporting surfaces, and if these are horizontal the re-actions must be simply vertical.

If the bearing surfaces are inclined, say, to the "rake" of the girders, as shown at $E F$, there will be a sliding force acting upon the girder. The load carried to the pier E will act in a vertical direction such as $k\,i$, intersecting the surfaces of contact at a point g. Let $g\,i$ equal, to any convenient scale, the load upon the pier E. Then complete the parallelogram of forces by drawing through the point g a straight line at right angles to the supporting surfaces, and terminating it where it meets a line $i\,h$ drawn from the point i parallel to those surfaces, and from h draw upwards the vertical line $h\,m$. Then in the parallelogram $g\,i\,h\,m$, $g\,m$ represents the sliding effort on the pier, and $g\,h$ the normal pressure upon the bearing-surface of the same, which will tend to overturn the pier inwards. This form of support must always be avoided.

Horizontal bearings being supplied, the moments of stress will be calculated in exactly the same way as those for a horizontal girder, the values of x being measured horizontally, and the effective span l being the horizontal distance between the supporting piers C and D. In ordinary building work the rollers shown in the figure are not used, as the spans of the girders are not large enough to require such special arrangements to allow for contraction and expansion under alterations of temperature, but I have here shown them as being advisable to use in the model referred to above.

Girders of this form are required to support "ramps" in warehouses, on which trucks are run from one level to another, also for the inclined horse-runs in buildings which have stables upon floors above the ground level.

In Fig. 16, A, B, and C show three forms of staircase carriages, and the strength of these and the stresses upon them are calculated in the same way as that given above — so long as the bearing-surfaces are horizontal, the values of x and l are measured horizontally.

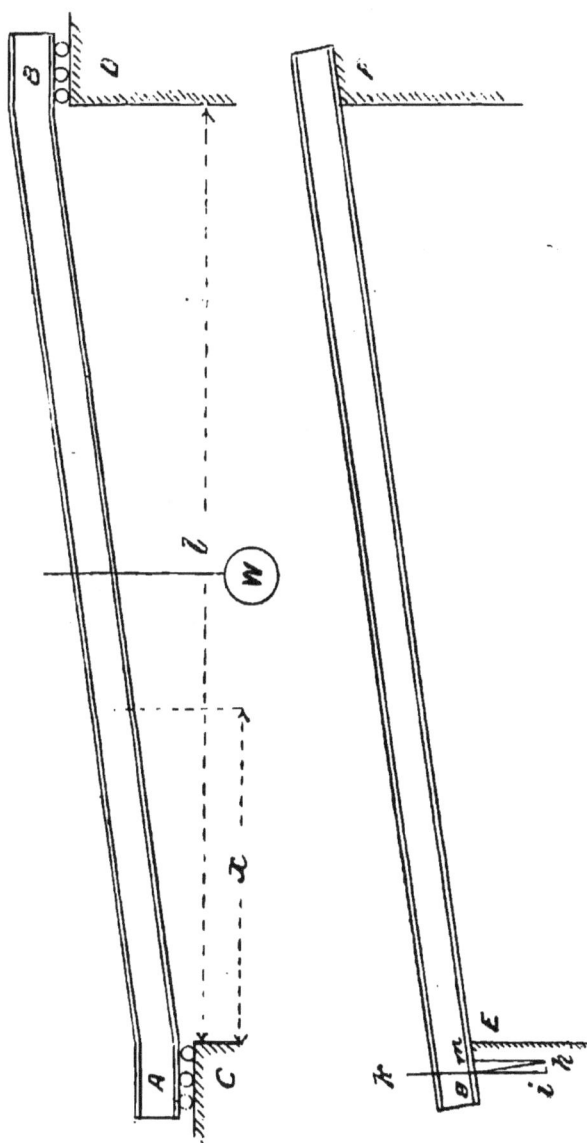

Fig. 15.

Having determined the stresses upon the flanges for a full equally-distributed load, we have covered all stresses that may be brought upon any part of the flanges by a partial load of the same intensity per lineal foot, but this is not the case in regard to the web. It will constantly happen that a main floor-girder, for instance, is only partially loaded, and this occurrence will bring shearing stresses upon the web that are not covered by those appearing under a maximum uniformly-distributed load. I shall assume that we have a floor girder of 20 feet span, loaded with one ton per lineal foot, and plot the curve of stresses at every foot under a retreating load.

Let *A B*, Fig. 17, represent the horizontal line of the web, divided, as shown, into twenty parts, with vertical lines drawn down from each division, on which to plot the ordinates of the curve.

When the girder is fully loaded, the reactions, and therefore the shearing stresses, are equal over each point of support, one half of the weight being carried by each, and

$$R = R' = \frac{w \cdot l}{2} = \frac{1 \times 20}{2} = 10 \text{ tons.}$$

This is plotted off to scale and shown by the ordinate *A o*; under this load the stress on the centre of the web = 0, as half the load passes each way to the respective points of support. Now suppose the load to move towards *B*, until the part 1 — *B* alone is fully loaded, then the reaction *R* will be—

$$= \frac{w \times (1 - - - B) \times 9\cdot 5}{l} = \frac{1 \times 19 \times 9\cdot 5}{20} = 9\cdot 025 \text{ tons.}$$

The distance 1—*B* gives the length over which the load extends, and its centre of gravity is 9·5 feet from the support

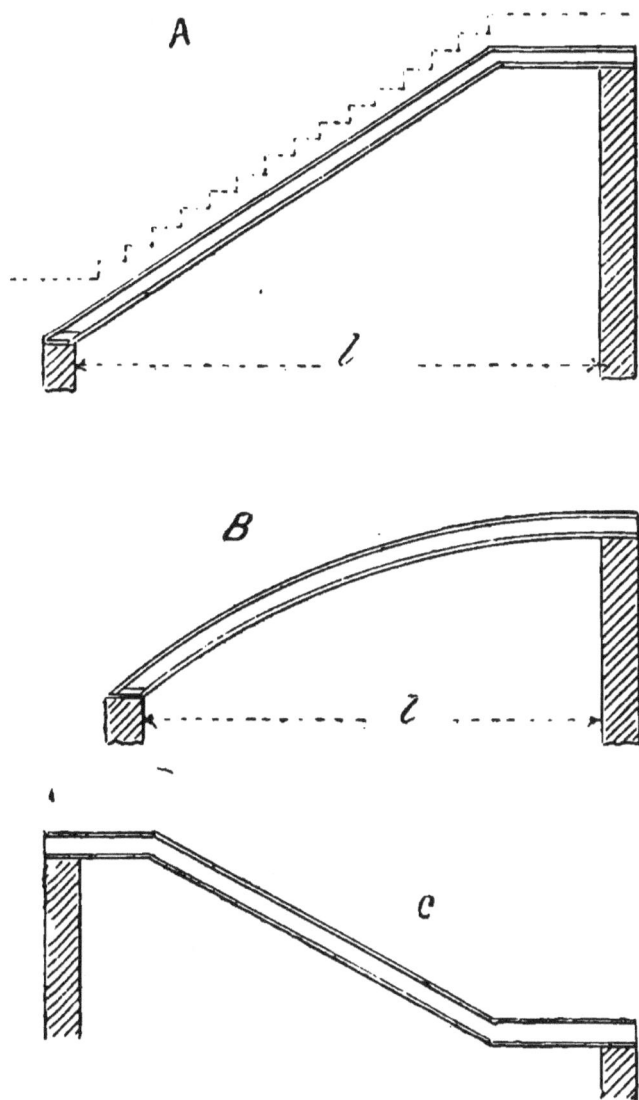

Fig. 16.

B, and it acts upon the support A with this intensity, while the reaction has a distance of $l = 20$ feet about the point B, and by this we divide the downward moment to find the reaction.

Now, as there is no vertical force acting between A and 1, this shearing force will be constant to that point, and its value is plotted in the ordinate 1 . . . 1.

Proceeding in the same way with further consecutive retrogressions of load to the centre of the span, we obtain the following ordinates :—

$$2 . . 2 = \frac{1 \times 18 \times 9}{20} = 8\text{·}100 \text{ tons.}$$

$$3 . . 3 = \frac{1 \times 17 \times 8\text{·}5}{20} = 7\text{·}225 \quad \text{,,}$$

$$4 . . 4 = \frac{1 \times 16 \times 8}{20} = 6\text{·}400 \quad \text{,,}$$

$$5 . . 5 = \frac{1 \times 15 \times 7\text{·}5}{20} = 5\text{·}624 \quad \text{,,}$$

$$6 . . 6 = \frac{1 \times 14 \times 7}{20} = 4\text{·}900 \quad \text{,,}$$

$$7 . . 7 = \frac{1 \times 13 \times 6\text{·}5}{20} = 4\text{·}225 \quad \text{,,}$$

$$8 . . 8 = \frac{1 \times 12 \times 6}{20} = 3\text{·}600 \quad \text{,,}$$

$$9 . . 9 = \frac{1 \times 11 \times 5\text{·}5}{20} = 3\text{·}025 \quad \text{,,}$$

$$10 . . 10 = \frac{1 \times 10 \times 5}{20} = 2\text{·}500 \quad \text{,,}$$

There is no need to carry this curve any further, as it would have continually diminishing ordinates, if the load had been moving towards A; a curve of maximum shearing stress would have been described for the other half of the web, similar to that shown by the line 1, 2, 3 . . . 10; and the strength of each half of the web must therefore be suited to resist the stresses thus found.

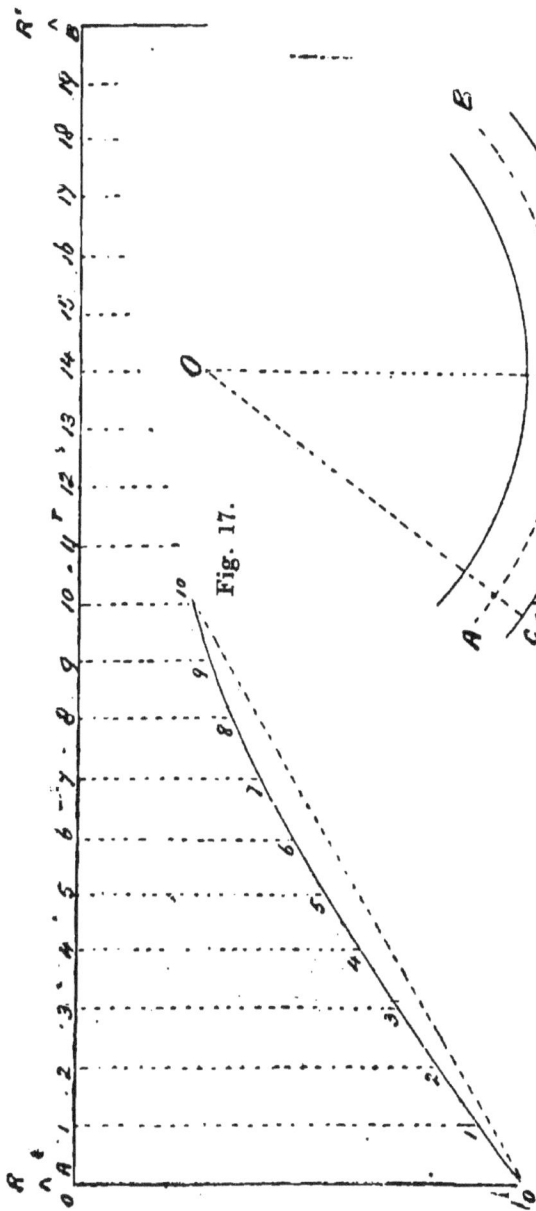

Fig. 17.

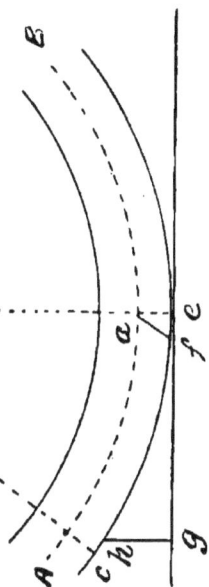

Fig. 18.

In practice it will be sufficient to find the stresses $A o$ and 10 . . . 10, and draw a straight line from o to the lower end of 10 . . . 10, for the curve is between the straight line —shown dotted—and the line $A B$, so that we cannot fall short of strength, and at the same time the difference between the ordinates to the curve and those to the straight line does not show any extravagant margin. When a load comes upon a girder it must necessarily deflect it to an extent proportional to the intensity or magnitude of such load. In Fig. 18, $A B$ represents part of a bent beam, the dotted line showing the neutral surface and O being the centre of curvature. From a draw a parallel to $b c$, then $e f$ is the outside extension in the length $a b$. By similar triangles $O a$ is to $a b$ as $a e$ is to $e f$. Let $\frac{d}{2} =$ the depth $e a$, $E =$ modulus of elasticity, $s =$ stress per sectional square inch at e, and $R =$ radius of curvature at that point,

$$\frac{O a}{a b} = \frac{a e}{e f}; \text{ therefore } R = \frac{\frac{d}{2}}{\frac{s}{E}} = \frac{d E}{2 s}$$

If the stress is uniform throughout the length of the girder, the radius of curvature is constant, and the deflection at any point is found by the ordinary formula, thus: Let $e g$ be a tangent to the circle at e, then to find $g h$ we have,

$$g h = \frac{\overline{e g}^2}{2 R}$$

Applying this to girders which have uniform stress through-out the flanges, where $l =$ span of girder, $d =$ depth, c and $t =$ the stresses on top and bottom flanges respectively in tons per sectional square inch ; $e\,g = \dfrac{l}{2}$, and $g\,h = D =$ deflection of girder at centre ; then

$$D = \frac{\overline{e\,g}^{2}}{2R} = \frac{\dfrac{l^{2}}{4}}{2R} \; ;$$

$2s = c + t$, therefore

$$2R = \frac{2\,d\,E}{c + t}$$

and,

$$D = \frac{\dfrac{l^{2}}{4}}{\dfrac{2\,d\,E}{c + t}} = \frac{l^{2}\,(c + t)}{8\,d\,E}$$

In practice it is seldom that such conditions can be found, and if the sectional area and moment of resistance vary, so will the radius of curvature ; so it is advisable to determine under what laws the deflection varies and fill in constants obtained by experiment.

If we take the equation $R = \dfrac{d\,E}{2\,s}$ and include all the terms except the load and given dimensions in the term a, to be found by experiment, we get,

$$R = a\,.\,\frac{d}{s} \; ; \; D = \frac{l^{2}}{8\,R} = \frac{l^{2}}{\dfrac{8\,a\,d}{s}} = \frac{l^{2}\,s}{8\,a\,d} = a'\,\frac{l^{2}\,s}{d}$$

But s at the centre varies as $\dfrac{W\,l}{b\,d^{2}}$ for rectangular beams, or

$$= a''\,\frac{W\,l}{b\,d^{2}}$$

therefore,

$$D = a' \times \frac{l^{2}}{d} \times a'' \times \frac{W\,l}{b\,d^{3}} = a'''\,.\,\frac{W\,l^{3}}{b\,d^{3}}$$

in which the value of a''' must be filled in from experiment.

When the section is not a solid rectangle $b d^3$ is to be replaced by $b d^3 - b_1 d_1^3 \ldots b_n d_n^3$, as in the determination of moments of resistance; let this quantity be represented by m.

For iron of good quality we find the following formulæ—Let $l =$ span in feet; $W =$ load in tons; b and $d =$ breadth and depth in inches; and $D =$ deflection in inches.

CAST IRON.—Girder loaded in the centre,

$$D = \frac{W l^3}{14 m}$$

With the load uniformly distributed,

$$D = \frac{W l^3}{22 \cdot 4 m}$$

Cantilever fixed at one end and loaded at the free end,

$$D = \frac{8 W l^3}{7 m}$$

The same with uniformly distributed load,

$$D = \frac{3 W l^3}{7 m}$$

WROUGHT IRON.—Girder loaded in the centre,

$$D = \frac{W l^3}{28 \cdot m}$$

With the load uniformly distributed,

$$D = \frac{W l^3}{44 \cdot 8 m}$$

Cantilever loaded at the free end,

$$D = \frac{4 W l^3}{7 \cdot m}$$

The same with uniformly distributed load,

$$D = \frac{3\,Wl^3}{14\,m}$$

For rolled girders supported at both ends and loaded centrally,

$$D = \frac{Wl^3}{18\,m}$$

With uniformly distributed load,

$$D = \frac{Wl^3}{29\,m}$$

For single web riveted girders, loaded at the centre,

$$D = \frac{Wl^3}{16\,m}$$

With uniformly distributed load,

$$D = \frac{Wl^3}{25\,.\,6m}$$

For MILD STEEL the deflections may be taken as about 0·7 of those for wrought iron; but all these figures are only approximate. The deflections vary inversely as the modulus of elasticity of the material. The following are some of the results taken from practice:—

Modulus of Elasticity.

Cast iron rectangular bars .	. 6,785 tons to 8,434 tons.
Square and circular tubes .	. 5,453 ,,
Rolled iron girders . .	. 7,304 ,, ,, 9,630 ,,
Single web plate girders .	. 7,000 ,,
Double ,, ,, .	. 10,500 ,,
Conway (tubular) bridge .	. 8,372 ,,
Rolled English steel girders .	. 10,000 ,, ,, 12,000 ,,

The variations in these figures are so wide that we cannot calculate with accuracy the deflection of any given girder unless its modulus of elasticity is first determined.

In order to prevent girders from deflecting below the horizontal line drawn between their points of support, it is usual to make them slightly curved upwards; this is termed making them with a "camber," for which, in built girders, the common practice is to allow one inch rise in the centre for every forty feet of span; for other forms this allowance should be varied inversely as the modulus of elasticity of the material.

In public buildings of considerable magnitude, in the construction of which girders of long span are required, plate or other solid webbed girders are unsightly, and if the depth of the girder is over four feet, one of the forms of open webbed girders will also be more economical.

CHAPTER IV.

TRIANGULAR AND LATTICE WEBBED GIRDERS.

A SIMPLE form of triangular webbed girder is illustrated by the diagram $A\,B$, Fig. 19. The web consists of a number of equilateral triangles, of which the top and bottom flanges form the bases. Although, regarded as a girder, the structure as a whole is under transverse stress, all its parts are under direct stress. Let a load W be applied at the centre of the span, then half will be carried by each point of support, A and B.

We can deal in the simplest way with the determination of stresses by applying the principle of the parallelogram of forces. From the central apex 1, let fall a vertical line to the centre of gravity of the weight W, and upon this line mark off from 1 the distance 1 a, to represent to any convenient scale the weight W; then using the bars 1, 2, as sides, complete the parallelogram 1 $b\,a\,c$; then 1 b and 1 c will represent the compressive stresses upon the bars 1, 2, on each side of the centre of the span. The web-bars in compression are termed " struts," those in tension being known as "ties." The actual relation of the stress to the load will be that of 1 b to 1 a; therefore

$$W \times \frac{1 \cdot b}{1 \cdot a}$$

would equal the thrust upon bar 1, 2. Join the points $b\,c$,

H

and the line b c will bisect the line 1 a, so that $1\,h = \frac{1}{2}\,W$. Now $\frac{1}{2}\,W$ is the load which is carried by the bar 1, 2. Call this load $= L$, then the stress upon the bar 1, 2

$$= \frac{W}{2} \times \frac{1 \cdot b}{1 \cdot h} = L \times \frac{1 \cdot b}{1 \cdot h}$$

But the triangle 1, b, h, is similar to the triangle 1, 2, i; therefore $\dfrac{1.b}{1.h} = \dfrac{1,2}{1.i}$. As 1, 2, is the length of the strut and 1, the depth of the girder, it follows that the load passing through the bar being given, the stress upon the bar will equal such load multiplied by the length of the bar (strut or tie) and divided by the depth of the girder. This is a simple rule which can easily be remembered. In the case in question the whole triangle 1, 2, 2, is equilateral, and its base being horizontal and the line 1 i vertical, the triangle 1, 2, i, is a right-angled triangle. By the qualities of right-angled triangles (Euc., bk. 1, prop. 47) the square upon the hypothenuse 1, 2, is equal to the sum of the squares upon the base 2 i and the perpendicular 1 i. 1, 2, 2 being an equilateral triangle, 1, 2 $= 2 \times 2\,i$; if we call $C =$ compressive stress upon the strut 1, 2, then,

$$C = L \times \frac{1,2,}{1 \cdot i} = L \times \frac{2 \times (2\,i)}{(1 \cdot i)}$$

But,

$$(1 \cdot i)^2 = (1,2)^2 - (2,\ i)^2; \text{ and } (1,\ i) = \sqrt{(1,2)^2 - 2(\ i)^2}$$

As, however, $(1,2) = 2 \times (2\,i)$;

$$(1,\ i) = \sqrt{\{2 \times (2 \cdot i)\}^2 - (2\,i)^2} = \sqrt{4 \times (2\,i)^2 - (2 \cdot i)^2}$$
$$= \sqrt{3 \times (2\,i)^2} = 1.732 \times (2 \cdot i)$$

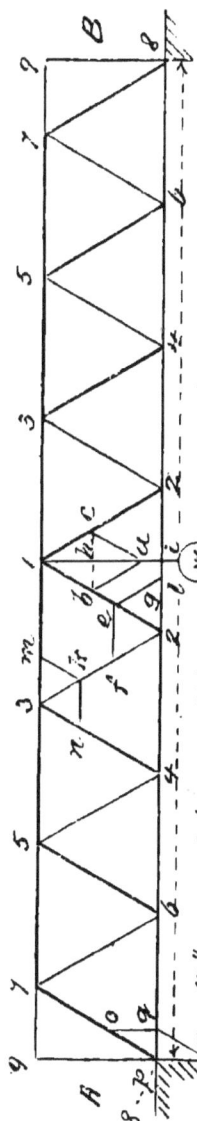

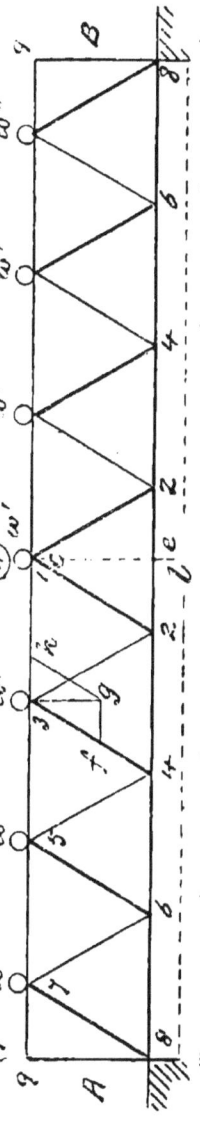

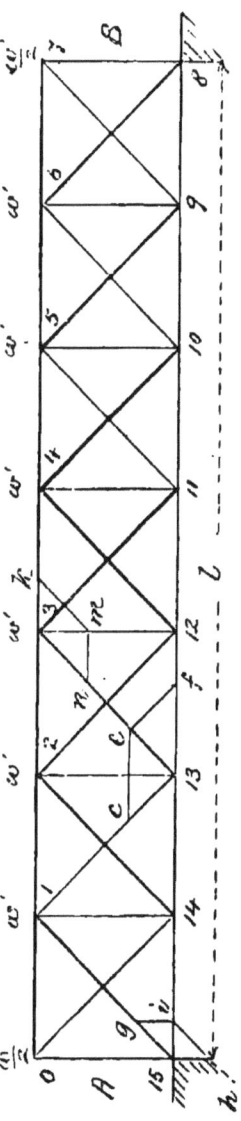

Fig. 19.

Fig. 20.

Fig. 21.

And replacing (1 . i) in the above equation by this value,

$$C = L \times \frac{2 \times (2, i)}{1 \cdot 732 \times (2, i)} = 1 \cdot 154 . L$$

The load, therefore, being known upon any bar at the angle of 60 degrees to the horizon, the stress upon the bar is found by multiplying such load by 1·154. The angles of web-bars are almost invariably either 60 degrees or 45 degrees with the horizon; so a knowledge of the factors for these two types will usually suffice for all cases. Girders with the bars at the latter angle I shall deal with subsequently.

Returning now to the diagram A B, we find on the bar 1, 2, a thrust 1, b; this will cause tension upon the bars 2, 3, and 2, 2; make 2 e = 1 b, and complete the parallelogram 2, f, b, g; then 2 f and 2 g represent the tensile stresses upon the bars 2, 3, and 2, 2. As, however, all the triangles made up by lines drawn parallel to those that form the girder must be similar to them, these also are equilateral, and the lines 2 e, 2 f, and 2 g are all equal, and the tension upon the bars 2, 3, and 2, 2, are each equal to the compression upon the strut 1, 2. The tensile stress on bar 2, 3, will bring compression upon the bars 2, 4, and 2, 1. Make 3 k = 2 f (which = 2 e = 1 b), and complete the parallelogram 3, n, k, m. Then the compressive stress 3 n and 3 m are equal to the tensile stress 3 k (which = 1 b). This may be repeated to the end of the girder with the same results. On every web-bar there is a stress of the same intensity, but alternating in character, and at every junction of the web-bars with the flanges an equal stress is brought upon the latter; those on the top flange come from web-ties and produce compressive stresses, and those upon the bottom flange are caused by struts and create tensile stresses. The stress brought upon the bottom flange by the last strut will be one-half the strut stress, as it is resolved vertically upon the pier (as shown by the parallelogram $o\,p\,r\,q$), and horizontally upon the flange, instead of upon an inclined tie and the flange. The vertical pressure $p\,r$ on the pier will be equal to the load L, which $= \dfrac{W}{2}$.

Now, although the stresses upon the web-bars are constant throughout, it is obvious that those upon the flanges accumulate. Take the top flange, and there are three apices, at which they occur. On the bar 7, 5, there is a stress $= 1\cdot154\,L$; at apex 5, another equal stress is added, making $2\cdot308\,L$; and at apex 3, a similar addition brings the maximum stress upon the top flange to $3\cdot462\,L$. On the bottom flange the stress upon the bar 8, 6, is $\dfrac{1}{2} \times 1\cdot154\,L = 0\cdot577\,L$; on 6, 4, it becomes $= 0\cdot577\,L \times 1\cdot154\,L = 1\cdot731\,L$; on 4, 2, the stress is $2\cdot885\,L$; and on 2, 2, it is $= 4\cdot039\,L$; adding $1\cdot154\,L$ in each case.

A comparison of the central flange-stress in this structure with one of the same span and depth, but having a plate-web, will be instructive.

Let the span be 70 feet, then every side of a triangle will be 10 feet, and the depth 1 i will be,

$$d = \sqrt{(1,2)^2 - (2,i)^2} = \sqrt{10^2 - 5^2} = \sqrt{75} = 8\cdot66 \text{ feet}$$

Referring to $A\,B$, Fig. 9, and the equation for moment of stress under a concentrated load, we find that

$$M = Wy - \frac{Wy^2}{l}$$

WM B STEVENS
MEMORIAL LIBRARY
MANAYUNK

In the present example the load W is in the centre, and, therefore, $y = \dfrac{l}{2}$; replacing y by this value, we get,

$$M = W \times \frac{l}{2} - W \times \frac{l^2}{4\,l} = \frac{W \cdot l}{4}$$

The moment of resistance also is,

$$M = (S \times T) \times \frac{d}{2}$$

Assuming the total resistances of the two flanges to be equal, $S = T$ and

$$M = 2\,T \times \frac{d}{2} = T\,d$$

As the moments of stress and resistance must be equal, therefore,

$$T\,d = \frac{Wl}{4}$$

In our present case $W = 2 \times L$,

$$T\,d = \frac{Wl}{4} = \frac{2 \cdot L \cdot l}{4}$$

Whence,

$$T = \frac{L \cdot l}{3\,d} = \frac{L \times 70}{2 \times 8 \cdot 66} = 4 \cdot 041\ L$$

The slight discrepancy is due to dropping decimals after $1 \cdot 154$, which would run out to $1 \cdot 15415$, &c.; but three places of decimals are quite enough for the questions now under consideration.

This comparison of figures shows that the same flange-stress is found by each method of calculating when the central stress is under consideration; at any other point this equivalence of stress will not obtain, for in the plate-webbed girder the increments of flange-stress occur at close intervals, such as four inches, being in fact the pitch of the rivets, while in the triangular-webbed girder the intervals are long, in the case taken ten feet.   This at once suggests that a plate-girder might be treated as a lattice-girder with its web-bars touching each other, and that a solid rolled girder might be regarded as having a web of inclined bars indefinitely narrow touching each other.   This method works out to the same result as that otherwise obtained, but for such mathematical refinements I have not space, and perhaps my readers would not have the inclination to follow them.

The bars 9, 7, and 9, 8, are not necessary to the construction of the girder, unless it carries its load distributed over the top flange, but under a central load the continuation of the top flange may be required, so that the ends may be built into walls.

Fig. 20 is a diagram of a girder similar to that shown above, but it is assumed to have a uniformly distributed load = w tons per lineal foot of span, such as a wall might put upon it.   This continuous load will put a transverse stress upon the top flange between the points of junction with the web-bars, and will cause that flange to take the duty of a short-span continuous girder, in addition to resisting the horizontal stress brought upon it by the ties in the web.   At present, however, I shall deal only with the direct stresses.

The load upon the top flange may be considered as carried at the apices of the triangles, and the load w' on each apex (except the end ones) will be = w multiplied by the base of a triangle, as 5, 3, and 3, 1.   On the end apices the load will be three-quarters of this, because half the load upon the bar 7, 9, will be carried by the upright 9, 8.

Let the vertical line $c\,e$ be drawn from the apex e through the depth of the girder, then the loads on each side of this line will be symmetrical, and therefore the stress also, so one-half of the girder may be regarded as carrying half the load directly to the support upon which it rests. In the previous example it is noticeable that all bars, in the web, which incline downwards towards the nearest support, are struts, and those inclining downwards towards the centre of the span are ties. This may be taken as an axiom in all cases where the construction and the load are both symmetrical, but not otherwise.

Following the stresses up in this way, those from the central apex load w' will be $1\cdot154 \times \dfrac{w'}{2}$ upon all the web-bars, and these will cause increments of stress of equal intensity at every apex. The next apex load, over 3, goes entirely to the support A, and therefore will send through the bars 3, 4; 4, 5; 5, 6; 6, 7; and 7, 8; stresses of $1\cdot154.w'$; the load over apex 5 will send equal stresses through the bars 5, 6; 6, 7; and 7, 8; the load upon apex 7 will cause a thrust upon bar 7, 8; $= \frac{3}{4} \times 1\cdot154.w' = 0\cdot866\ w'$.

An examination of these results will show that the stresses upon the web-bars increase in direct ratio to the distance from the centre of the span up to the last bar; using — to show tension, and × for compression.

The stress upon bar .   1, 2 $= +0\cdot577w'$.
,,   ,,   . 2, 3 $= -0\cdot577w'$.
,,   ,,   . 3, 4 $= +0\cdot577w' + 1\cdot154w' = +1\cdot731w'$.
,,   ,,   . 4, 5 $= -1\cdot731w'$.
,,   ,,   . 5, 6 $= +1\cdot731w' + 1\cdot154w' = +2\cdot885w'$.
,,   ,,   . 6, 7 $= -2\cdot885w'$.
,,   ,,   . 7, 8 $= +2\cdot885w' + 0\cdot866w' = +3\cdot751w'$.

Until the end apex is reached the stresses run in the

ratios 1, 3, 5 ; so it is only necessary to remember this, and for a uniform load, all the web stresses up to the end strut may be easily calculated from that upon the first one at the centre.

Upon the top flange at each apex there is, in addition to the stress brought upon it by the tie, a stress from the load immediately over it, as shown by the parallelogram 3, f, g, h; where, if $3\ g = w'$, $3\ h$ will equal the increment of horizontal stress upon bar 3, 1. From previous calculations we find $h\ g = 1\cdot154 \times (3\ g)$; but $3\ h = \frac{1}{2}\ h\ g$. Therefore $(3\ h) = 0\cdot577 \times (3\ g) = 0\cdot577\ w'$.

The stresses upon the flanges may now be summarised; those upon the upper flange will all be in compression.

The stress upon bar . 7, 5 = $+ 2\cdot885w' + 0\cdot577w' = + 3\cdot462w'$.
,,   ,,   . 5, 3 = $+ 3\cdot462w' + 1\cdot731w' + 0\cdot577w' = + 5\cdot77$.
,,   ,,   . 3, 1 = $+ 5\cdot774w' + 0\cdot577w' + 0\cdot577w' = + 6\cdot924$.

Upon the bottom flange the stresses are,

The stress upon bar . 8, 6 = $+ \frac{1}{2} \times 3\cdot751w' = - 1\cdot875w'$.
,,   ,,   . 6, 4 = $- 1\cdot875w' - 2\cdot885w' = - 4\cdot76w'$.
,,   ,,   . 4, 2 = $- 4\cdot76w' + 1\cdot731w' = - 6\cdot491w'$.
,,   ,,   . 2, 2 = $- 6\cdot491w' - 0\cdot577w' = - 7\cdot068w'$.

To compare this flange-stress with that found by the expression for the plate-girder, it will be convenient to ignore the action of the upright 9, 8, and consider w'' of the same value as w'.

For the central stress upon the flange of a plate-girder we have

$$T = \frac{Wl}{8 \cdot d} = \frac{W \times 70}{8 \times 8\cdot66}$$

But $W = 7 \cdot w'$, therefore

$$T = \frac{7 \times w' \times 70}{8 \times 8\cdot66} = - 7\cdot07\ w'$$

The slight excess is accounted for by taking w'' as equal to w'.

Such counter-checks as this are of great value in practice, as, although the two results do not exactly agree, any wide variation will indicate some clerical error; and if a man has to check his own work, he can only do it with confidence by using two methods to reach one result.

I will now take the case of a "trellis" girder, Fig. 21, which is assumed to carry a floor with a varying load. This will involve two sets of calculations; one for the dead weight of the floor, and the other to find the maximum stresses brought upon the web-bars by the varying load.

$A B$ shows a "trellis" girder of seven bays—or panels as they are sometimes called. The ties and struts are all placed at an angle of 45 degrees to the horizon, and at every point of junction there is a vertical bar to distribute the load equally between the two series of triangles which form the web. These vertical bars are not necessary to the construction, and are commonly omitted in small girders; but in moderate-sized girders they are so formed as to act as vertical stiffeners. When the stiffeners are not used the summit loads w' come alternately upon each series of triangles. From the angles of the bars to the horizon being 45 degrees, it follows that each panel is a square, of which the ratio of the diagonal line to one side is as $\sqrt{2} : 1$. That is, as 1·414 is to 1. This, therefore, is the ratio of the stress upon any bar to the load it carries.

I will take an actual example in this and work it out fully. Let the girder shown be one of a pair carrying a gallery or bridge from one side of a railway-station to the other across the railway lines and platforms; let the span be 40 feet, and the width in the clear between the girders 12 feet; then each main girder will have to carry half the floor area, which will be $40 \times 6 = 240$ square feet. For the weight of passengers 120 lbs. per square foot must be allowed. We shall never get a load equal to this, but an allowance is made for the jar and vibration due to hurrying crowds of people ; the probable load will not exceed 70 lbs. per square foot. If the floor is made of iron plates covered by concrete and asphalte four inches thick, its weight may be taken at 35 lbs. per square foot ; then the total distributed load will be, for each main girder, $240 \times \{120 + 35\} = 37,200$ lbs. $= 16\cdot6$ tons, which is equal to $0\cdot415$ tons per lineal foot of span. The weight of the girder itself will be about $0\cdot035$ tons per foot run for this load and ratio of depth to span. This brings the total load per lineal foot to $0\cdot415 + 0\cdot035 = 0\cdot45$ tons, of which $0\cdot321$ ton is live or moving load, and $0\cdot129$ is permanent or dead load. The maximum stresses upon the flanges occur under a full load, and with that I shall first deal.

The distance from apex to apex is evidently $40 \div 7 = 5\cdot714$ feet, which is also the depth of the girder. The load $w' = 5\cdot714 \times 0\cdot45 = 2\cdot571$ tons. This is divided by the uprights between the two series of triangles, giving the increment of weight at every apex as $2\cdot571 \div 2 = 1\cdot285$ tons. The increment of *stress* upon the web-bar will therefore be $= 1\cdot285 \times 1\cdot414 = 1\cdot817$ tons.

In the previous case there was an apex at the centre (c) of the span, so a half load went each way ; but in the present example this does not, and each load goes in its entirety to the nearest point of support. Under a uniform or any symmetrical load there will be no stress at all upon the bars 3, 11, and 4, 12 ; but on the others, under a full load the stresses will be, in tons,

The stress upon bar . 3, $13 = +1 \cdot 817$.

,,   ,,   . 13, $1 = -\{1 \cdot 817 + 1 \cdot 817\} = -3 \cdot 634$.

,,   ,,   . 1, $15 = +\{3 \cdot 634 + 1 \cdot 817\} = +5 \cdot 451$.

,,   ,,   . 12, $2 = +1 \cdot 817$.

,,   ,,   . 2, $14 = +\{1 \cdot 817 + 1 \cdot 817\} = +3 \cdot 634$.

,,   ,,   . 14, $0 = -\{3 + 3 \cdot 634 + 1 \cdot 8173\} = -5 \cdot 451$.

These stresses, it will be noticed, increase as the numbers 1, 2, 3, &c.

The relation of the flange-stress to that on the web-bar is shown by the parallelogram c, 13, f, e; in which e 13 is made equal to the stress upon the bar 3, 13; when 13 f will represent the flange-stress to which it gives rise. The triangle 13, e, f is similar to the triangle 13, 3, 11; and is half a square in which the diagonal represents the flange-stress when one side is equal to the web-bar stress, therefore, from the ratio of the diagonal to the side of a square, the flange-stress will be equal to the stress on the web-bar multiplied by $1 \cdot 414$. The last strut and tie, however, will only bring half this stress upon the flange, as shown by the parallelogram h, i, g, 15; in which the diagonal represents the web-stress, and one side of the square that on the flange, which involves dividing by $1 \cdot 414$ instead of multiplying by it. At every apex there is also another stress put upon the flange, as shown by the parallelogram 3, n, m, k. If 3 m represents the load $\frac{w'}{2}$, then 3 k will be the stress put by it immediately on the bar 3, 4. As 3 m and 3 k are sides of the same square, they are equal, and this increment of stress is equal to the load. The same will occur at the other apices. The stresses will be of the same intensities on both flanges, but those on the top are compressive, and those on the bottom flange are tensile. The following are the stresses, in tons, on the flanges :—

The stress upon bars   0, 1 ; and 15, 14 $= 5\cdot451 + 1\cdot414 = 3\cdot855$.

    ,,   ,,   1, 2 ; and 14, 13 $= 3\cdot855 + \{3\cdot634 \times 1\cdot414\} + 1\cdot285 = 10\cdot278$.

    ,,   ,,   2, 3 ; and 13, 12 $= 10\cdot278 + \{1\cdot817 \times 1\cdot414\} + 1\cdot285 = 14\cdot132$.

    ,,   ,,   3, 4 ; and 12, 11 $= 14\cdot132 + 1\cdot185 = 15\cdot417$.

We can check this result also by the principle of moments. The load at 0 is carried directly on to the pier A by the upright 0, 15, and does not affect the stresses upon the structure between the supports.

The reaction of the pier A that does affect the structure is equal to $3\ w' = 3 \times 2\cdot571 = 7\cdot713$ tons for a point at the centre ; this acts at the distance of half the span, or 20 feet, so the upward moment will be $7\cdot713 \times 20 = 154\cdot26$ foot-tons. The three loads between the pier and the centre will exert downward moments equal in all to $2\cdot571 + 14\cdot285 + 2\cdot571 + 8\cdot571 + 2\cdot571 + 2\cdot857 = 66\cdot108$ foot-tons. The difference between these moments is $154\cdot26 - 66\cdot108 = 88\cdot152$ foot-tons. This divided by the depth, $5\cdot714$ feet, gives the direct stress at the centre of either flange :

$$\frac{88\cdot152}{5\cdot714} = 15\cdot42 \text{ tons,}$$

which corroborates the previous calculation. The stresses upon the other half of the girder will be symmetrical with those just determined.

I will now take the stresses upon the web-bars, due to the dead load. The dead load per lineal foot of span is $0\cdot129$ tons, and the load per apex $= 0\cdot129 \times 5\cdot714 = 0\cdot737$ ton. The increment of load on each web-bar is half this, or $0\cdot368$ ton, and the increment of stress $= 0\cdot368 \times 1\cdot414 = 0\cdot52$ ton. As other stresses will subsequently be added to those now about to be tabulated, the signs $+$ and $-$ must be used to signify compression and extension respectively ; the intensities of the stresses upon each side of the centre of the span are symmetrical, therefore :

The stress upon bars . 3, 11 ; and 4, 12 = 0.
,,   ,,   . 3, 13 ; and 4, 10 = + 0·52.
,,   ,,   . 12, 2 ; and 11, 5 = − 0·52.
,,   ,,   . 2, 14 ; and 5, 9 = + 0·52 × 2 = + 1·04.
,,   ,,   . 13, 1 ; and 10, 6 = − 1·04.
,,   ,,   . 1, 15 ; and 6, 8 = + 0·52 × 3 = + 1·56.
,,   ,,   ·. 14, 0 ; and 9, 7 = − 1·56.

The live load is 0·321 ton per lineal foot, and the corresponding load per apex $= 0.321 \times 5.714 = 1.834$ tons. The increment of stress will be $= \dfrac{1.834}{2} \times 1.414 = 1.597$ tons.

I shall consider the load as first covering the whole length of the girder and then gradually receding, and calculate the stresses upon the web-bars, commencing from the end A from which the load is supposed to move towards B. With the full load, there will be three apex loads to be carried by the bars 1, 15, and 14, 0 ; that is, $3 \times 1.597 = 4.791$, this being tension upon 14, 0 ; and compression on 1, 15. Now let the load move back from apex 1 ; then the centre of gravity of the four remaining apex loads will be at apex 4, as one is over it and two on each side ; the proportion of load, therefore, passing to the pier A will be, by the principle of moments :

$$= 5 \times 1.834 \times \frac{3 \text{ bays}}{7 \text{ bays}} = 4.073 \text{ tons,}$$

from which the stress upon bars 2, 14 ; and 1, 15 $=$

$$+ \frac{4.073}{2} \times 1.414 = + 2.879,$$

and the stress upon bars 13, 1 ; and 14, 0 $= - 2.879$.

Taking the load back another bay and measuring by bays

as before, we find the centre of gravity of the remaining loads midway between the apices 4 and 5; therefore, the load upon A will be

$$= 4 \times 1\cdot834 \times \frac{2\cdot5}{7} = 2\cdot62 \text{ tons};$$

the increments of stress will be:

on bars 3, 13; 2, 14; and 1, 15 $= + \dfrac{2\cdot62}{2} \times 1\cdot414$

$$= + 1\cdot852$$

„   12, 2; 13, 1; and 14, 0 $= -1\cdot852$

Going back another bay, the centre of gravity of the remaining loads is at apex 5, and the load passing to A is:

$$= 3 \times 1\cdot834 \times \frac{2}{7} = 1\cdot572 \text{ tons};$$

the increments of stress will be:

on bars 4, 12; 3, 13; 2, 14; and 1, 15 $= + \dfrac{1\cdot572}{2} \times 1\cdot414$

$$= + 1\cdot111$$

„   11, 3; 12, 2; 13, 1; and 14, 0 $= -1\cdot111.$

Here we at once observe that there are stresses upon the bars 4, 12, and 11, 3, which under a uniformly distributed load suffer no stress at all.

At the next step the centre of gravity of load recedes to midway between apices 5 and 6, and the load on A is:—

$$= 2 \times 1\cdot834 \times \frac{1\cdot5}{7} = 0\cdot786 \text{ ton};$$

the increments of stress will be:

on bars 5, 11; 4, 12; 3 13; 2, 14; and 1, 15 $= + \dfrac{0\cdot786}{2}$

$$\times 1\cdot414 = + 0\cdot556$$

„   10, 4; 11, 3; 12, 2; 13, 1; and 14, 0 $= -0\cdot556$

The last movement before the disappearance of the live load leaves one apex load at 6, and of this there passes to A

$$1\cdot834 \times \frac{1}{7} = 0\cdot262 \text{ ton,}$$

from which the stresses are,

on bars 6, 10 ; 5, 11 ; 4, 12 ; 3, 13 ; 2, 14 ; and 1, 15

$$= + \frac{0\cdot262}{2} \times 1\cdot414 = + 0\cdot185$$

„   9, 5 ; 10, 4 ; 11, 3 ; 12, 2 ; 13, 1 ; and 14, 0

$$= - 0\cdot185$$

It seems advisable here to point out how it is that a maximum load gives maximum stresses upon the flanges but not upon all the bars of the web—it is this, that the characters of the stresses upon the flanges do not change, the top flange is always in compression and the bottom flange always in tension ; but it is found that in the web-bars the stress produced by a partial load on any particular bar is not always of the same sign as that accruing under a maximum load.   As both series of triangles are under equal stresses, through the equalising uprights distributing the loads between them, only one series need be considered.   I will take that one which runs from 0 to 8.   The bar 14, 0 ; must always be in tension, because no stress can be brought upon it except at the lower end, and its maximum stress is under a full load.

Bar 14, 2, with the first set back of live load, has a stress from live load $= + 2\cdot879$, and from dead load $+ 1\cdot04$, total $2\cdot879 + 1\cdot04 = + 3\cdot919$ tons.   Under the full load the stress was $+ 3\cdot634$ tons, so the varying load stress must be taken in determining the strength of this bar although it remains in compression.

Proceeding in the same way, bar 2, 12, has a stress from the live load $= -1.852$ tons, and from the dead load $- 0.52$; total $- 1.852 - 0.52 = - 2.372$ tons. Under the full load the stress was $- 1.817$ tons, so here again the varying load stress must be used.

Bar 12, 4, has a stress from the live load $= + 0.556$ ton; from the dead load there is no stress, nor was there any from the full load.

Bar 4, 10, has a stress from the live load $= - 0.556$ ton, and from the dead load $+ 0.52$ (this being the same as that upon bar 3, 13); total $= - 0.556 + 0.52 = - 0.036$ ton, so that one stress nearly counterbalances the other.

Bar 10, 6, has a stress from the live load $=$ 0.185 ton, and from the dead load $- 1.04$; total $+ 0.185 - 1.04 = - 0.855$; so these bars behind apex 4 are to be calculated from the stresses accruing from a full load.

In the example under consideration the stresses are altogether so slight that their variations are not of importance in view of the sections which must in practice be used, but even in this case the investigation shows that the bars 3, 11; and 4, 12, must not be omitted. In heavily loaded girders of greater span, every difference will have to be considered in determining the sections.

It may fairly be said that there is always more satisfaction in designing a large and heavily loaded girder than a small one, because in the former there is a chance of approximating very closely to the calculated sectional areas, which does not offer in the latter case, but it happens that in railway station work a great number of these lightly loaded structures are required, and therefore it is necessary to deal with them.

I

Now, coming to the practical part of the question, the sections absolutely requisite are very small, especially in the flanges, as might be expected from the great depth of girder compared with the span.

This depth, however, is necessary to afford protective fencings on each side of the bridge, for it would be wasteful to put in shallow girders where it would be necessary to mount parapets upon their tops.

The maximum stress upon the flanges—that at the centre—is only 15·417 tons ; and this would require in the top flange a sectional area

$$= \frac{15 \cdot 417}{4} = 3 \cdot 854 \text{ square inches.}$$

This flange could be easily made of one tee-iron, but before deciding this matter the end struts and ties must be dealt with, in order to ascertain what rivets will be required to connect them with the flanges ; for if tee-iron is used each rivet connecting a trellis-bar with it will be in single shear, whereas by using two angle-irons the rivets may each be brought into double shear.

The stresses upon the end trellis-bars are, at the maximum, 5·451 tons. This in tension would only require a nett sectional area

$$= \frac{5 \cdot 451}{5} = 1 \cdot 09 \text{ square inches.}$$

But in compression, where the bending element occurs, a larger sectional area is needed.

For the present purpose it is sufficient to determine the rivet area required in the connections of the end struts and ties with the flanges. This will be 5·451 ÷ 4 = 1·383 square inches. A rivet ¾ inch in diameter is quite

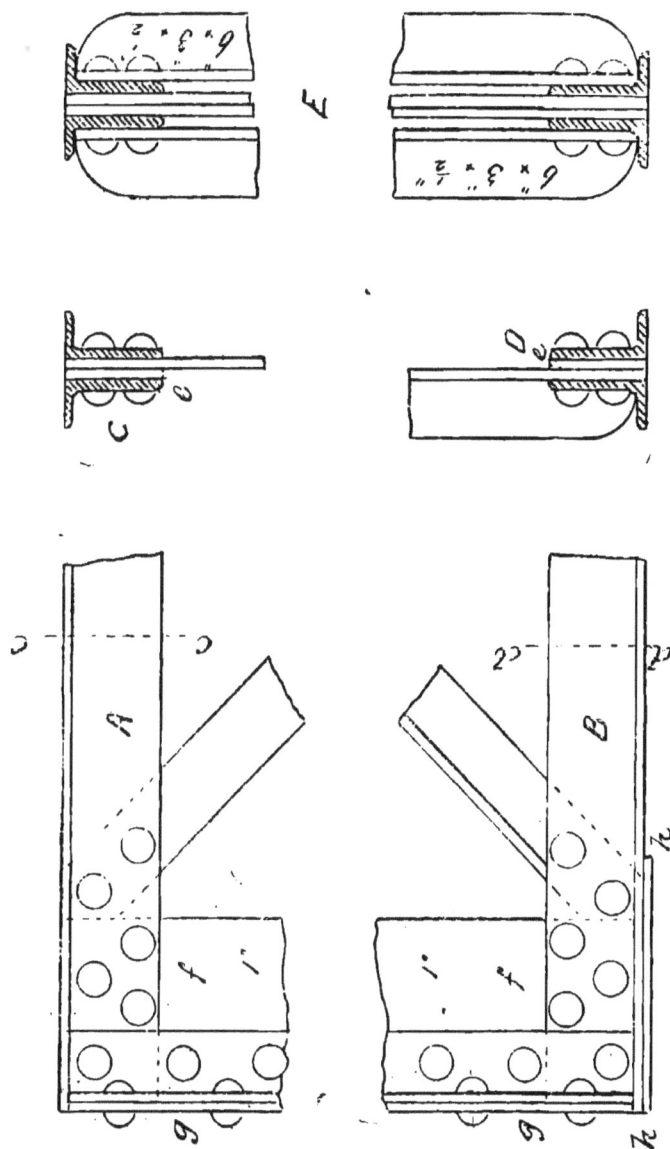

Fig. 22.

small enough to use in work of this class; the number
of rivet areas required will be 1·383 ÷ 0·44 = 3·14; that
is, four rivet sections are required.

Considering the length of the strut, about two tons
per sectional square inch is all that may probably be
allowed, so this may be taken in the first instance to determine its sectional area, which will be 5·451 ÷ 2 =
2·725 square inches. An angle-iron 3 in. × 3 in. × ½ in.
will give the required sectional area, and one limb may
be cut away where it joins the flange, as there will be no
bending stress. The tie bar will require a sectional area
= 5·451 ÷ 5 = 1·09 sq. inches nett. A 3 in. × ½ in. bar
will give, with loss by one ¾ in. rivet, 2·25 × ½ = 1·125
sq. inches nett; and these sections being adopted the strut
and tie will each be three inches wide, and their ends half
an inch thick. This thickness will be used throughout the
girder; then where the ends of ties and struts meet at the
flanges, they will cross and pass between two angle-irons;
the uprights, which also act as stiffeners, being riveted outside the angle-irons.

In Fig. 22, details of the top and bottom end joints are
shown at A and B. The flanges of the girder are each
formed of two angle-irons 2 in. × 3½ in. × ½ in. thick
placed back to back, with the web-bars between them; C
and D show vertical sections on the lines c c and d d; as
the end bars are not crossed by others, the space between
the angle-irons is made up by packing e e.

The ends of the flanges are connected together by a web
plate f f, one inch thick, or of two half-inch plates, and the
end of the girder is finished off by a plate g g, 5 inches
wide, connected with the web plate by angle-irons 3 in. ×
3 in. × ½ in., riveted to both as shown. In order to
stiffen the girder vertically, the upright bars at the junc-

tious of the web bars with the flanges may consist each of two tee-irons 6 in. × 3 in. × ½ in. riveted, one on each side, and back to back, outside the main angle-irons, as shown in the flange section *E*.

The bars in the centre panel may be in tension under one variation of load, and in compression under another, and they should therefore both be made of angle-irons to secure rigidity; the other bars inclining downwards towards the centre of the span may be flat. The struts which incline upwards from the centre will of course be angle-irons. Where the ties and struts cross they should be riveted together to resist lateral vibration.

If the load is distributed equally along the top flange of the girder then that flange will act not only to resist the direct stress of compression, but also the compressive and tensile stresses due to the load between any two successive apices, and the whole flange fully loaded will form a continuous girder, so that each separate span may be regarded as having its ends fixed; under which circumstances the maximum bending moment will be $= \dfrac{Wl}{12} = \dfrac{w \cdot l^2}{12}$ over any apex. *l* in this expression is the distance between any two successive apices.

If the element is already under transverse stress and receives a direct compressive force, this will increase the compressive stress on the part of the section above the neutral axis and reduce that on the part below, and it will there partly counterbalance the tensile element of the transverse stress.

The neutral axis of any section is found, so long as the limit of elasticity is not closely approached, to practically coincide with the centre of gravity, and we shall see that, on account of the form of flange used, the compressive

stress does not reach its working maximum when the tensile stress does. A simple tee-section, such as that shown
in Fig. 23, will illustrate my meaning. This section is
assumed to be 6 in. × 6 in. × ⅜ in. thick. This section
may be regarded as comprising the three rectangles $a\,c\,f\,k$,
$k\,h\,i\,m$, and $m\,g\,e\,b$. If we calculate the moments of area
of these rectangles about the line $a\,b$ as an axis and then
divide their sum by the total area, the quotient will be the
distance from the axis $a\,b$ to a line $o\,o$, passing through the
centre of gravity of the whole section. The rectangles
being symmetrical figures, the centre of gravity of each
one will coincide with its centre of figure; thus for the
rectangle $k\,h\,i\,m$ the centre of gravity will be 3 inches
from the line $a\,b$, and for each of the other rectangles the
distance will be ⅜ in. = 0·375 in. The sum then of the
moments of area will be thus made up :

For rectangle $k\,h\,i\,m$.   .   .   .   . $6 \times 0·75 \times 3 = 13·500$
 ,, rectangles $a\,c\,f\,k$ and $m\,g\,e\,b$. $2 \times 2·625 \times 0·75 \times 0·375 =$ 1·476
                                                                14·976

The total area of the section is $6 \times 0·75 + 2 \times 2·625 \times 0·75$
$= 8·437$. The distance of the centre of gravity (and
neutral axis) from the line $a\,b$ will be,

$$\frac{14·976}{8·437} = 1·775 \text{ inches.}$$

This will therefore be the maximum distance of metal
under compressive stress from the neutral axis ; and the
maximum distance of metal on the tension side from the
neutral axis will be

$$= 6 - 1·775 = 4·225 \text{ inches.}$$

If, then, the maximum working tensile stress is 5 tons per

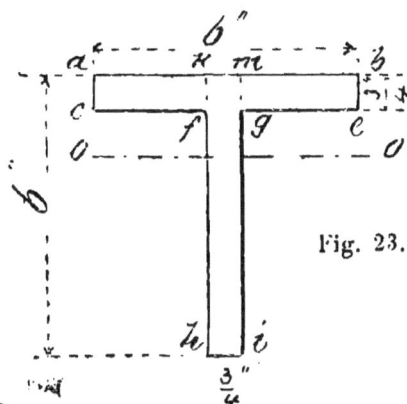

Fig. 23.

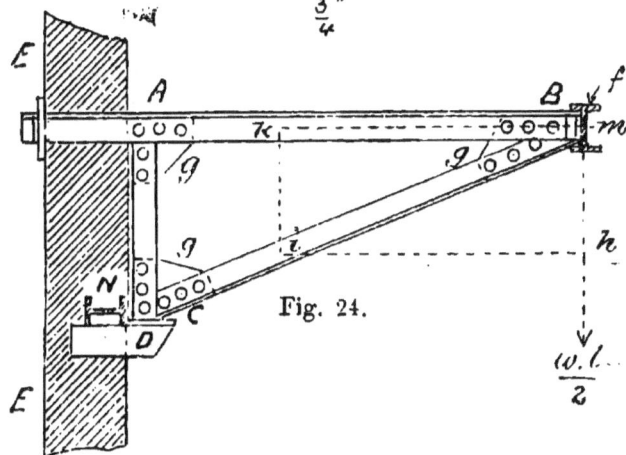

Fig. 24.

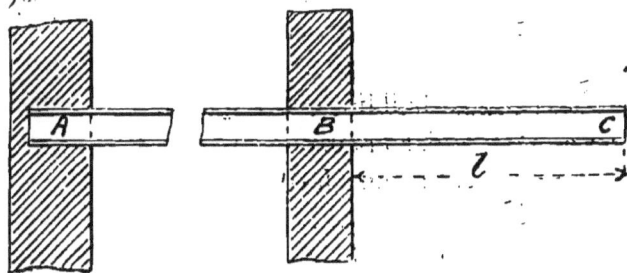

Fig. 25.

sectional square inch, the maximum compressive stress
will be

$$= 5 \times \frac{1 \cdot 775}{4 \cdot 225} = 2 \cdot 1 \text{ tons per square inch.}$$

So the metal above the neutral axis can still take 2 tons
per sectional square inch of direct stress. Below the neutral
axis there will be the tensile stress to deduct from the
direct compression before any compression comes upon the
material. It is most convenient to calculate the stresses
from transverse load first, and then adjust the section
to meet the direct stresses as well. These calculations
must be separately made for each particular case.

If the load were upon the bottom flange, similar results
ensue, but the characters of the stresses are reversed
throughout ; the tensile stress is less than the compressive under the transverse load, and the direct stress is
tensile.

When the apices are close together—two or three feet
apart—this transverse stress may be neglected as the margin
of strength allowed for direct stress will cover it, but it is
always best to bring the load on to the girder at the junction of the web-bars with the flanges if practicable ; this
can readily be done in the case of a girder carrying a
bridge floor, but not with one under a wall, and for this
reason a plate girder is preferable to carry a wall.

It frequently happens that galleries are required running
along the walls of buildings, and where there is sufficient
headway, a triangular bracket of the form shown at Fig.
24 is as convenient as any form of cantilever that can be
adopted; it is very rigid, and does not allow any perceptible
drop at the outer end. It consists of a horizontal member,
$A\ B$, which must be tied into the supporting wall $E\ E$ by
extending it backwards into the wall and building in, or,

better, taking it right through the wall and cleating it to a flat plate e, with angle-iron cleats ; the plate e then bears against the wall and cannot be drawn through it. The outer end of the horizontal member is supported by a strut, $B\,C$, of which the foot, C, rests upon a corbel, D, built into the wall. This foot is connected to the top member by an upright $A\,C$, and so a complete triangular frame is formed. As shown the three principal members consist each of two angle-irons placed back to back and connected together by joint-plates g, riveted between them. The ends of the brackets should be cleated by angle-irons to a facia girder f, running the length of the gallery.

Let $l =$ the distance from the wall to the facia girder f; and $w =$ the load per lineal foot, then the total load will be $w\,l$, and of this half will be taken at A and half at B ; the former may be borne wholly by the wall at A, or it may pass partly or entirely through the upright $A\,C$ to the corbel D. To find the direct stresses let fall from the intersection of the centre lines of the members $A\,B$ and $B\,C$ a vertical line and on it mark off $m\,h = \dfrac{w\,l}{2}$. This load produces two stresses, one in tension upon $A\,B$ and the other compression upon the strut $B\,C$. Complete the parallelogram $m\,h\,i\,k$; then $m\,k$ will represent the tension upon $A\,B$, and $m\,i$ will represent the thrust upon the strut $B\,C$. $ki = m\,h = \dfrac{w\,l}{2}$. The triangle $C\,A\,B$ is similar to the triangle $i\,k\,m$, therefore the distribution of stresses may be found from a diagram formed by the centre lines of the bracket itself. Taking the letters A, B, and C to apply to the intersections of these centre lines, the tension upon $A\,B$ will be

$$= \frac{w\,.\,l}{2} \times \frac{A\,B}{A\,C}$$

the compressive stress upon $B\,C$

$$= \frac{w\,.\,l}{2} \times \frac{B\,C}{A\,C}$$

Let $A\,B = 8$ feet, which is also the clear width of the gallery, and $A\,C = 3$ feet; and let the brackets occur at distances apart of ten feet throughout the length of the gallery. From the principles of right-angled triangles, the length of the strut will be :

$$B\,C = \sqrt{(A\,B)^2 + (A\,C)^2} = \sqrt{8^2 + 3^2} = 8\!\cdot\!544 \text{ feet.}$$

If this gallery is provided as a means for operatives to leave a factory after work, or as an escape in case of fire, it will be liable to the heaviest load which a crowd of human beings can bring upon it; this is about 1 cwt. per superficial foot, but in order to allow a margin for the vibration caused by a moving load it may be taken as $1\frac{1}{2}$ cwt. per superficial foot; then the load upon each bracket will be :

$$= 8 \times 10 \times 1\tfrac{1}{2} \text{ cwt.} = 120 \text{ cwt.} = 6 \text{ tons,}$$

and therefore,

$$\frac{w\,l}{2} = \frac{6}{2} = 3 \text{ tons.}$$

The stress upon $A\,B$

$$= 3 \times \frac{8}{3} = 8 \text{ tons.}$$

and that upon $B\,C$

$$= 3 \times \frac{8\!\cdot\!544}{3} = 8\!\cdot\!544 \text{ tons.}$$

It will be convenient, for practical reasons, to keep all the angle-irons in the frame of the same size, and therefore the size of the largest will be calculated. This will obviously be the strut $B\,C$, which besides having the greatest stress has also to be calculated as a stanchion in reference to its ratio of width to length.

If $r =$ the ratio of length to width of an angle channel or tee-iron, the breaking stress per sectional square inch, acting in compression, is :

$$= 19 + \left\{ 1 + \frac{r^2}{900} \right\}$$

Let the angle-irons in the brackets be 3 in. ; then 3 in. × 3 in. $= 0\cdot25$ ft. $=$ the width, and

$$r = \frac{8\cdot544}{0\cdot25} = 34\cdot176 \; ; \; r^2 = 1168,$$

then the breaking or crippling stress per sectional square inch will be

$$= 19 + \left\{ 1 + \frac{1168}{900} \right\} = 8\cdot268 \text{ tons.}$$

Taking 4 as a factor of safety, the working stress per square inch is $8\cdot268 \div 4 = 2\cdot067$ tons per square inch ; the thrust upon the strut is $8\cdot544$ tons, and therefore its gross area must be $8\cdot544 \div 2\cdot067 = 4\cdot13$ square inches. If the two angle-irons are each $\frac{3}{8}$ in. thick, the sum of their gross areas will be :

$$2 \times \left\{ 3 + (3 - \tfrac{3}{8}) \right\} \times \tfrac{3}{8} = 4\cdot218 \text{ square inches,}$$

which covers the required sectional area.

The tensile stress upon $A B$ is 8 tons; therefore, taking 5 tons per sectional square inch as the working stress, the nett sectional area required will be $8 \div 5 = 1\cdot6$ square inches. There will be a rivet hole $\frac{3}{4}$ inch diameter in one limb of each angle-iron—if there are holes in the horizontal limb to be attached to the floor by bolts to the brackets these may be arranged to alternate with those in the vertical limbs, so that there will be only one hole in any vertical section of one angle-iron, then the nett sectional area of the two angle-irons will be :

$$2 \times \{3 + (3 - \tfrac{3}{8} - \tfrac{3}{4})\} \times \tfrac{3}{8} = 3\cdot663 \text{ square inches.}$$

So there is plenty of sectional area to spare, and holes may be put wherever desired in the horizontal limbs. The upright $A\,C$ is shorter than the strut $B\,C$ and therefore will be stronger, and it cannot in any case have a greater load than $\dfrac{w \cdot l}{2}$, and that will only accrue when the tail end of the member $A\,B$ does not rest upon the wall at all.

It is observed that there is tension amounting to 8 tons upon the member $A\,B$; this will also be the pull upon the wall $E\,E$; there will be an equal thrust against the wall at C, so there will be an effort to break the wall off at C, or, if it is strong enough to resist this, a further effort to pull the wall over, by reason of the overhanging weight.

It is to be assumed that there will be a floor within the building at the same level as the gallery, and in this case a tie rod can be taken through such floor, and the plate e made to bear against an internal, or perhaps a back wall, and so secure the front wall from the danger of being pulled out.

In some cases where the stress upon the ties has been very considerable, they have been carried back to a girder fixed across the building, so as to bear its ends against the side walls which run parallel to the ties, and this adds very much to the expense.

There yet remains the thrust at C to be met, of which the tendency will be to push the wall in and so cause a general collapse. If the internal walls running at right angles to the wall $E\,E$ are not too far apart, a rolled girder N may be built into the wall $E\,E$ in the position shown, so that its back flange will bear against the edges of the internal walls, which will then act as buttresses to resist the thrusts from the struts of the external brackets.

If these means of stiffening or relieving the front wall are not available, the gallery must be carried upon canti-levers, $B\,C$, Fig. 25, which are formed by the ends of girders, $A\,B\,C$, of which the back ends are built into walls sufficiently far back to give the requisite hold. If there is a floor carried upon the part $A\,B$, this will also help to balance the load on $B\,C$, and vertical stress only will come upon the wall. Under a uniform load upon the canti-lever $B\,C, = W$, the centre of gravity will be distant $\dfrac{l}{2}$ from B, and therefore the moment of stress at that point will be

$$M = \frac{W\,l}{2}$$

With a load W at the free end C, the moment of stress would be $M = W\,l$.

CHAPTER V.

MANUFACTURE—CONNECTIONS.

IT would be quite out of question for an architectural pupil or student to go through a practical course of construction in the foundry and girder yard—the only way to obtain a thorough knowledge of constructional ironwork—but it is highly necessary that he should have a general idea of the processes carried on there, and this I will now endeavour to impart to my readers. Beginning with the foundry we find that the castings with which we shall have to deal are made to certain models or "patterns" made of wood of a slightly larger size than the required work; this is to allow for shrinkage in cooling, which, in cast iron, amounts to one-tenth of an inch per foot (a column 10 feet high would require a pattern 10 feet 1 inch long). The pattern-maker is not called upon to calculate this difference, as pattern-makers' special rules are made which give the excess of length in the reading; thus, if a length of one foot and one-tenth of an inch is divided into twelve parts and each part again divided into eighths and sixteenths, we shall get a rule with all pattern-makers' measurements on it. Of course the drawing supplied to the shop should have every dimension figured clearly upon it, otherwise the workman will have to measure off it with a standard rule or scale, and between that and his contraction rule he may make a mistake.

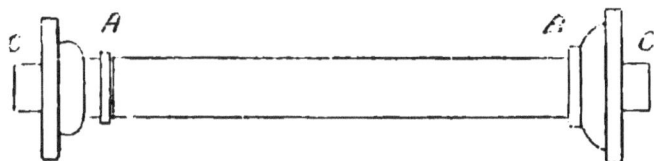

Fig. 26.

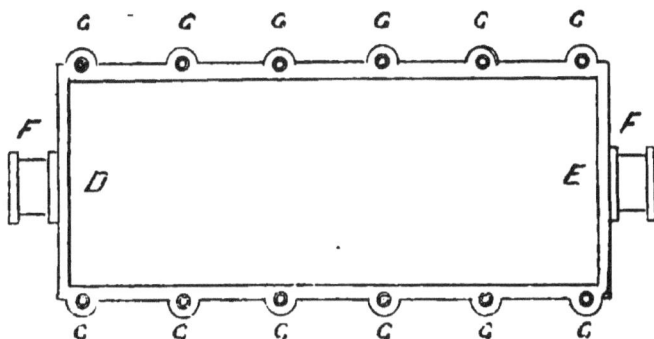

Fig. 27.

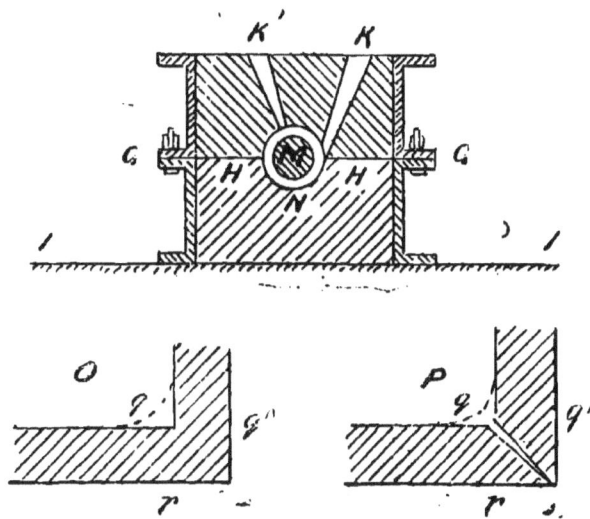

Fig. 28.

The general process of casting is as follows :—Two boxes without either top or bottom, called "flasks," are so made that one will fit to the top of the other, being kept in exact juxtaposition by pins in the sides of one flask which enter holes in lugs on the sides of the other.

Fig. 26 shows a pattern of a column, $A B$; this is carefully turned in a lathe and the surface smoothed off so that the moulder's sand shall not adhere to it. At each end is a round projection called a core print, which makes a recess in the sand to receive the end of the core which is inserted to form the hollow in the column.

A plan of a flask is shown at Fig. 27. It consists of a rectangular frame $D E$, with journals or trunnions $F F$ at its ends, to allow of its being easily turned over on end supports without jarring or shaking about. It may have ribs inside to give a hold for the sand when the casting is a large one. Along the sides of the flask are cast lugs G, which are fitted with pins in one box and corresponding holes in that which pairs with it. This box is set on the foundry floor and filled with green sand, smoothed off at the top surface. A cavity is scooped in it and the pattern $A B$, Fig. 26, imbedded to one half its depth. Another flask is fitted to this, and filled in with sand which is carefully pressed down around the upper half of the pattern so as to receive an exact impression of its form ; the impression in the lower box will necessarily be defective as the pattern was merely pressed down in a roughly scooped cavity. The flasks are now turned over so that the under one comes on the top ; this is taken off and the sand knocked out, when it is replaced and carefully filled, so that in it also is an exact impression of half the pattern. This time, however, certain pegs K, K', Fig. 28, are placed in the sand to form channels for influx of the molten metal, and others for the escape of the air from the mould and also of the gases driven out of the sand by the heat. Great care should be exercised to drive out all bubbles, which if left in form honeycomb castings and render the work useless. Fig. 28 shows a transverse section of the flasks when ready to receive the metal in the annular space N, between the sand in the boxes and the solid core M, which is supported in the core prints at each end of the mould; the whole rests upon the floor II. The "git" K should extend to the bottom of the mould for the inflow, and gits K' be placed at the top for the outlets. Further facilities for the escape of gases from the sand heated by the casting may be afforded by piercing it with stiff wires to within a short distance of the cavity. The cores are made of tough loam mixed with straw wound upon a bar and turned true by causing the bar to revolve while a scraper is applied to its surface ; they are dried in ovens before being placed in the moulds. If the cores are very long they may be supported in the mould by broad-headed nails which rest upon the bottom of the mould, and then they will also require other nails on the top to prevent the cores from being floated up by the superior gravity of the metal. When the moulding work has been finished over the pattern the top flask is lifted, and the pattern tapped to loosen it from the sand below, it is then lifted out, and the surfaces of the mould repaired with trowels if anywhere scarred or broken. The adhesion of the sand between the two boxes on the line HH is prevented by sprinkling the lower part of the mould with coal dust or parting-sand, before putting in the green sand in the upper flask. The pattern having been removed, the core is put in place and the upper flask replaced and secured by cotters passing through slots in the pins carried by the lugs G, and the mould is then ready as shown in Fig. 28.

A column with a plain cap is a very simple matter to mould, but when the cap carries a complicated stilting the matter is not so easy, and special cores may be requisite in some cases. These will not be cores necessarily passing right through the casting, but may be small pieces to form undercuts or cavities, the part of the mould corresponding to which would come away with the pattern when it is drawn out of the sand. The designer should use all diligence to avoid where possible the occurrence of these undercuts and try to so shape his work that the pattern may draw clear out of the sand. In some intricate cases the patterns are made in separate pieces and held together by pins, so that they may be drawn out of the mould piecemeal. When special cores are used the pattern-maker prepares wooden core-boxes, in which they receive their required forms respectively.

All the re-entering angles must be rounded off, for any internal corners will be very likely to cause cracks in the casting. The reason of this is shown at O and P, Fig. 28. At O the metal is supposed to have just set; then as it cools and therefore contracts the thicknesses, $q\,q'$ and $q\,r$ will decrease and tend to draw open the inner angle as shown at P, thus starting a crack in the direction of $q\,s$. If, however, the corner is filled in as shown by the dotted lines, the contraction will only cause a sharpening of the curvature. In small castings these inside angles may be filled up with a composition made of yellow wax and Burgundy pitch, but for patterns of any size, fillets of wood properly hollowed out to form round corners are used.

In any casting where there is no obstruction, the air and gas bubbles will rise freely to the top, and therefore it is desirable to cast columns in an upright position; as when lying horizontally or at an inclination to the horizon, the core prevents the bubbles from rising vertically, and in the horizontal position those that form after the mould is full are very apt to collect in the top part of the mould and there remain, thereby causing one side of the column to be honeycombed, and therefore much weaker than the other side. This is a very serious defect, because not only is there a loss of direct strength but also an increased tendency to curvature from one side compressing more than the other, and thus augmenting the transverse stress. It should therefore be insisted that some inclination shall be given to the flasks, and as much as possible should be stipulated for; in point of fact, it will often be the best policy to pay a better price and get the columns made at some foundry where there are facilities for casting columns vertically.

Another point of practical importance is the variation in thickness in castings. If there is a wide difference in the thickness of the different parts of a casting, it must be allowed plenty of time to cool gradually; this, however, may or may not be given to it; it is best to avoid great and especially sudden changes of thickness, even if a larger quantity of material is used than is absolutely required by the exigencies of the case.

Where appearance need not be studied, as in warehouses and depôts, a decided preference is given to stanchions over columns, as defects are easily detected in the former, but difficult of location in the latter.

The defect that may be regarded as peculiar to hollow columns is inequality of thickness, and it will be a first-class foundry in which this does not occur to some extent, if the columns are not cast vertically. The simplest way of finding out whether the thickness is uniform is by drilling small holes in the column and measuring the thickness at each hole; but this is objectionable, it must weaken the column, though of course the holes are plugged up again. The difference of strength can be tested by putting the column under transverse stress and noting its deflection; then turning it a quarter round and again noting the deflection under an equal load, and so through the two remaining quadrants of the circle. If the column is of equal thickness all round and the metal homogeneous and free from blow holes, the four deflections should be equal; if they are not equal an observation of their differences will indicate the weak side. The objection to this system of testing is its expense.

Rectangular stanchions are open to the objection that they present sharp corners, and therefore are inconvenient in railway depôts and other buildings where carts and vans have to be moved about, and equally so in the show-rooms of large warehouses. In the large depôts and other places where economy is not the first object columns would be used, and precautions taken to secure good materials and workmanship; but in other circumstances stanchions are frequently used and encased in some material which gives them the outward appearance of columns. Unfortunately the only common section of stanchion which lends itself to this circular enclosure is the cross shaped, which I certainly consider the worst section to use. The section F, Fig. 1, is an improvement upon this, but as it cannot be got into the same space it has not been adopted. I have never seen or heard of its being actually used.

I may mention here that wrought iron and steel cross-shaped stanchions have been used in this connection, but the adaptability of these cannot be considered while dealing with cast iron.

This matter of encasing is not confined to stanchions, and in some very important buildings cast-iron columns have been encased in marble, the result being highly satisfactory.

Where such an excellent finish cannot be afforded, stanchions filled in with fine concrete and surrounded with some artificial imitation of granite or marble may suffice.

When cast-iron columns or stanchions are encased it of course adds to the diameter very materially—say, four inches—and therefore in such places as banks and show-rooms it is best to use solid cast-iron columns; although being of less diameter than hollow ones, they will require greater sectional area, we may partly set this off by the reduced amount of encasing material required, and in a solid column no varying thickness of metal can occur to give rise to bending stresses.

As a matter of fact cast-iron columns and stanchions are almost invariably made much stronger than is necessary; they are calculated for loads that never come upon them, and often absurdly high factors of safety are taken. A factor of safety that makes the working stress one-fourth of the ultimate resistance of the material is sufficient for all kinds of builder's iron work.

The production of a girder in cast iron is in all respects similar to the process used for a stanchion, and the same precautions as to inside corners and varying thicknesses must be observed.

As soon as the casting has sufficiently cooled the flask is opened and it is lifted out and the necks which were formed in the git passages knocked off, and the face of the casting where they were joined on trued up with a cold metal chisel and a file. The casting takes on a coating of partly vitrified sand, which is of a protective character and only requires to be smoothed in rough places by rubbing off with hard coke.

It sometimes happens that a projecting lug is accidentally knocked off a casting, and in such cases a process called burning on has been frequently used for replacing it. This consists in making a mould for the part to be renewed to fit the part of the casting treated, which is then heated to a temperature approaching the melting point, and the lug or other part is then cast on to it, and the whole allowed to cool very gradually. This operation may be successful, but I should never like to trust it when the part so burnt on has to resist any stress, and moreover the partial heating of the main body of the casting must put some internal stress upon it, and under this it will sometimes crack. Probably the safest method of applying this principle would be to reunite the broken part by electricity used as in electric welding—so called—which is done by placing the surfaces to be joined in juxtaposition and passing a powerful current of electricity through them; the break of continuity in the metal places so high a resistance in the course of the current that the temperature rapidly rises and the surfaces are fused together, as soon as this occurs the resistance ceases and the temperature therefore falls. I have not, however, heard of this process being applied to cast iron.

In order that columns and stanchions may have true bearings, the ends, that is the cap and base, should be accurately faced up in a turning lathe, for their strength is very materially affected by the solidity or otherwise of their bearing upon their base stones and that of the load upon their caps, and if the columns run in vertical tiers, one upon the top of another, it is indispensable that their ends shall be turned true and parallel to each other; and in order to insure such parallelism the column or stanchion should be centred in a lathe, with two slide rests carrying tools by which both ends are faced simultaneously.

The best form of joint between columns is shown in Fig. 29, which is a vertical section taken through the centre line of the columns, *B* is the top of the lower, and *A* the bottom

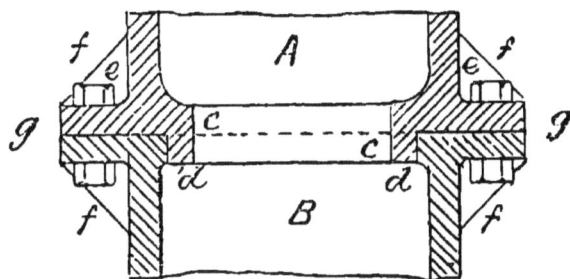

Fig. 29.

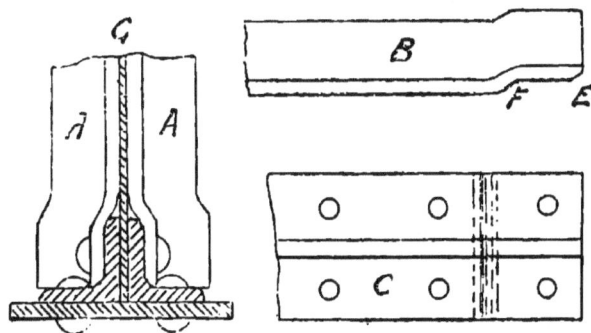

Fig. 30.

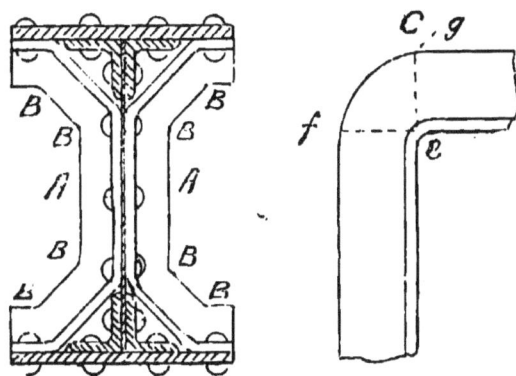

Fig. 31.

of the upper column. The top part of column B at d is accurately bored out to receive a spigot end $c\ c$, on the upper column which is turned to fit it, the meeting faces $g\ g$ are also accurately turned so that the two ends shall fit together without the slightest shake, and they are held together by bolts $e\ e$, which ought to be turned and fitted into drilled holes, the bolt and nut threads being cleanly cut, then a very solid and reliable joint is made. The cap and base may be stiffened by brackets $f\ f\ f\ f$ between the bolt holes. The great demand for cheapness unfortunately precludes, in most instances, the adoption of this true joint, and the ends are merely faced and held together by unturned bolts placed in holes made much larger than is necessary in order to insure the bolts going in with facility, and it is impossible that this can make a really steady joint, and one calculated to resist the effects of constant vibration.

The surfaces of castings which are to be bored, turned, or otherwise machined, should be indicated by a light tint on the drawing running along those parts so that the pattern-maker may know where to increase the thickness to allow for the metal cut away in finishing the work.

In no case must a casting have recesses or pockets in it in which water can accumulate, otherwise a frost may cause its destruction ; some years since a large iron viaduct was damaged to the extent of some thousands of pounds from this cause ; so if pockets are necessary, let holes or passages be made at their lower ends to allow water to drain out should any find its way into them.

The faces of the iron should come hard together at the joints, there should be nothing put between them as is sometimes done to save the expense of facing the metal, and especially is lead to be avoided, because in the presence of moisture the lead and iron form a galvanic couple, the action of which results in the destruction of the latter. When the column bases are anchored to their bed stones by holding-down bolts, these should be run in with sulphur, which does not attack the iron.

Iron is a highly oxidizable material, so much so that if it be prepared in a very finely comminuted state and scattered into the air, it immediately bursts into flame, and even ordinary iron filings thrown into a flame are consumed with bright scintillations. The sand skin on a casting is a great protection so long as it is continuous, but if there are exposed parts the metal will rust, the rust seeming to eat into it. The reason of this is easily explained by chemical affinities, but as I have not space to discuss these here the statement of the fact must suffice.

Having thus far dealt with foundry work the manipulation of the malleable metals now comes under review. Rolled iron and steel cannot in all qualities be twisted and bent and forged into varied shapes with equal facility. Hard cindery wrought iron will not bear working at all, and mild steel is best bent cold by degrees and annealed between the successive bendings. Steel will not bear working at high temperatures, and for that reason is very difficult to weld. Most of the operations of the forge in connection with constructional iron and steel work refer to cranking, bending, and joggling the ends of rolled bars of various sections, and of these processes I shall now give an outline, premising, however, that it is best to avoid all smith-work, if possible, especially upon rolled joists and girders; the less done to this class of material after it leaves the rolls the better, and this being known and admitted, the designer may often by a little forethought dispense with a good deal of labour, which may vitiate the strength and durability of the work.

In Fig. 30, *A A* represent the lower ends of two tee-iron

stiffeners joggled at their ends over the bottom angle-irons of a built-up girder; B is a side, and C a front elevation of the end removed from the girder.

If a considerable number of bars are to be joggled in exactly the same way, cast-iron dies should be made and fitted to a hydraulic press, and between these the work of joggling is quickly and accurately done, and the bars being all worked in the same dies will necessarily be exactly alike. If, however, there are only a few to be made they are generally drawn over a block by a smith, but it is impossible for him to make as perfect a job as the machine does. Channel and angle-iron and steel also require joggling in some positions, but these sections are both more troublesome to handle than tee-iron. It will be readily understood that the web part of the tee-iron is the only part in which difficulty will occur, but in this section it is reduced to a minimum as the web is supported equally on both sides by the table. The vertical limbs of the channel bar and that of an angle-iron being only stayed on one side will tend to yield laterally in one direction or the other when being bent by the hammer. When these sections are joggled in dies these prevent deviation of the vertical limbs. In point of economy there is a very great saving in power in using the press over the hammer, and it does not draw out the bars operated upon as does the hammer.

The ends of the bars, to make neat work, should be cut off square, or splayed to meet the parts of the work upon which they abut. In small work this is commonly done with a chipping chisel and hand hammer, but more satisfactory work is made by cutting the end true with a circular saw. The back edge of the end of the bar at E is to be rounded off to fit into the root of the main angle-iron, and the curve at F formed in an easy sweep to pass round

the upper edge of the same, and lie snugly against the web
G. As the duty of the tee-iron is to stiffen the web, it is
obvious that their ends should be carefully fitted so as to
rest steadily against the flanges.

In Fig. 31 is shown a cross section of a plate girder with
wide flanges, the edges of which are supported by the web
stiffeners *A A*, which for this purpose are cranked at the
points *B B*, &c. The making of these cranks will require a
vee-shaped piece, cut out of the web, and its parts, after
bending, welded together, unless the surplus metal can be
taken up in thickening the web, which will be much better
than cutting the bar, and the greater thickness will stiffen
the cranks, and if the bar were cut we could never be quite
sure of having a sound weld when it is joined up again. Here,
also, clean, sharp work may be effected under the press,
which can never be got with the hammer, as the latter has
to work in detail on the bend, while the former takes the
whole at once. If it is necessary to crank a tee or angle-
bar with the web outwards, as shown at *C*, it will be un-
avoidable to cut the web and then make it good by welding
in a piece shown by the triangle *f e g*, otherwise the web
would be drawn out to a thinness incompatible with the
required strength.

Bending to curves of large radius is done both hot and
cold. If a number of angles or tees are to be curved to the
same radius, one bar is first carefully curved by the hammer
or press, to form a template for the rest, which are brought
to a red heat, and bent round it successively. The template
bar must be kept cool with water to prevent its softening
and losing its accuracy of form. Cold bending is done in a
different way altogether. If a plate is to be curved longitudinally, this is done by passing it through a set of three
rollers, *A B B*, Fig. 32. These bending rolls, as they are

called, are so mounted in their frames that the level of the
bearings of the upper roll A can be varied in relation to
the bearings of the lower rolls B B. If the adjustment is
such that the bottom of the roll A is just the thickness of
the plate C above a straight line touching the tops of the
rolls B B, a plate passed through will remain straight, or if
it has any kinks in it they will be taken out, but if A is
depressed below this line, it will impart a curvature to the
plate depending upon the amount of the depression. The
curvature, unless it is very slight, will require the passing
of the plate several times through the rolls, and perhaps
may need annealing during the process. If the plate is to be
curved in the direction of its width it is best done in dies.

For curving tee, channel, and other bars cold, a machine is
used, the acting parts of which are shown in diagram at
D D, and E. D and D are adjustable stops, and which
may be rigidly fixed in any required position, and midway
between them is a block E, which is moved backwards and
forwards by an eccentric or crank on the driving shaft of
the machine. The stops D D may be kept so far back that
the block E will not in its travel touch a bar C resting
against them, but if they are advanced this block will
impinge upon the bar and bend it, and by moving the bar
along between the strokes of the block E, it will be uniformly
bent along its length, or any part of it that may be required; or by varying the adjustment the curvature of the
bar may be varied, so that by means of this machine
parabolic, elliptic, and other arcs can be obtained, as well
as the circular ones. The accuracy of the bending is tested
by applying a wooden template, made to the true curve, to
the bar from time to time during the process of bending.

In the above manipulations the work turned out is what
it appears to be—a joggled end is an end with its full section

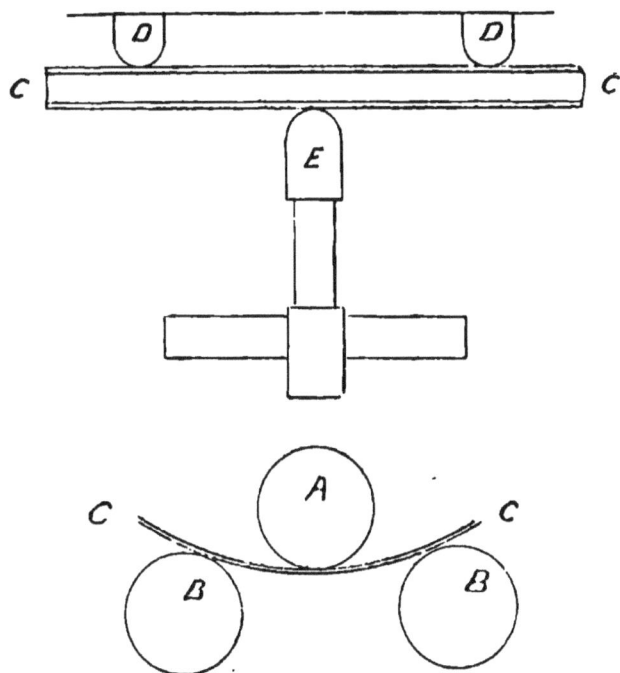

Fig. 32

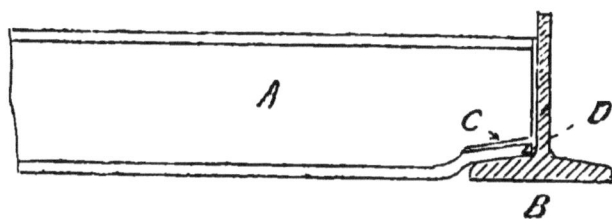

Fig. 33.

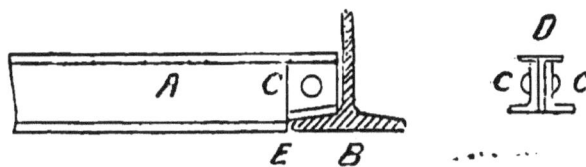

Fig. 34.

of metal preserved intact throughout the joggle; but when we look at the practice which obtains in connection with the labours on rolled girders for builder's work, we find some things that are not so, and conspicuous among them is the so-called "forged-joggled end." Now any work which is weaker than it appears to be is a source of danger, and I do not know of any other class of work in which the permanent severance of the lower flange from the web, just over the point of support, would be tolerated, if that flange is to be regarded as conveying the load to its bearing.

In Fig. 33 is shown a side elevation of the detail alluded to. It is necessary to carry the end of a rolled girder A upon the bottom flange B of a main girder in such a manner that the undersides of the two girders shall be flush—and this in buildings is a constant requirement. There is a difficulty in joggling up the flange of a rolled girder, because the web will be sure to buckle, therefore a piece must be cut out of the web. This piece is slotted out close to the bottom flange, and to the necessary height shown by the line C, and the flange is then hammered up to it, as shown at D, but no connection exists at this part between the web and the flange.

The strength of this hammered-up flange cannot be regarded as of any value, it merely acts as a packing to prevent the web of girder A from cutting into the flange of B. If this flange were welded to the web it would make an excellent job; but I doubt if this is ever done, nor is it likely to be while those in charge of the structures in which such connections occur are satisfied with the *appearance* of strength and solidity. The least objectionable method is to cut away the bottom flange of A altogether, and let its web rest directly upon the flange of B, and if the load is such that the edge of the flange does not afford sufficient bearing surface, rivet an angle bracket on each side of it, as shown in Fig. 34. Here the lower flange is cut away from the point E to the end of girder A, and on each side of the web there are riveted angle-iron brackets C, as shown in the end elevation at D, the horizontal limbs of these brackets take their bearings upon the bottom flange B of the main girder.

The mere supporting of the end of one girder upon the top or bottom flange of another will only be sufficient when there is some lateral stay to prevent the end of the first girder from slipping sideways upon the other. Such a stay is provided by floor joists on each side, or by a filling in of solid concrete, or other form of fire-resisting floor, but if there is not such a stay, on both sides, the supported girder end must be kept from lateral movement by cleating its web to that of the main girder, as shown in Fig. 35. A is the supported girder shown in horizontal section, with its web secured to that of girder B by an angle-iron cleat C which is fastened to A by a rivet D, and to B by a bolt E, the latter being necessary for connecting the work in the course of erection. The same parts of this joint are shown in elevation by A', B', C', D', E''.

If a connection is required to take the load from one girder to another, the single cleat is of no use whatever, for the girder A would swing round upon the bolt, or, if the cleat were deep enough to take two bolts, would put such a twisting stress upon the cleat bolts as would throw it out of its proper position, and very likely bring the whole load upon the flange of the girder B. In such a case, then, the girders should be double cleated together, as shown in the connection of girders F and G, where cleats are riveted upon each side of girder F by rivets, and fastened to girder G by bolts I. Even if this is a slovenly made joint, it may be assumed that the slip on each side of girder F will be alike, and so no twist will come upon it, though the holding capacity of the bolt is questionable.

Wherever a connection can be made entirely with rivets this should be done, for the rivets must fit their holes if they are hot enough for making heads upon them, and if not a loose rivet is easily detected. In builder's iron-work the bolts very seldom fit the holes, and for that reason the joints made with them must be rickety, but as they are soon covered up and hidden from sight this condition escapes observation, unless the clerk of the works has some knowledge of what iron-work construction should be.

At one time riveting was done entirely by hand, the rivets were placed on a plate with holes in it, through which their stems hung over a fire until they were white-hot, the heads being protected from such intense heat by the supporting plate. The rivets were taken as wanted, one by one, and pushed through the hole in the plates to be riveted together, the head was held close up by pressing a hammer against it, and the white-hot end first hammered down by light blows, and then completed by holding a die or "snap" over it, and striking this with a hammer to give the rivet head a shapely form. In boiler work this was usually conical; in girder-work, either hemispherical, or cheese-shaped, both of which are stronger than the conical head, if there is any longitudinal stress upon the body of the rivet. As the rivet cools it draws the plates strongly together, and makes very solid work if the hammering up and snapping of the head has not taken long enough to allow the rivet to cool. The first improvement on hand riveting was the introduction of steam riveters, and these have now been displaced by hydraulic riveting machines. These last have the great advantage of being easily made in

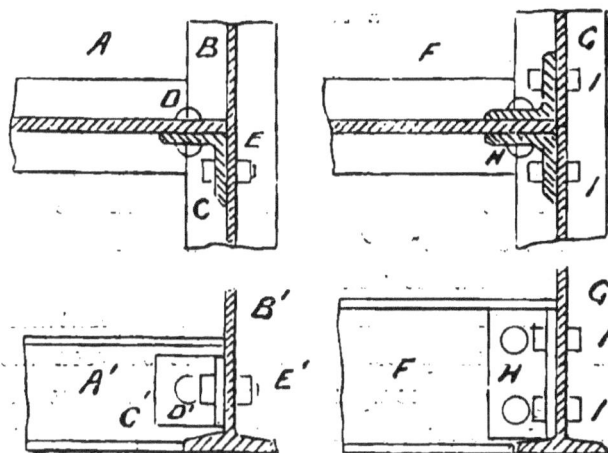

Fig. 35.

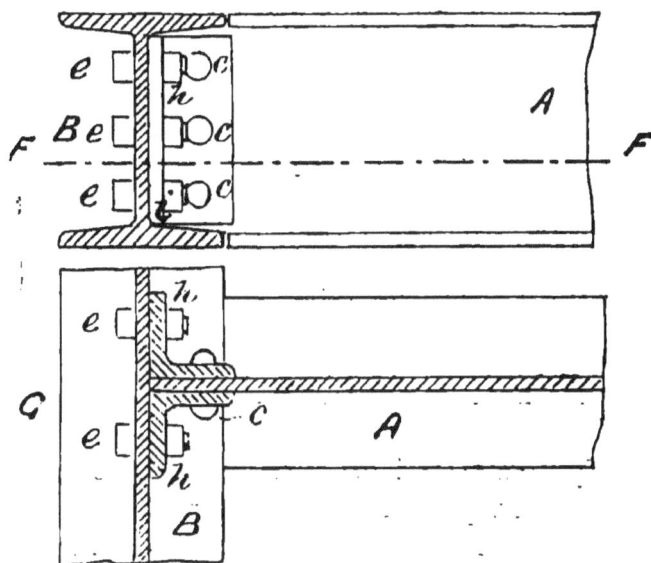

Fig. 36.

L

very portable forms, and they are also very readily handled. One may be hung on a crane and be run along the flange of a girder in process of construction with considerable expedition, the water-pressure being supplied by a flexible hose. By those machines the rivets are noiselessly and instantaneously closed, and so exert a greater contractile force upon the plates than do rivets which are closed by hand. One end served by close riveting is the exclusion of moisture from between the plates where it would cause deterioration of strength through rusting. Another is the prevention of the plates buckling apart between the rivets under longitudinal compressive stress; but lines of rivets must not be put too close together, as the plate is weakened by punching or drilling holes close together, and it may be taken, as a rule, that the rivet pitch should not be less than six diameters of the rivet; thus the distance from centre to centre of $\frac{5}{8}$-inch rivets should not be less than $3\frac{3}{4}$ inches; that of $\frac{3}{4}$-inch rivets, $4\frac{1}{2}$ inches; of $\frac{7}{8}$-inch rivets, $5\frac{1}{4}$ inches; and of 1-inch rivets, 6 inches.

The pressure which is put on a $\frac{3}{4}$-inch rivet head by a hydraulic riveter is about twenty tons, so the hot metal is thoroughly pressed into the rivet-hole, which it is very important that it should fill. The length of rivet body which should be allowed beyond the thickness of plates through which it is passed should be $1\frac{3}{4}$ diameters for hydraulic riveting, though $1\frac{1}{2}$ diameters will be enough for hand riveting. If a collar should be formed round the base of the head of the rivet, through excess of material, it should be left on, as if it is cut away the tool used for that purpose may cut into the plate, and so weaken it.

The mode adopted for marking and perforating the plates and bars for riveting is of great importance. Where ordinary single punching machines only are available, wooden templates, the sizes of the plates—or frames fitting to their sizes will do—and wherever rivet-holes are required in the iron, corresponding holes are drilled in the wooden frames. These frames are then clamped to the plates and bars, and a wooden stump, which fits the holes easily, is dipped in white lead, and successively pressed into each hole, making a white ring on the iron in the exact position of the required rivet-hole. The plate is slung in chains and worked along by the puncher's mates, while he, sitting in front of the machine, brings each mark exactly under the punch as it descends ; a practised hand will do good work in this way at the rate of fifteen holes a minute. A more certain method consists in making conical depressions in the iron by means of a centre punch placed in the template holes and struck with a hammer, and then putting the plate through a machine with a nipple punch. In this the punch has in its centre a small point or " nipple," which, falling into the countersink made by the centre-punch, secures the correct central position of the hole when punched. Similar accuracy can be secured without the preliminary process of marking the iron if the plate is put on the carriage of an automatic feed punching machine, which moves the work forward the exact distance required after each stroke of the punch ; but this, of course, is only available for lines of rivet holes of uniform pitch. A few machines have been made which would punch holes to different patterns by having a row of punches, any or all of which could be thrown out of action as each stroke was made ; it was worked on the principle of the Jacquard loom, but is too expensive in proportion to its utility to be widely adopted.

The violent action of a punch necessarily strains the metal immediately round a hole, and for this reason it is often specified that holes should be punched, say ⅛-inch less in diameter than the finished size, and the holes broached out to the full size by a broaching machine ; by this means the damaged material is removed, and that which remains is sound, and, moreover, the hole is left cylindrical, which is not the case when it is simply punched, as then it is larger on the under than on the upper side of the plate. This will be evident to any one who examines the "burrs" which accumulate under the punching machines.

Holes drilled out of the solid are to be preferred to punched, or punched and broached holes, but the sharp arrises should be taken off their edges by an obtuse broaching tool.

If the holes have to be drilled singly the expense is greatly increased, and accuracy of pitch is not assured, but multiple drilling-machines are made, which drill a number of holes at once, and as one man can attend to several of these machines, economy as well as accuracy is ensured. Special drills are used for this purpose, very stiff so that they shall not run out of line on encountering some difference of hardness in the metal operated upon. Twist-drills are very satisfactory for this purpose. One very great advantage obtainable by the use of drilling machines is that several plates may be clamped together in the positions they are required to occupy in the completed work, and the holes drilled through all of them at one operation, thus securing a truly cylindrical hole throughout the whole thickness of a flange for the reception of the rivets.

If the plates—assuming there are three or more in the thickness of a flange—are punched, they must be punched separately, and if they are not broached out afterwards the holes cannot be cylindrical through the whole thickness of plates when put together, for each hole has some taper in it. In such a case the rivet is closed with rings around it —assuming it to be hot enough to fill the hole—and these rings form sharp angles in the rivet bodies, which tend to aid shearing stress; if the rivet is too cool to be forced into all the irregularities of the hole, it will not obtain its full bearing, and, under moving loads, will soon become loose. Even when the holes after punching are broached out, there is always an uncertainty, if they are worked separately, about the holes coinciding throughout when a number of plates are put together, and if they do not the rivets will be partly sheared in closing.

It will be seen from these remarks that the preparation of plates and angle-irons for riveting together is a matter of superlative importance to the security of the work, and the same remark applies to bolted work where any stress comes upon the bolts.

I feel it necessary to point out particularly the require-ments of riveted, and, more especially, bolted connections which must be made during the erection of a structure on account of the ramshackle way in which a good deal of this work is done in connection with builders' ironwork, some of it is absolutely disheartening to a man who has been accustomed to engineers' work. Nearly all of this results from want of forethought; nothing should be left to the men, for every labourer thinks he can bolt two iron girders together, and do it well, when very likely, if the bolts served him, or which he picks up, are too long for the job, instead of laying them aside and getting others, he will fill up the spaces with three or four, or more, washers under the nut, which, combined with the usual ill-fit of the bolt to the hole, renders the connection practically useless; you might as well tie the things together with a piece of wire.

Washers placed round bolts are very necessary in certain positions; if the case necessitates the bolt-head resting against the inner or sloping side of a rolled joist flange, a taper washer is required between the bolt-head and the flange to give a fair bearing to the former, and so prevent a side wrench upon the body of the bolt. If the nut comes upon the slope of the flange, a similar washer should be used under that. Washers in other classes of work are put under the nuts to prevent their cutting into the metal as they are screwed down; but the nature of the work with which we are dealing is not such as to require this precaution. While on this subject, I may point out a very common blunder, which is to screw down the nuts as tight as the operative can do it, under the impression that thereby the work is made more stable. This idea may lead to accidents, as a screw thread affords the means of enormously increasing the force used in tightening up a nut, as a little calculation will show. Suppose we take a 12-inch spanner to tighten up a nut upon a 1-inch bolt. The pitch of the thread on that bolt will be $\frac{1}{8}$th of an inch, and the force exerted on the nut in a direction parallel to the axis of the bolt will be to the force applied to the end of the spanner, as the path described by the latter in one revolution is to $\frac{1}{8}$-inch; therefore the multiplication of force is—

$$= 3\cdot1416 \times 24 \div \tfrac{1}{8} = 603\cdot187.$$

Twenty-four inches is the diameter of the circular path described by the end of the spanner. If now 20 lbs. pressure be applied at the end of the spanner, the longitudinal force upon the bolt will be 603 × 20 = 12,060 lbs. = 5·43 tons nearly. At the bottom of the thread the bolt is about $\frac{3}{4}$-inch in diameter, which gives a sectional area of 0·44 square inch, so the stress per sectional square inch would be $5·43 \div 0·44 = 12·34$ tons, which would probably exceed the limit of elasticity of the material, and therefore permanently stretch and injure it. If the nut and bolt on which it is screwed are quite dry, however, the friction of the nut would stop it when about 2 tons stress was reached, with only 20 lbs. on the spanner, but if the parts were oily a pressure of nearly 5 tons would be obtainable; but a man can put 50 lbs. or more on the spanner, so this is no guarantee of safety. It may be taken that when the nut has an all-round solid bearing upon the surface against which it is screwed that it is tight enough. Where one girder has to support the end of another, there are two correct positions for the latter; in the first, the supported end rests upon and reaches across the top flange of the supporting girder, so that its weight is equally distributed over the width of the latter, and therefore causes no tendency to twist or rock under varying loads. In the other position, the web of the supported girder A, Fig. 36, is cleated to that of the supporting girder B, which is shown in section. The girders taken for illustration are each rolled steel girders 16 inches deep and 6 inches wide on the flanges. No weight should be allowed to come direct upon the bottom flange, it should all be taken from the web of girder A to that of girder B, by angle-iron cleats h, arranged as shown in the horizontal section C taken on the line $F F$.

The web of the girder A is to be so cut that it cannot bear upon the bottom flange of girder B when this can be avoided, but this necessitates the use of properly fitting bolts $e e e$, to connect the angle-cleats h with the web of girder B, and therefore the end of the web is commonly cut to rest with more or less equality of bearing on the lower flange of this girder, regardless of the fact that a twisting stress upon the root i of a rolled girder is about the most destructive to which it can be exposed. If there happen to be two girders resting on the lower flange of girder B, and their loads are equal and constant, the twisting stress alluded to will not occur, but if the loads are unequal and are not constant but varying, there will be an alternating stress upon the root i of a most trying description.

The proper attachment of the cleats to the girder A is easy enough, as they can be riveted on with rivets $c\ c\ c$, but in work small enough to be served by rolled girders there is seldom room enough to rivet the cleats to the girder B on the site, and if there is room, there is seldom enough of it to warrant the introduction of riveting plant on the premises ; though certainly more might be done in this direction in towns in which high-pressure hydraulic mains are laid whence power to actuate hydraulic riveters is available.

While upon this subject I cannot pass on without referring to the great waste of money that occurs in the drilling of holes on the site in buildings. A hole that will not cost a penny to punch in the yard will cost sixpence to drill in the building, and then the work will be ill-done. If the architect will only decide upon what he wants, and make his plans absolutely complete before he begins building, there need be none of this waste ; the position of every hole can be determined, and all of them can be punched or drilled in the iron yard with the certainty of a proper fit when they are put together on the site. In railway depôts and large carriers' buildings this system has been used with perfect success, and there is no reason why it should not apply equally well in the execution of constructional iron-work for smaller buildings, if they are designed *before* they are built.

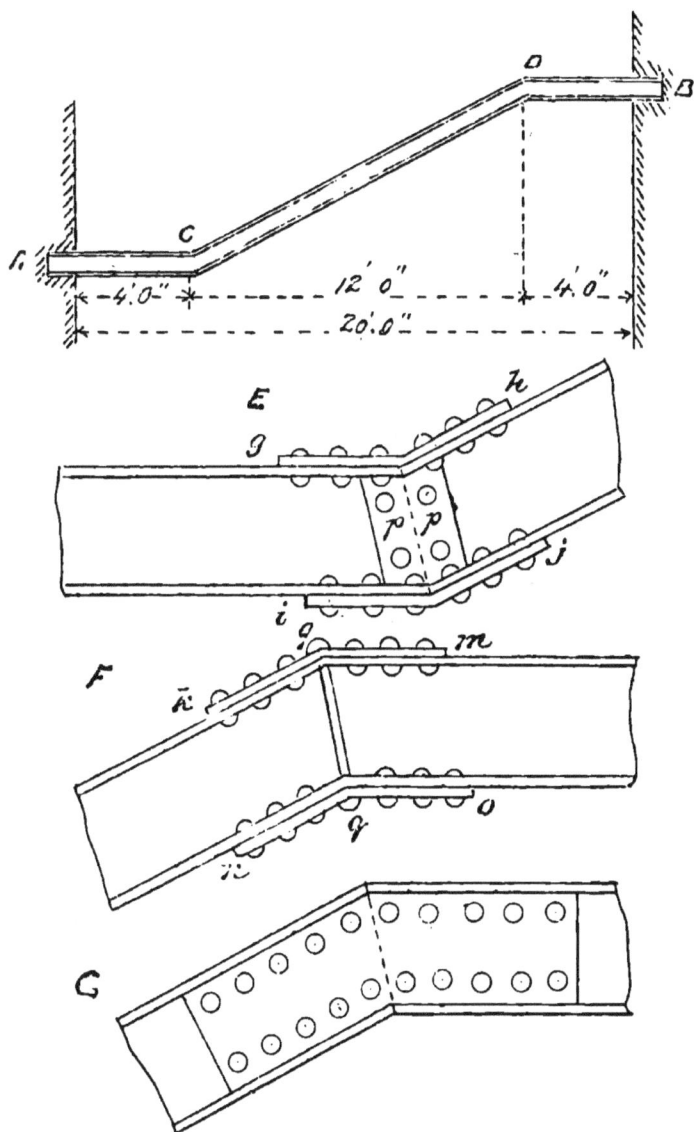

Fig. 37.

The most curious connections suggested in building work are to be found in sketches of wrought-iron and steel stair-carriages. Anything that is required in this way can be readily made in cast iron or mild cast steel, and these I have constantly used with satisfaction myself; but the idea of the strength and safety of rolled joists for this purpose has gained so strong a hold that the difficulties of making cranked and twisted joints are quite 'overlooked in view of the fact that the run of the stair-carriage is a rolled joist.

The question of stair-carriages involves some curious reflections; it is trite to say that in some case they are wanted, or they are not wanted, but I have known them to be ordered by a district surveyor to be fitted up, after the stairs were completed, each stair being pinned into the wall and quite capable of carrying its load as a cantilever. In such a case it does not matter much what is put up, so that it does not put additional load on the stairs it appears to support.

We have, however, here to deal with cases in which the stairs are supported by their carriages, and to consider the connections necessary when we are tied to the use of rolled joists.

In the general run of work it is not often that a joist larger than an 8-in. × 4-in. steel is required as a staircase carriage, and the difficulty of making any satisfactory cranked joint in these I will now point out.

As in other cases this matter is most easily dealt with by taking an example, and in so doing I shall assume that the carriage does actually take half the weight of the staircase and its living load. A side elevation of such a stair-carriage is shown by $A\ B$, Fig. 37. $A\ C$ is horizontal to carry the lower landing, and also $D\ B$ which supports the upper landing. The part $C\ D$ which supports one end of each step is necessarily inclined, but its effective span—as previously explained—is the horizontal distance between C and D, 12 feet in the example taken; the total span of the carriage is 20 feet. Now let the staircase be 4 feet wide in the clear, then when it is crowded with people the total load between the walls A and B will be,

$$20 \times 4 \times 2 \text{ cwt.} = 160 \text{ cwt.} = 8 \text{ tons.}$$

The load of 2 cwt. per superficial foot will cover the weight of the people as well as that of the stairs, even allowing the latter to be of concrete. One-half of the load 4 tons comes upon the stair-carriage, and reference to a table of strength shows that if German joists are used, one 8 in. $\times$ 5 in. will be requisite, as this is rather a long span. To make sharp bends as shown at C and D is impossible, and therefore the steel joist must be cut and jointed at those places. These joints are shown in detail at E and F. The moment of stress will be the same for both joints. The whole load on the cranked carriage is 4 tons, therefore the load per foot horizontal is, $\frac{4}{20} = 0\cdot2$ ton. The distance of either C or D from the nearest point of support is 4 feet, therefore the moment of stress required is,

$$M = \frac{w \cdot x''}{2}\left\{x - l\right\} = \frac{0\cdot2 \times 4}{2}\left\{4 - 20\right\} = -\ 6\cdot4 \text{ foot-tons,}$$

the minus sign showing that there is tension on the lower flange and compression upon the upper. If the stress can be picked up by flange cover plates, the stress to be taken at each flange will be equal to the moment of stress (M)

divided by the depth of the carriage, which is 8 in. $= 0\cdot66$ feet, therefore this stress will be,

$$S = \frac{6\cdot4}{0\cdot66} = 9\cdot6 \text{ tons.}$$

The number of rivets on each side of the joint required to meet this stress will be—the rivets being $\frac{3}{4}$ inches diameter, and their areas therefore $0\cdot44$ square inch—

$$= \frac{9\cdot6}{0\cdot44 \times 4} = 6 \text{ rivets,}$$

the safe shearing stress being taken at 4 tons per sectional square inch.   If a 4-inch rivet pitch is used, and two rows of rivets, the length of cover on each side of the joint will be 3 rivets $\times$ 4 pitch $= 12$ inches, and the cover plate must therefore be 2 feet long.   We have now to see how the action of the cover plates is affected by the joint being cranked.   At the joint E the top cover $g\,h$ is in compression, and this will press it against the top flange of the girder so that it will do its work without putting any longitudinal stress upon the rivets.   The cover plate on the tension flange will also be pressed against it, so that in this position the cover plates both act properly.   The net sectional area of the plate $i\,j$ must not be less than,

$$\frac{9\cdot6}{5} = 1\cdot92 \text{ square inches.}$$

The flange width is 5 inches, so this would also be the width of the cover plate; deducting two $\frac{3}{4}$-inch rivets, the effective width left is $5 - 2 \times \frac{3}{4} = 3\cdot5$ inches, and therefore the requisite thickness of the cover will be,

$$\frac{1\cdot92}{3\cdot5} = 0\cdot55 \text{ inch,}$$

so the plate will be made $\frac{9}{16}$-inch thick.

The gross area of the top cover plate must be,

$$\frac{9 \cdot 6}{4} = 2 \cdot 4 \text{ square inches.}$$

The necessary thickness will be $2 \cdot 4 \div 5 = 0 \cdot 48$ inches, so ½ inch will be sufficient for this, but it is best to keep these plates both of the same thickness, so that if they are misplaced—which may happen if the workmen are careless—there shall in any case be the requisite strength of cover. Some people like to hide the joint in the web by riveting the plates $p\,p$ on each side of it, but this is quite unnecessary as—if there is no encasing or other covering—the paint will cover it.

Now in regard to the joint F the conditions of the cover plates are reversed, compressive stress will tend to spring the plate $k\,m$ away from the girder flange and so bring a pull upon the heads of the rivets, and tension will have the same effect upon the bottom plate $n\,o$. A long rivet $g\,g$ on each side of the girder web would certainly serve to tie the cover plates, which would then require to be longer to allow a pitch distance on each side of the tie rivet to the next ordinary rivet.

The consideration of this joint, in which the cover plates do not seem thoroughly reliable, brings us to the question of the strength of a fish-plate joint in such a position. In this mode of construction, the moment of shearing resistance of the rivets is necessarily much less than that of an equal number of rivets in the flanges of the girder. The rivets must be kept in sufficiently to let their heads clear the inner sides of the flanges, which in an 8 by 5-inch girder would be about 1¾ inches from the outside of the flanges, so that the effective depth for the moment of resistance of the rivet areas would be,

$$8 = 1\tfrac{3}{4} \times 2 = 4\tfrac{1}{2} \text{ inches.}$$

The direct shearing stress upon the upper and lower rows of rivets will be as $4\frac{1}{2}$ in. $= 0\cdot376$ ft., and the moment of stress is 6·4 tons.

$$S = \frac{6.4}{0.375} = 17\cdot0\dot{6} \text{ tons.}$$

As there will be a plate on each side of the web of the girder, the rivets will be in double shear, and the number of rivets required on each side of the joint will therefore be —in each row—

$$\frac{17\cdot0\dot{6}}{4 \times 0\cdot44 \times 2} = 5 \text{ rivets.}$$

The arrangement of the rivets is shown at C, Fig. 37. If the rivets are pitched 3 inches apart the total length of the joint plate, measured along one of the lines of rivets, will be 2 feet 6 inches.

At the joint itself the joint plates will act as rectangular beams under transverse stress. If $b =$ thickness in inches of each plate, and $d =$ depth in inches, the moment of resistance for working stress will be,

$$M = \frac{s.b.d^2}{b} = \frac{4 \times 2b \times 7^2}{6} = 65\cdot\dot{3}\, b.$$

This is taking the depth of the joint plates as 7 inches, and the direct working resistance of the metal as 4 tons per sectional square inch; the formula gives the resistance in inch-tons; in foot-tons it will be $65\cdot\dot{3}\, b \div 12 = 5\cdot\dot{4}\, b$. Equating this with the moment of stress

$$5\cdot\dot{4}\, b = 6\cdot4, \text{ therefore } b = 1\cdot18 \text{ inches.}$$

Therefore each joint plate should be $1\frac{1}{4}$ inch thick to give the proper transverse strength at the joint, and this is assuming the work is riveted. If bolts are to be used the result is not so good—or rather, I should say, is worse—and I should certainly prefer a cast carriage to one joined up by fish plates bolted on each side of the web.

On account probably of the distrust of cast iron, columns and stanchions of that material have been in many cases rejected in favour of substitutes built up of wrought iron and steel. This may very reasonably be done in railway depôts where there are chances of very heavy blows and where very heavy loads occur which require supports of sufficient size to allow of free manipulation ; but it is a mistake to use wrought iron or steel for the stanchions of ordinary commercial buildings, as it is difficult to make satisfactory connections between the shafts and caps or the shafts and bases, or to adapt the shafts to carry intermediate girders, such as may readily be taken on a bracket on the side of a cast-iron stanchion, or have its end supported in a pocket cast in the stanchion.

In storage warehouses where appearances are of no consequence, wrought columns and stanchions of large size can be used, and sections of metal suitable for building them up are specially rolled.

In Fig. 58 A shows a stiff form of column made of segmental bars with longitudinal flanges b connected by rivets c. The longitudinal flanges add much to the strength of the column by resisting collapse. The shaft is fastened to the base by strong angle-iron brackets $e\ e$, &c., and the cap may be attached by similar means; but if the bases extend far outside the limits of the shaft the angle-iron should not be used, but the tee-iron brackets as shown in plan at B and in vertical section at C.

The simplest way to get over the difficulty of a large base to a wrought shaft is to make it of cast iron or cast steel and make good the joint with iron cement. The casting is made with a recess of the same shape as the shaft, but with sufficient play to allow the cement to be readily put in and caulked down ; the socket can of course be made high enough to allow of brackets being cast on it to properly spread the load over the whole base.

In commercial and other public buildings architects often wish to keep the columns small, and yet there must be something allowed outside the iron column for ornamentation, such as producing the appearance of a marble or granite column. We are then perhaps rigorously cut down to a diameter of 6 or 7 inches, and if the architect insists on built steel stanchions we must rivet four angle-irons together, which at the larger size could only be $3\frac{1}{2}$ in. by $3\frac{1}{2}$ in. by $\frac{1}{2}$ in. thick. Such an arrangement is shown at D, Fig. 38, where $f f f f$ are sections of the angle steels resting on a base plate $g g g g$. The difficulty of attachment is obvious, there is a bare 3 inches clear on any limb of either angle-iron, and if a bracket is riveted on to one it blocks the surface of the return limb of that bar, by taking up another $\frac{1}{2}$ inch as shown at h, so that only four brackets could be used in such a case. Now suppose this stanchion to be 14 feet high, and made of the best English iron or of mild steel, then its breaking resistance per sectional square inch will be,

$$= \frac{20}{1 + \dfrac{r^2}{900}}$$

when $r =$ ratio of length to diameter which in this case is, 14 ft. $\times$ 12 $\div$ 7 ins. $=$ 24, therefore the breaking stress is

$$= \frac{20}{1 + \dfrac{(24)^2}{900}} = 12\cdot2 \text{ tons nearly.}$$

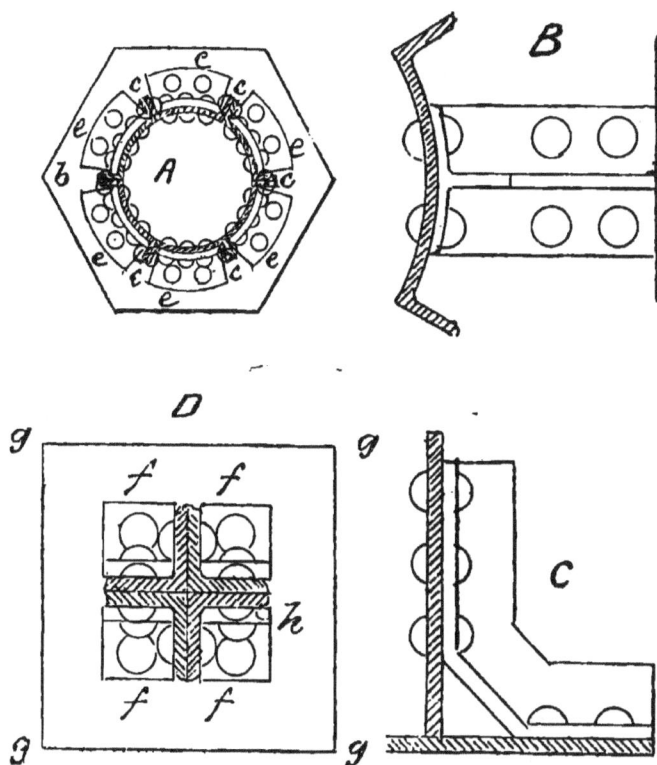

Fig. 38.

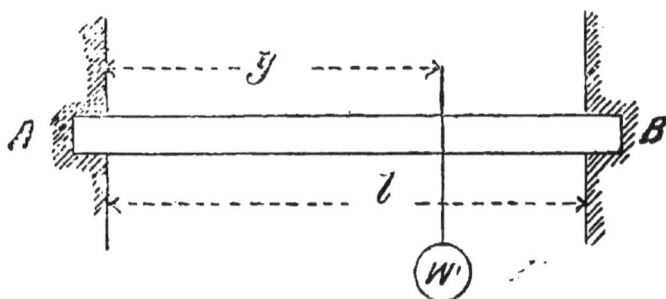

Fig. 39.

M

With 6 as a factor of safety this gives as the working strength $12 \cdot 2 \div 6 =$ say 2 tons per sectional square inch.

The sectional area of the stanchion is,

$$4 \times \{3\tfrac{1}{2} + 3\tfrac{1}{2} - \tfrac{1}{2}\} \times \tfrac{1}{2} = 26 \text{ sq. ins.}$$

Such a stanchion would therefore safely support $26 \times 2 = 52$ tons.

Now let us consider what sized base will be required to carry this load on to the foundation; by this I do not mean the subsoil or rock, as the case may be, for a pier may be extended by footings all round until the required area is covered, but to the bedstone upon which the base of the stanchion rests. How much pressure may be put upon the bedstone will, of course, depend upon its quality; assuming that a good quality of limestone is used, it may be loaded safely to 7 tons per square foot; but as it has to spread the load it must have a true bed on the masonry or brickwork below, and if it is 1 foot thick, should not extend more than 9 inches beyond the baseplate of the column, all round. The area of iron baseplate required will therefore be $52 \div 7 = 7 \cdot 43$ square feet, which would be furnished by a base $2 \cdot 72$, say 2 feet 9 inches wide. So wide a base could not be fitted with sufficient brackets from such a shaft to fairly distribute the load, if the base is made of wrought metal.

A steel cylinder 7 inches in diameter with an angle steel ring 6 in. $\times$ 6 in. $\times$ $\tfrac{1}{2}$ in. riveted on as a connection to the base, and a smaller one to connect the shaft with the cap, would make a much better and neater job than the built-up stanchion, and cost less.

We are not tied down to having small steel columns, for Messrs. Russell & Sons, of Wednesbury, roll solid steel tubes up to 20 inches in diameter and $\tfrac{3}{4}$ inch thick, and these may be made round, square, octagonal, hexagonal, or of other forms, if required. The relative strengths of different materials in long columns are—calling cast iron = 1,000 : wrought iron 1,745 ; cast steel 2,518 ; and the relative strengths for flat and round ends are : both ends rounded, 1 ; one end flat and firmly fixed, 2 ; both ends flat and firmly fixed, 3.

As the crippling resistance of a solid rolled tube will certainly not be less than that of a riveted one, we shall be safe in using for the ultimate strength of mild steel hollow columns per square inch of sectional area,

$$S = 16 \div \{1 + 0.00022\, r^2\};$$

and for hollow square columns,

$$S = 16 \div \{1 + 0.00017\, r^2\}.$$

CHAPTER VI

LOADS AND STRESSES IN BUILDINGS.

In no matter is there more uncertainty than is to be found in the question of loads in the structural ironwork of buildings, except in certain classes of work in which the uses of the structures are known and clearly defined.

Floors of dwelling-houses, from the smallest artisans' tenements to the most aristocratic flats, are now carried by iron or steel joists, especially when the filling in of the floor is executed in fire-resisting material, such as concrete, or combinations of concrete with terra-cotta or fireclay tubes or lintels, and the use of these floors has for many years been increasing; first adopted for important public and state buildings, its use has gradually but continuously been extended to every class.

If a fire-resisting floor is to be used, the architect should decide upon the system to be adopted, before the structural ironwork is designed, for the respective weights and modes of construction of these floors vary widely, and also the stresses brought by them upon the supporting joists. These floors may generally be divided into two classes : those that put thrust, or lateral stress, upon the walls or girders carrying them, and those which merely bring vertical load upon their supports.

If any lateral stress occurs in the floor it must be taken

up by ties, or provided for by the solidity of the walls, or the added strength of piers or buttresses. The fact that different floors, made to attain the same end, may produce such different stresses upon the walls, is sufficient argument in itself that the architect should design his floors in detail, as well as in general plan and section. The practice of giving block plans to the various makers of special floors, in competition, is altogether wrong, for a building suited to carry one kind of floor may be quite unfitted for another, which may be cheaper ; the trouble comes afterwards. I do not intend to deal with the relative values of fire-resisting floors, further than their construction affects the nature of the stresses upon their supports, and the convenience of fixing them.

Live, moving, or useful loads will first be considered ; the dead load of the structure itself is easily and absolutely ascertainable. Let us take a room crowded with people of something over the average weight; we cannot get more than six twelve-stone men on a square yard of surface, unless pressure is applied ; that gives just 112 lbs. per superficial foot for live load. If the floor of an assembly room were packed to this extent, the occupants could only move with difficulty, so vibration need not be considered ; but such a packing as this is practically impossible, except, perhaps, in corridors of theatres and railway stations. The weight of soldiers in close marching order is about 70 lbs. per superficial foot. Some remarkable data as to the weight of men that may be concentrated on a small surface have been published, some reading as high as 150 lbs. per foot super ; but this coming from experiments in a monastery, it may be that a number of exceptionally tall men were crowded into a small apartment ; such men, from the rigor of their lives, would necessarily be hard, and weigh more in proportion to their heights than those who live a freer life, and occupy more space.

There comes in the question of allowance for vibration, but this ought not to be mixed up with that of actual weight. I do not believe that 112 lbs. per superficial foot is ever reached in any public building by a moving crowd, and therefore that may be taken as the maximum useful load.

In private houses the floor load is necessarily much less, for if reception and ball-rooms are excluded, we shall seldom find one person to the square yard, and the whole family, with guests included, will not average 56 lbs. to the square foot. Furniture must also be taken into consideration, but this—in the central spaces—does not exceed the live load it displaces; heavy furniture, such as bookcases and pianos, are usually close to the walls, and, therefore, the loads they bring upon the joists act so closely to the points of support, that the transverse stress, as compared with uniformly distributed load, is very slight; what it actually is may be calculated, as will be shown subsequently. If a library is arranged to have bookcases in the centre of the floor, as well as at the sides, a special case arises, which must be dealt with in determining the sizes of the joists.

In bedrooms in private houses, the weight of furniture and inhabitants will seldom *average* 28 lbs. per square foot, but heavy pieces of furniture may cause local loads, that must be considered in determining the section of the joists; and it must be remembered that such heavy loads may be moved from one part of the room to another, for purposes of cleaning, temporarily, or for a longer period, to suit the varying convenience of the occupant, therefore, all the joists in that room must be made strong enough to support the heaviest piece of furniture it is intended to receive, and also to take the weight in the centre of their spans.

In drill halls and rooms used for rythmical athletic practice, vibration in its most destructive form will occur, but whether it has any serious effects will depend upon the intervals of impact upon the floor. Every elastic body has a certain period of vibration, and if the impacts producing vibration have the same interval, every successive impact will increase the distance or amplitude of the vibration, and this may be continued until the limit of safe deflection has been passed, after which a collapse is imminent. We should, therefore, study to prevent any possibility of synchronism between the vibrations of the floors and the movements of those practising on them. For this purpose, the quickest return available is that to be adopted, and that will be found in the most rigid joist, and the rigidity increases, for a given span, as the depth of the girder; hence the requisite strength should be obtained preferably from depth of girder, rather than from breadth and thickness of the flanges. The effects of dancing or drilling on a floor is not similar to the dropping of weights upon it, for the elasticity of the muscles causes the weight to come gradually upon the floor, and the energy of the action is in a great measure expended in generating heat in the bodies of those engaged in it. If a man walks on to a girder until he reaches the centre of the span, he will cause it to deflect gradually until it has attained its maximum deflection on his arrival at the centre, that is, its maximum under the steady load of his body; if he now jumps, the ensuing deflection of the girder will be momentarily increased, but the still load to produce the same effect cannot be determined unless the deflection of the girder is measured.

Work—in a mechanical sense—is the product of a force

multiplied by the distance through which it acts, and work is done in deflecting a girder, and the effect of a jump, or other impact, will be equal to that of a weight gradually applied to produce the same deflection.

If a load W brought gradually upon the centre of a girder causes a deflection equal to y, then the work done by the load is $W \times y$. If the same load is dropped from a height h on to the girder, and it causes a maximum deflection $= y'$, then the work done will be equal to $W \times \{h + y'\}$. The deflection of a given girder is in direct simple ratio to the load producing it; if 5 cwt. will deflect a bar through half an inch, 10 cwt. will deflect the same bar one inch, but the work done is evidently in ratio to the square of the load, as when the deflection is doubled both the load and the distance through which it acts are doubled; therefore the weights to produce different deflections will vary as the square-roots of the work expended in causing them.

If a load of 100 lbs. produces a deflection of one inch on a given bar, the work done in producing this deflection is 100 inch-lbs. Now suppose the weight to be let fall from a height of five inches, it will have accumulated work equal to 500 inch-lbs. before it touches the girder, and before it comes to rest this work must be expended in causing deflection, in addition to the 100 inch-lbs. that would act in gradual application, or the whole work is 600 inch-lbs.

The weight which, gradually applied, would produce a deflection equal to the falling weight, would be,

$$W' = \sqrt{\frac{600}{100}} \times W = 2 \cdot 449 \, . \, W$$

In the acts of marching and athletic posturing there is no actual fall of the load, but a shifting of position, though a skipping exercise would involve an actual descent of load, though, as explained previously, this would not be suddenly applied in its whole intensity. It may therefore be taken that the movement will not do more than produce twice the deflection which would accrue from the same load applied gradually. The load to be dealt with will not be the full weight of 112 lbs. per superficial foot, as room must be given for action. Now in close marching order soldiers will not average more than 80 lbs. per square foot of floor room, and scholars in gymnasia would certainly be lighter than this; so a live load of $1\frac{1}{2}$ cwt. is sufficient to take for drill-rooms and training-schools. Of course the dead weight of the floor must be added to this for the total load.

In storehouses, such as granaries and other buildings designed for specific purposes, there is no difficulty in determining the maximum useful load upon the floors, as the weights of the materials are known, or are easily ascertainable, and the same is the case with libraries and shops of various kinds. In churches and other places of worship we may be satisfied with taking the seating capacity of the building, and the same with assembly-rooms and places of entertainment, where the seats are fixed; but where this is not the case, allowance must be made for a closely-packed crowd, and in buildings which may be occasionally used for political purposes the effects of violent stamping, which can be managed even in a crowded room, must not be disregarded; in such cases it is advisable to calculate for a double load.

The continuous running of machinery in factories and mills cannot be regarded as causing a potential increase of load, but it certainly calls into existence tremors which will probably extend throughout the building, and these will exert a racking action upon the connections of the girders and stanchions or columns used in the structure. To reduce this to a minimum the running machinery should nowhere be directly connected to any girder or stanchion which forms part of the constructional iron or steel work of the building. I know this cannot always be avoided, especially in regard to shafting, but the evil effects must not be overlooked.

The dead weight of the floor must be allowed for, and that of any partitions which may be carried upon it. If the partitions run parallel to the joists, an extra strong joist, or two joists side by side, may be placed immediately under each partition, and so take the weight of the partition alone, and not communicate to it the vibration of the floor. This, however, will not always happen, and when partitions occur at right angles to each other, some of them must necessarily cross the floor joists.

Now as these joists are of uniform section throughout their lengths, and as in the tables previously mentioned the strengths are given for uniformly distributed loads, it is necessary to find, in cases where other loads are concerned, what uniformly distributed load would give the same maximum moment of stress. A central load causes twice the stress that the same weight would, if distributed, bring upon the joist; so if there is a four-ton central load on a joist, one fitted to carry eight tons uniformly distributed must be chosen. We want, however, a general expression for the equivalent load uniformly distributed to any local load.

Let $A B$, Fig. 39, represent a joist or trimmer, which carries a load $= W$ at a point distant y feet from the point of support A, and let l equal the clear span. The maximum moment of stress will be at the point of application of the load, and will be,

$$M = W' \times \frac{l - y}{l} \times y,$$

in which equation $W' \times \frac{l - y}{l}$ is equal to the reaction of the support A, and y is the distance at which this reaction acts to produce the moment of stress. We have shown that under a uniform load the maximum moment of stress is,

$$M = \frac{W \cdot l}{8}$$

If $W =$ the distributed load causing it; therefore, equating these, we find,

$$\frac{W \cdot l}{8} = W' \times \frac{l - y}{l} \times y;$$

therefore,

$$W = 8 \, W' \times \frac{l \, y - y^2}{l^2} = 8 \, W' \times \left\{ \frac{y}{l} - \left(\frac{y}{l} \right)^2 \right\}$$

We will apply this to the case shown in the figure, in which $y = 10$ feet, and $l = 14$ feet, and take W' to equal 6 tons; then,

$$W = 8 \times 6 \left\{ \frac{10}{14} - \left(\frac{10}{14} \right)^2 \right\} = 9 \cdot 816 \text{ tons.}$$

If, however, a joist is subject to mixed loads, the finding of a single distributed load, giving the same maximum stress as all the actual loads together, will be more complex. Now let the same joist have, in addition to the local load W', a uniformly distributed load of 2·8 tons, that is, 0·2 ton per lineal foot of span. At any point in the span the moment of stress will be equal to the sum of the moments accruing from each load, and the question arises as to whether the maximum will be under the local load or at the centre of the span. From formulæ already found, the moment of stress under the local load, but due to the equally distributed load, will be,

$$M = \frac{w \cdot x}{2} \left\{ x - l \right\} = \frac{0 \cdot 2 \times 10}{2} \left\{ 10 - 14 \right\} = -4 \text{ ft.-tons.}$$

The moment from the local load will be, designating the reaction minus as usual,

$$M = -W' \times \frac{l - y}{l} \times y = -6 \times \frac{4}{14} \times 10$$
$$= -17 \cdot 14 \text{ ft.-tons.}$$

Total under local load $= -17 \cdot 14 - 4 = -21 \cdot 14$ ft.-tons. At the centre the moment from the distributed load is,

$$M = -\frac{w \cdot l^2}{8} = -\frac{0 \cdot 2 \times 14 \times 14}{8} = -4 \cdot 9 \text{ ft.-tons,}$$

and the moment at the centre of the span of the reaction at *A*, due to the local load, will be,

$$M = -W' \times \frac{l - y}{l} \times \frac{l}{2} = -6 \times \frac{4}{14} \times 7 = -12 \text{ ft.-tons.}$$

Total at centre of span $= -12 - 4 \cdot 9 = -16 \cdot 9$ ft.-tons. So the maximum moment in this case occurs under the local load, and the uniformly distributed weight to produce the same stress is thus found :—

$$\frac{W \cdot l}{8} = 21 \cdot 14 \; ; \quad \frac{W \times 14}{8} = 21 \cdot 14 \; ; \quad W = 12 \cdot 08 \text{ tons.}$$

If there are more loads than two, the moments of stress

must be summed under each of the local loads, as well as at the centre of the span, to find the maximum, which is then to be equated with $\dfrac{W \cdot l}{8}$ to find the equivalent equally distributed load.

The amount of load that a partition will bring upon a wall will of course depend upon its construction and the material of which it is composed. A partition formed in breeze concrete six inches thick will weigh 42 lbs. per superficial foot of face; so if this were running across joists placed two feet apart, and the room is ten feet high, the local load on each joist immediately under the partition will be,

$$= 2 \text{ ft.} \times 10 \text{ ft.} \times 42 \text{ lbs.} = 840 \text{ lbs.} = 0 \cdot 38 \text{ ton.}$$

If the partition runs along a joist of say 16 feet span, the equally distributed load upon it will be,

$$= 16 \text{ ft.} \times 10 \text{ ft.} \times 42 \text{ lbs.} = 6720 \text{ lbs.} = 3 \text{ tons.}$$

A 4½-inch brick partition weighs about one-third of a hundredweight to the superficial foot, but as it must be plastered upon both sides, its weight when finished will come up to that of the breeze concrete partition.

When the breeze partition or one of brick has set, it will be almost, if not quite, self-supporting, provided there are no doorways in it, but if one occurs the continuity of the work is destroyed, and its whole weight will always be upon the supporting joist, and as alterations to buildings frequently involve the cutting of new doorways, it is not safe to take it for granted that a partition made complete in the first instance will always remain so.

Framed wood partitions filled in with concrete or brick may be trussed so as to carry their own loads clear of the floor, but of course these must not afterwards be cut about. If the expense is not in the way, a framed iron or steel partition may be made, but this will generally cost more than a joist strong enough to carry an ordinary partition, because the load is so light that framed work cannot be kept down to anything like the theoretical weight required.

It may make this question of loads clearer if I take an example, and work them out for the whole building, assuming that some kind of concrete flooring is to be used, and the constructional work to comprise steel girders and cast-iron columns and stanchions.

Fig. 40 shows a front elevation and Fig. 41 a plan or horizontal section of a warehouse building, which is assumed to be sixty feet square inside, and the whole of the ground-floor front left open, with the exception of the two columns *A A*, which help to support the superstructure; further in, on the ground floor, are other columns *B B*, to take the ends of main girders *C C C*, which carry the floors. No internal walls are carried up; so the columns *B B*, on the ground floor, must carry others in tier above them to support the main girders of the upper floors and the roof. Whether stanchions will be required upon the columns *A A* will depend upon whether the brickwork or masonry, as the case may be, is strong enough to carry the upper loads, and the same contingency is to be looked for in regard to the back wall.

Each floor will be required to carry a useful load of, say, 3 cwt. per superficial foot of area; the dead weight may be taken as half a cwt., making a total load of 3·5 cwt. per foot super. If the roof is also formed by a concrete flat, the total load upon that may be taken as 1·5 cwt. per foot super.

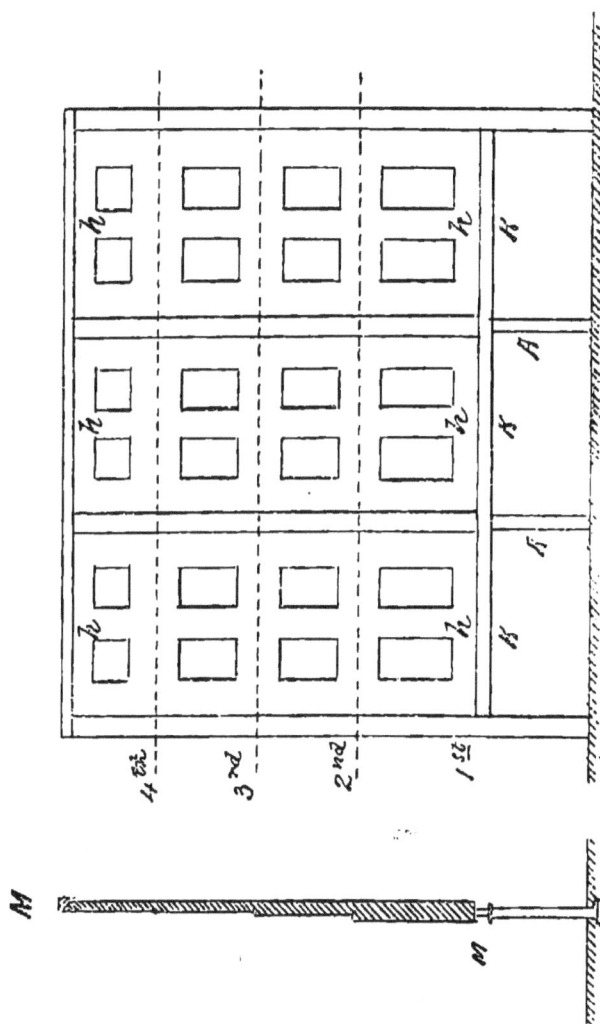

Fig. 40.

Lines drawn through the centres of the columns $B\,B$, parallel to the front and back walls, will divide the floor area into nine equal portions, each 20 feet square, so the floor surface carried by one main girder C will be 400 square feet, as shown by the dotted square $g\,g$. Therefore the load upon each will be,

$$= 400 \text{ sq. ft.} \times 3\cdot5 \text{ cwt.} = 1400 \text{ cwt.} = 70 \text{ tons.}$$

I shall assume that English steel joists of material guaranteed to stand 32 tons per sectional square inch in tension, before rupture, are specified to be used, the factor of safety being 4, so that the maximum working load does not exceed one-fourth of the estimated breaking weight. The effective span is less than 20 feet by the width across the column-cap, so it may be taken as 19 feet, which will allow a little margin.

Referring to a manufacturer's table of strengths, I find I must use two rolled girders side by side, each being 18 inches deep by 7 inches wide on the flange, and weighing 84 lbs. per foot run ; this would give a strength of 80 tons working load, but two girders of the next size smaller stock sections would only give a working strength of 51·2 tons ; so the loss by using excess of metal cannot be avoided, unless a compound girder is used, and then the extra cost per ton incurred in riveting on the plates to the top and bottom flanges would in most cases exceed that due to the excess of weight in the plain girders.

If, however, there is not sufficient height to allow of the use of such deep girders—they would be required under every floor—compound girders must be resorted to.

The main girders need not project downwards their whole depth below the floor, as the floor joists may run on to angle-irons riveted to the main-girder webs, or they may

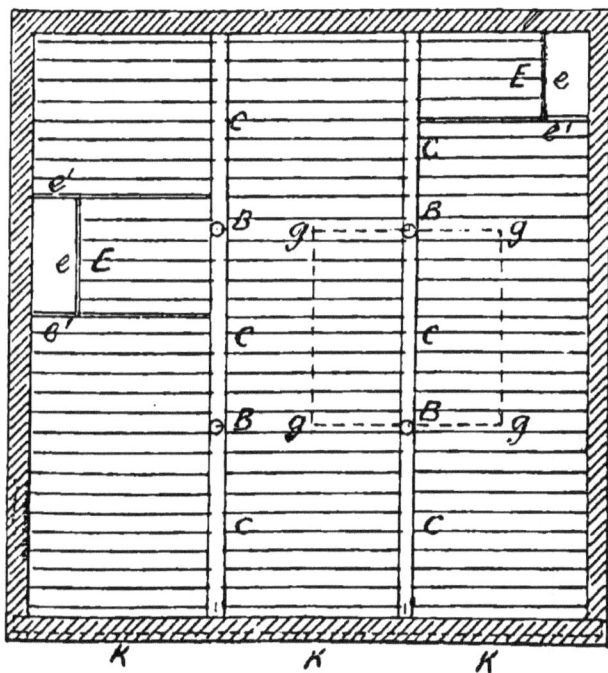

Fig. 41.

rest upon the bottom flanges of the main girders, if these are not high enough to project upwards through the floor finish.

We will now see what size floor joists we shall require, assuming they are to be spaced at two feet centres. These joists are filled in with concrete, and although this packing, if properly made, will tend to prevent buckling and lateral displacement, it can add nothing to the direct carrying strength of the joist, it can take no part of the longitudinal flange-stresses upon the joist, and must be regarded merely as a slab to carry the intervening load on to the joists.

In regard to longitudinal transverse strengths, an inelastic material cannot act in concert with one which is elastic.

Respecting concrete and artificial stone of various kinds no formula has yet been evolved by which their transverse strengths may be determined. The molecular changes which occur under stress are not known, and altogether we are in darkness on the subject. If a floor of steel joists and solid concrete deflects under a load, it is most likely that the concrete will crack at intervals on the under side, and so be effectively broken up into short bearers between the joists, if it was originally good enough to become solid.

In determining the size of the floor-joist, we must then ignore the concrete, and use one capable of carrying the whole superposable load. Taking the distances from centre to centre of the main girders as the joist-spans, these will be 20 feet, and the load on each will therefore be,

$$= 20 \text{ ft.} \times 2 \text{ ft.} \times 3\cdot5 \text{ cwt.} = 140 \text{ cwt.} = 7 \text{ tons.}$$

The nearest stock section above this strength is 8 inches

deep by 6 inches wide, and weighs 36 lbs. per foot run; the working strength of this joist is equal to a load of 7·6 tons, but the next below it is only fitted to take a load of 6·4 tons.

The sections here taken are those kept in stock, but if the quantity of joists required is sufficient for a special rolling order, nearer approximations can be obtained. A lighter, 8 inches × 6 inches, can be rolled which affords a carrying strength of 7·3 tons; but the question of time comes in here. We cannot rely upon getting a parcel of steel rolled in less than five or six weeks; it is not that it takes this time to roll the joists, but each order must take its turn, unless it is sufficiently large to pay for changing the rolls and putting in those required for the particular section.

There need be no difficulty in this matter, if the plans are properly drawn and dimensioned, and the builder is bound to follow these dimensions, the steel-work can be ordered before the building is commenced; but if the plans are in so unsettled a state that we must wait until the walls are up before the sizes of the rooms can be ascertained, there will not be time for getting the floor-joists rolled, and the extra expense of buying out of stock must be incurred.

Let us now see what we can do with a compound girder, assuming that the whole thickness of the floor is not to exceed 16 inches from the top of the concrete, flush with the tops of the floor-joists, down to the undersides of the main girders.

Two rolled girders, 12 inches deep by 5 inches wide and 19 feet span, will carry together 25·47 tons; this, deducted from the total load of 70 tons, leaves $70 - 25·47 = 44·53$ tons to be taken by plates riveted to the flanges of the

rolled girders. The effective depth between the plates will
be one foot, so the stress on each flange at the centre
will be,

$$S = \frac{W \times l}{8 \times d} = \frac{44\cdot53 \times 19}{8 \times 1} = 105\cdot75 \text{ tons.}$$

At a working load of 8 tons per sectional square inch the
required sectional area of plates will be,

$$\frac{105\cdot75}{8} = 13\cdot22 \text{ square inches.}$$

There is, however, to be taken into account the loss by
rivet-holes. These will be $\frac{3}{4}$ inch in diameter, and there
will be two rows in the top flange and two in the bottom.
The central thickness of the flanges of the 12-in. $\times$ 5-in.
girder is 0·6 in., so the loss of sectional area in the
flanges of the rolled girders will be, either top or bottom,
$= 2 \times 0\cdot75 \times 0\cdot6 = 0\cdot9$ square inches; therefore the effective sectional area of plates required will be $13\cdot22 + 0\cdot9 =$
$14\cdot12$ square inches. If two plates 16 inches wide by $\frac{1}{2}$ inch
thick are used for each flange, then, deducting $1\frac{1}{2}$ inches
from the width for loss by rivet-holes in the plates, the
remaining sectional area $= 2 . \{16 - 1\cdot5\} \frac{1}{2} = 14\cdot5$ square
inches. The depth of the main girder over the plates will
be 14 inches, and this will leave one inch top and bottom for
rivet-heads and irregularities. The floor-joists would require to have their ends notched on the top to clear the
flanges and the main girders, and would rest upon angle-
irons riveted to the webs of the latter.

In some forms of floor this notching may be avoided.
Thus, if the floor is to be finished with boards carried on
small sleeper-joists on the concrete, the floor-joists could

then be carried on the bottom flanges of the main girders; this would bring their top surfaces to within $3\frac{1}{4}$ inches of the top of the main girder, and sleeper-joists 4 inches deep would allow the boards to clear them.

The roof-girders carry each the same area—400 feet—as the floor main girders, but with a lighter load. The load upon each will be,

$$= 400 \text{ sq. ft.} \times 1\cdot5 \text{ cwt.} = 600 \text{ cwt.} = 30 \text{ tons.}$$

We cannot find anything less than two rolled girders, 12 in. $\times$ 6 in. at 57 lbs. per foot run, for this purpose, and this has a working load of 34 tons.

The load upon one joist will be,

$$= 20 \text{ ft.} \times 2 \text{ ft.} \times 1\cdot5 \text{ cwt.} = 60 \text{ cwt.} = 3 \text{ tons.}$$

A 5-in. $\times$ 5-in. joist at $25\frac{1}{4}$ lbs. is the least to take this load, and that is not deep enough for the span; but we can use 8-in. $\times$ 4-in. joists, which are much stronger and more rigid, and only 25 lbs. per foot run.

Where the supports stand clear, columns will not only look better but also be more convenient than stanchions, as they have no corners to knock against, and if carts or trolleys are being run about amongst them, they have no flanges to be chipped by things running against them. I shall therefore take B B B B to be columns; they will be bedded one foot below the ground-floor level. From the ground to the first-floor level is marked 12 feet in height, to this is to be added one foot down to the bottom of the base, making 13 feet, from which is to be deducted the total thickness of the floor to give the height of the column. If the compound girder is used the rivets must be flush on the underside where the ends rest on the columns; the depth of main girder—above which the floor-joists do not rise—is 14 inches, and an ordinary allowance for finish is two inches, making the height from the top of the column to the finished floor level 16 inches, therefore the absolute height of the column itself will be,

$$= 13 \text{ ft.} - 1 \text{ ft. } 4 \text{ in.} = 11 \text{ ft. } 8 \text{ in.}$$

Columns of this height and in such a position should be made not less than 12 inches external diameter. Bringing the length into inches and dividing by the diameter, the ratio of the former to the latter is found to be,

$$r = \frac{11 \times 12 + 8}{12} = 11\dot{\cdot}6.$$

By the formula given in a previous chapter the breaking resistance of this column, per sectional square inch, will be,

$$= \frac{36}{1 + \dfrac{(11\cdot6)^2}{400}} = 26\cdot865 \text{ tons per square inch.}$$

To allow for possible imperfections in the castings a higher factor of safety will be taken for the columns than that used for the rolled girders and floor joists, and one-sixth the breaking weight will be taken as the safe working stress ; this will be,

$$\frac{26\cdot865}{6} = 4\cdot477 \text{ tons per sectional square inch.}$$

Now what is the load on each column B on the ground floor ?   It has to carry two ends of girders C, at each level, that is a load equal to the total load on one entire girder at each level.   The roof girder has a maximum load of 30 tons, and the girder load at each of the four floor levels will be 70 tons; so the total on the ground floor column will be $30 + 4 \times 70 = 310$ tons. Dividing this by the safe working strength the required sectional area is found,

$$= \frac{310}{4 \cdot 477} = 70 \text{ square inches (nearly)}.$$

We must now find the internal diameter of column to give this area in the thickness of the metal, which will be the difference between the areas corresponding to the inside and outside diameters. Taking a table of areas of circles it is found that the area of a circle 12 inches diameter is a trifle over 113 square inches, therefore the inside area will be $113 - 70 = 43$ square inches. $42 \cdot 718$ square inches $=$ the area of a circle $7\frac{3}{8}$ inches in diameter, and this will be taken for the inside diameter, and the thickness of metal in the column will be

$$\frac{12 - 7\frac{3}{8}}{2} = 2\frac{5}{16} \text{ inches,}$$

which would in practice be made $2\frac{1}{4}$ inches.

As we ascend the column load decreases. On each of these on the first floor over B the load is reduced to $30 + 3 \times 70 = 240$ tons. The heights of the columns will now correspond to the distances of the floor levels apart, so those on the first floor will be 13 feet high; keeping these also 12 inches in diameter $r = 13$, and the breaking strength per sectional square inch will be

$$= \frac{36}{1 + \dfrac{(13)^2}{400}} = 25 \cdot 31 \text{ tons,}$$

and the working resistance $25 \cdot 31 \div 6 = 4 \cdot 22$ tons (nearly).

The sectional area required for the floor bearing columns on the first floor is, therefore,

$$\frac{240}{4\cdot 22} = 57 \text{ square inches (nearly).}$$

Proceeding, as before, the area of the hollow part of this column should be $113 - 57 = 56$ square inches, which corresponds to a diameter of $8\frac{1}{2}$ inches (nearly), which makes the thickness of this column $\{12 - 8\cdot 5\} \div 2 = 1\frac{3}{4}$ inches.

The load on the columns on the second floor will be $30 + 2 \times 70 = 170$ tons : and the height being 11 feet we may reduce the diameter to 10 inches; then $r = 11 \times 12 \div 10 = 13\cdot 2$. The breaking stress of these per square inch will be,

$$= \frac{36}{1 + \dfrac{(13\cdot 2)^2}{400}} \quad 25\cdot 087 \text{ tons,}$$

and the working stress $= 25\cdot 087 \div 6 = 4\cdot 18$ tons per square inch.

The sectional area required will be,

$$= \frac{170}{4\cdot 18} = 41 \text{ square inches (nearly).}$$

The area corresponding to 10 inches diameter is $78\cdot 54$ square inches ; the internal area of the column must not be more than $78\cdot 54 - 41 = 37\cdot 54$ square inches, which is slightly in excess of that of a circle $6\frac{7}{8}$ inches diameter ; this would be made $6\frac{3}{4}$ inches, and the thickness of metal $1\frac{1}{8}$ inches.

On the third-floor columns the load will be reduced to $30 + 70 = 100$ tons, and the column diameter may be reduced to 8 inches; the height of the column is again

11 feet, so $r = 11 \times 12 \div 8 = 16\cdot5$. The breaking stress per sectional square inch will be

$$= \frac{36}{1 + \dfrac{(16\cdot5)^2}{400}} = 21\cdot43 \text{ tons,}$$

and the working stress $= 21\cdot43 \div 6 = 3\cdot57$ tons per square inch. The sectional area required will be

$$= \frac{100}{3\cdot57} = 28 \text{ square inches.}$$

The area corresponding to 8 inches diameter is $50\cdot26$ square inches, so the internal area of this column will be $50\cdot26 - 28 = 22\cdot26$ square inches. The next convenient diameter below that corresponding to this area is $5\frac{1}{4}$ inches, which leaves a thickness of metal $= \{8 - 5\frac{1}{4}\} \div 2 = 1\frac{3}{8}$ inches.

On the fourth floor the column has only to carry the roof load of 30 tons; we may make it 6 inches in diameter. The height from the top floor to the underside of the roof girders is 9 feet, so $r = 9 \times 12 \div 6 = 18$; and the breaking stress per sectional square inch will be,

$$= \frac{36}{1 + \dfrac{(18)^2}{400}} = 19\cdot89 \text{ tons,}$$

and the working stress $= 19\cdot89 \div 6 = 3\cdot31$ tons per square inch. The sectional area required will be

$$= \frac{30}{3\cdot31} = 9\cdot07 \text{ square inches.}$$

The area of a circle 6 inches in diameter is $28\cdot27$ square inches; therefore the internal area of a fourth-floor column will be $28\cdot27 - 9\cdot07 = 19\cdot2$ square inches, which corresponds to a diameter a.little more than $4\frac{7}{8}$ inches, this would give a thickness of metal $= \{6 - 4\frac{7}{8}\} \div 2 = \frac{9}{16}$ inch; but I should never have a column less than $\frac{3}{4}$ inch in a building of this character.

We now come to the iron and steel work required under and in the front wall of the building; at the first-floor level the wall is carried upon girders resting upon end piers of masonry, and on the two columns A A.

The determination of the load in this part of the structure requires careful consideration, it is simply folly to average such a load per foot of surface or per lineal foot of girder; it often happens that after one storey is passed the whole wall load comes upon the piers over the ground-floor stanchions, each succeeding storey having the outer walls carried by a lintel across the windows.

In the example I have taken there are two windows in each bay, and as the steel joists run parallel to the front wall, there is no floor load to be carried upon the girders $K K K$. The wall having to carry only its own load above the girders $K K K$, lintels will not be required, the ordinary arching over the windows being sufficient. The floor loads will come from the main girders directly over the columns A A. These will be dealt with after the loads upon the girders K.

At M M is shown a vertical section of the front wall on any of the lines h h. The thickness of the wall from the first to the second floor is 2 feet $10\frac{1}{2}$ inches, and at each floor above this is reduced by half a brick, or $4\frac{1}{2}$ inches. The side walls on the ground floor are 2 feet 3 inches thick, so if the caps of the columns A are 18 inches square, the clear span of either side girder will be 20 feet less 9 inches $=$ 19 feet 3 inches; and that of the centre girder 20 feet less 18 inches $=$ 18 feet

6 inches. The section will be settled for the larger span, and kept the same throughout the three. On the first floor the windows are shown 8 feet high by 5 feet wide, they are placed symmetrically in regard to the span, and the space between them is 3 feet. The weight of the windows themselves is so slight in comparison with that of the brickwork that it may be neglected and the whole window area deducted from the wall, especially as there will be a deduction from the brickwork due to splays in the window recesses, which, however, will also be neglected. The load on the central part of the wall between the windows will, of course, come as a central load upon girder K; the windows being 5 feet wide, the width of brickwork between the window and the end of the span of girder K will be (19 ft. 2 in. $-$ $\{5 + 2 + 5\})$ $\frac{1}{2}$ = 3 ft. $2\frac{1}{2}$ in., so each side load will be taken as acting at $3\cdot125 \div 2 = 1\cdot562$ feet from the commencement of the span. A square foot of face surface of wall $13\frac{1}{2}$ inches thick is taken as weighing 1 cwt., and upon this basis the wall loads will be determined.

If the first-floor windows start two feet from the floor level, there will be beneath them a uniformly distributed load upon K, which will be

$$= 19\cdot25 \times 2 \times \frac{22\cdot5}{13\cdot5} = 64\cdot16 \text{ cwt.} = 3\cdot208 \text{ tons.}$$

The 22·5 inches represents the thickness of wall above the girder up to the second floor, and 13·5 inches the thickness corresponding to 1 cwt. per superficial foot of face.

The central load will be the central pier plus half the brickwork over the windows :

Weight of central pier between 1st and 2nd floors	$3 \times 13 \times \dfrac{22 \cdot 5}{13 \cdot 5} = 65 \cdot 00$ cwt.
Weight of central pier between 2nd and 3rd floors	$4 \times 11 \times \dfrac{18}{13 \cdot 5} = 58 \cdot 66$,,
Weight of central pier between 3rd and 4th floors	$4 \times 11 \times \dfrac{13 \cdot 5}{13 \cdot 5} = 44 \cdot 00$,,
Weight of central pier between 4th floor and roof	$4 \times 9 \times \dfrac{9}{13 \cdot 5} = 24 \cdot 00$,,
Half brick-work over 1st-floor windows . .	$5 \times 3 \times \dfrac{22 \cdot 5}{13 \cdot 5} = 25 \cdot 00$,,
Half brick-work over 1st-floor windows . .	$4 \times 2 \times \dfrac{18}{13 \cdot 5} = 10 \cdot 66$,,
Half brick-work over 2nd-floor windows . .	$4 \times 3 \times \dfrac{18}{13 \cdot 5} = 16 \cdot 00$,,
Half brick-work over 2nd-floor windows . .	$4 \times 2 \times \dfrac{13 \cdot 5}{13 \cdot 5} = 8 \cdot 00$,,
Half brick-work over 3rd-floor windows . .	$4 \times 3 \times \dfrac{13 \cdot 5}{13 \cdot 5} = 12 \cdot 00$,,
Half brick-work over 3rd-floor windows . .	$4 \times 2 \times \dfrac{9}{13 \cdot 5} = 5 \cdot 33$,,
Half brick-work over 4th-floor windows . .	$4 \times 3 \times \dfrac{9}{13 \cdot 5} = 8 \cdot 00$,,

$$\overline{276 \cdot 65} \text{ ,, } = 13 \cdot 83 \text{ tons.}$$

This central load is partially distributed, that is, it is spread over the central three feet of the span of the girder, therefore the moment at the centre will be the difference between the moment of the reaction on one point of support and the moment of gravity of half the load about a point at the centre of the span, the pressure on, and therefore the reaction of each support, will be $= \dfrac{13 \cdot 83}{2} = 6 \cdot 915$ tons, and the moment of this about the centre will be $6 \cdot 915 \times \dfrac{19 \cdot 25}{2} = 66 \cdot 556$ foot-tons. The downward moment will equal half the central load multiplied by the horizontal distance (9 ins. $= 0 \cdot 75$ ft.) of its centre of gravity from the centre of the span, it will be therefore $= 6.915 \times 0.75 = 5.187$ foot-tons, and the resultant moment will be $= 66.556 - 5.187 = 61.369$ foot-tons. To find the equally distributed weight to produce an equivalent maximum stress we must equate this with the formula $\dfrac{Wl}{8}$; thus,

$$\frac{Wl}{8} = \frac{W \times 19.25}{8} = 61.369;$$

therefore, $\quad W = \dfrac{8 \times 61.369}{19.25} = 25.503$ tons.

The load on each end of the girder will be as follows :—

Weight of wall between 1st and 2nd floors	$3.125 \times 13 \times \dfrac{22.5}{13.5} = 67.70$ cwt.
Weight of wall between 2nd and 3rd floors	$3.625 \times 11 \times \dfrac{18}{13.5} = 53.16$,,
Weight of wall between 3rd and 4th floors	$3.625 \times 11 \times \dfrac{13.5}{13.5} = 39.87$,,
Weight of wall between 4th floor and roof	$3.625 \times 9 \times \dfrac{9}{13.5} = 21.75$,,
Half brick-work over one tier of windows	$\dfrac{84.99}{2} = 42.49$,,
	$\overline{224.97}$,, $= 11.24$ tons.

The maximum moment of stress from the end loads will be
$$= 11.24 \times 1.562 = 17.556 \text{ ft.-tons.}$$
The distributed load to produce the same maximum stress being $= W$,

$$\frac{Wl}{8} = \frac{W \times 19.25}{8} = 17.556;$$

$$W = \frac{8 \times 17.556}{19.25} = 7.29 \text{ tons.}$$

The total equally distributed load to produce a maximum stress on girder K, equivalent to that resulting from all the different loads upon it will be :

$$3.208 + 25.503 + 7.29 = 36 \text{ tons.}$$

The girder must, therefore, be designed strong enough to carry this load. We find that three rolled steel girders 10 in. × 5 in.—English—will give something more than the required strength, and the question arises whether these or a compound girder should be used.

The width of the wall in this case is $22\frac{1}{2}$ inches, so if the rolled girders are put close together there will be an over-hang on each side of $\{22\cdot5 - 15\} \div 2 = 3\cdot75$ ins. This may be taken up on a stone core laid on the girders, but then the overhang is greater than is advisable, so the girders may be placed a little apart, their distances asunder being fixed by hollow distance pieces through which pass the bolts for holding the girders together; these bolts are necessary, for without them the girders might spread apart under the load if their bearings were not exactly uniform.

Notwithstanding the bolting together of girders placed side by side it does not follow that the superincumbent load will be equally shared amongst them, and this is the strongest argument for compound girders in place of coupled or triplicate ones merely held together laterally by bolts.

In innumerable cases, two rolled girders connected by a plate riveted to their top flanges have been used to carry walls, the plate in this case sometimes serving in place of the stone core—of course no one with the slightest knowledge of the methods of determining the resistance of girders would put it on for strength, without also having an equal plate on the bottom flanges of the rolled girders—it does not distribute the load in such a way that each girder under it has half the load, unless it happens that the bedding of the girder ends is true, and that the bottom surfaces of the bottom flanges lie in the same plane, and this is hardly likely when each one is riveted on *one side only* of its web to the top plate; between the webs the flange edges are almost certain to be drawn away from the plate. There are means of preventing this spread, but as this form of construction should never be used I shall not point them out.

In the case under consideration the centre of gravity of the wall load, section $M M$, Fig. 40, is not exactly over the centre of the width of the girder K, on account of the diminishing thickness of the wall, and therefore if girder K consists of two rolled girders, or three, bolted together, the load upon the outer one will be greater than that upon the inner, and if they are all of equal strengths the outer one will have more deflection and so allow the wall to cant outwards, which is a very serious matter, especially if any floor loads are carried above, for it necessitates the introduction of wall ties, and a structure that is not normally stable is never made so by patching.

It is to be taken, then, that the outside walls of buildings should not be carried upon any but compound, or upon built-up girders, according to the span and load to be dealt with.

In the example, two 8 in. × 6 in. English steel girders taken at one-third the breaking strength would carry 20 tons, leaving 16 tons for the plates; this worked out as in previous examples, would require 10 inches nett sectional area in the flange plate. The mean thickness of the flange of the rolled joist is 0·6 inch, so the loss of area by two ¾-inch rivet holes is 0·75 × 0·6 × 2 = 0·9 in., therefore there must be 11 square inches nett in the plate; a plate 20 inches wide will have, after deducting two ¾-inch rivet-holes, a nett width of 18½ inches, so to get 11 square inches nett the thickness of plate must be:

$$= \frac{11}{18 \cdot 5} = 0 \cdot 594 \text{ inch,}$$

the nearest commercial size to which is ⅝ inch, which would therefore be used.

We now come to the spacing of the rolled joists in this girder: the wall has an overhang on each side of $\{22 \cdot 5 - 20\} \div 2 = 1\frac{1}{4}$ inch, outside the flange plate; the rolled girders should be brought with their flanges one inch inside the outer edges of the plate, so that the overhang of the plate does not exceed twice its thickness; the distance between the flanges of the rolled girders will then be six inches, but as there can be no distinctly *local* pressure here unless the wall splits longitudinally and vertically, there is no danger of depression occurring in the plates.

There is now to be something said about the riveting. There is a habit of riveting the plates to the rolled girders with the rivets zig-zag, which means that a twist shall be put upon the plate, or on the girder according to which first receives the load. With a uniformly distributed load the increments of stress come equally upon each side of the flange as the centre of the span is approached, and therefore it should be uniformly transferred to both halves of the plate riveted on, which it is not if the rivets are zigzag. This habit—a very bad one—of zig-zagging has probably been the outcome of the use of the long pitches; to save labour these long pitches are used regardless of what pitch is necessary—and to save appearances the rivets are zig-zagged to make the pitch *look* shorter.

I may seem prolix on such a matter as this, but if students in architecture will sufficiently interest themselves in the iron and steelwork of the buildings to which they have access during their erection, they will see for themselves that there is ample reason for calling attention to the very lax styles of construction that are permitted to pass and be buried in concrete and stucco.

The next step is to determine the loads upon the columns *A A*. From the front wall each one receives half the load of the girders, $K K$, on each side, so we will take the whole load on one ; that is, one central and two end loads :

$$= 13 \cdot 83 + 2 \times 11 \cdot 24 = 36 \cdot 31 \text{ tons.}$$

Taking other loads, there is first the brick pier immediately above the column, 44 feet high, 1 foot 6 inches wide, with an average thickness of 18 inches, which weighs $1\frac{1}{3}$ cwt. per square foot of surface the weight of this pier will be

$$= 44 \times 1 \cdot 5 \times 1\frac{1}{3} \text{ cwt.} = 88 \text{ cwt.} = 4 \cdot 4 \text{ tons.}$$

The load upon each floor girder C is 70 tons, and half of this comes over the column A from each of the four floors, in all $35 \times 4 = 140$ tons ; from the roof girder comes a load of $60 \div 2 = 30$ tons ; so the total load upon the column A is

$$36 \cdot 31 + 4 \cdot 4 + 140 + 30 = 210 \cdot 7 \text{ tons.}$$

Making these columns of the same diameter—twelve inches —as those under the floor, and the heights being the same, the working resistance will also be $4 \cdot 477$ tons per square inch, and therefore the required sectional area will be

$$\frac{210 \cdot 7}{4 \cdot 477} = 47 \text{ square inches.}$$

The area of a circle 12 inches diameter is $113 \cdot 09$ square inches, therefore the internal area of columns A must be $113 \cdot 09 - 47 = 66 \cdot 09$ square inches, which corresponds nearly to a diameter of $9\frac{1}{4}$ inches, giving a thickness of metal $= \frac{1}{2} \{12 - 9\frac{1}{4}\} = 1\frac{3}{8}$ in.

The next point to be settled is whether the brick pier over this column is strong enough to carry the upper floor loads, or if stanchions will be required ; we have done with the wall load now.

The total weight of the pier has to be taken, 4·4 tons, and the second, third, and fourth floor, and roof loads, making altogether :

$$4·4 + 35 \times 3 + 30 = 139·4 \text{ tons.}$$

The sectional area of the pier is only 2·25 feet and therefore quite insufficient to carry this load, or even the roof load—therefore stanchions must be carried up in tiers ; and in the same way stanchions must be used in the back walls to carry the floor and roof loads.

Having fully detailed the method to be followed in reducing the columns under the floors as they rise, it is unnecessary to occupy space in following it through again in connection with the wall stanchions ; but it may be pointed out that the back wall stanchions carry no wall load, as the wall rises from the foundations, and any doorways and windows in it are assumed to be arched over with arches of which the strength is sufficient to carry the wall above. Where there is not sufficient height for the construction of such arches, lintel girders must be used ; and this is generally found necessary in buildings in which the windows are carried close up to the ceiling in order to give the best ventilation in the way of releasing the hot air accumulating at the top of a room, which if left would, by reason of its impurity, sink down again in cooling and so render the air in the lower part of the room unfit for respiration.

At the staircase openings $E E$, it will be necessary to use trimming joists e, to carry the ends of the ordinary joists and to have joists at e', e', e', with additional strength to take the loads from the trimmers, as well as that from the parts of the floor resting upon them.

The method of dealing with these loads has already been shown.

CHAPTER VII.

GENERAL ARRANGEMENT OF BUILDINGS.

THERE are two very distinct classes of buildings, those
which have internal walls to act as ties, and those which
consist of external shells of brickwork filled in with floors
and staircases carried by constructional ironwork ; in the
latter case the architect often seeks to tie his walls together
with metal ties, which may be specially provided for the
duty, or the main girders or floor-joists may be made to
serve in this as well as in their load-carrying capacity ; in
the latter case angle-iron cleats are riveted to their ends to
give them a good wall hold. This arrangement is not,
however, without its disadvantages on account of variations
of length due to changes in temperature. The variation of
temperature during the year in England is, in exposed
localities, about 81 degrees Fahrenheit, and this corresponds
to a variation in length in wrought iron and mild steel of
about one inch in one hundred and fifty feet of length. If
no provision is made to allow for expansion and contraction,
it is evident that a stress must be put upon the end bearings
of the girders, and the magnitude of this stress will be
considerable ; a variation of 27 degrees Fahrenheit in cast
iron, or 13½ degrees in wrought iron, produces a change of
length equivalent to that accruing from a direct stress of one
ton per sectional square inch of the girder or joist, and if
the ends are solidly built into the brickwork, the stress
must come upon the walls, tending to push them out or pull

them inward, according to whether the temperature is rising or falling.

The structural iron work within a building will not be liable to such extremes of temperature as occur in the open air, but a variation of 54 degrees may be expected, and if the girders were fixed when either extreme of temperature obtained they would be capable of causing a stress of 4 tons per sectional square inch when the other extreme is reached.

Any ordinary or even extra strong wall must yield to this. In the example taken in Chapter VI. the main girders carrying the floors were 18 in. deep by 7 in. wide, used in pairs placed close together; the sectional area of one girder this size is 24·67 in., and therefore one pair would be capable of exerting a stress equal to 24·67 × 2 × 4 tons = 197·36 tons. A variation of temperature of five degrees would cause a thrust or pull of 18·3 tons, so it is very obvious that if the girders have their ends actually keyed into the walls the latter must yield and move with variations of temperature.

When expansion of the girders is occurring, it may be taken up by distortion or lateral bending, but during contraction an absolute pull will come upon the walls, which cannot be relieved from it, if the wall holds of the girder ends fit in the walls; for instance, if holes are sunk in stone templates to receive the girder cleats, and these are then run in with lead or cement, the girders cannot work their ends loose, and the wall will be constantly worked inward and outward with the changes of temperature. In the sixty feet run of the girders the extreme variation of length will be $\frac{60}{150} = 0\cdot4$ in., and taking this to be divided between the two walls, the movement of each would be one-fifth of an inch. If the cleats are built into the brick walls only, it is probable that they would soon work loose to this extent, and so cease disturbing the walls, though they would yet serve to prevent them from falling outwards from any other cause so long as the cleats hold.

If the metal joists and main girders are buried in concrete of a non-conducting character, extremes of temperature will fail to reach them except in the case of a long-continued conflagration, when, if the main girders become so hot as to yield under their load, the cleats on their ends will be a source of great danger, because through them the walls of the building will be pulled in and the structure will thus be caused to collapse.

In designing any kind of building we must consider what is likely to happen in case of the place taking fire. It cannot be said that any structure is absolutely fire-proof, stone will crack and fly to pieces, wrought iron and steel will become red-hot and bend, letting down the floors they are designed to carry, and cast-iron columns and stanchions will melt with the same result.

It is very evident that the best way to reach a practically fire-proof system of construction is to design one in which the metal elements of the structure are automatically cooled by draughts of air flowing continuously around them. There are several systems in which this is aimed at and carried out to a greater or less degree; so long as the fabric can be kept cool there is no fear of its failure, although a fire may be consuming its contents, but there is a limit to the time that concrete encasements will exclude the heat, and some concretes are not suited for the purpose at all, having combustible substances in their composition.

There is an old saying that heat goes upwards, and this is true, but not by any means the whole truth, for it also descends, therefore the top of a floor should have as much protection from heat as the underside; numbers of cases have occurred in which a fire has attacked the top floors of buildings and burnt them right out to the basements.

All the rooms in any large or important building should be so constructed, that anyone of them can be isolated from the others, and the staircases should also be contained between walls by which they are shut off from the rooms, so that if a fire break out it may be confined to the room in which it originates.

In floors where the joists are all of iron or steel, the cost will depend very materially upon the planning of the work, and in most structures, whether public or private, the observance of economy of cost is a prime consideration. Taking into consideration, in the first place, the different qualities of British and foreign rolled joists, I will point out by a few examples how differences of price compare with relative strengths. The figures quoted on both sides are taken from the published tables issued by firms of good repute, and refer to joists freely supported at both ends and uniformly loaded, the safe load being taken as one-third of the breaking weight. Market prices and special quotations vary in ways, and under certain conditions, which would make any observations based upon specified prices useless so far as general application of them is concerned, and I shall therefore show the amount of extra strength of joists in regard to equal weights, and from this the permissible extra cost may be ascertained.

In choosing sections, we are tied to those rolled by the manufacturers, we cannot specify just what we want; so it is no use arguing one way or the other that a certain section is ill-designed, they must be considered as they are. For convenience of calculation, the strengths will be taken for spans of ten feet.

Now, if two rolled joists are made from the same metal, and their proportions are exactly similar, then their strengths will be in exact proportion to their weights per foot run, so that the quotients found by dividing the safe loads by the weights would be equal ; from this it follows that by dividing the safe loads of joists by their weights per foot run we find a series of critical numbers which give the relative values of the joists, in fact, they show the carrying efficiencies per pound weight per foot run. Taking a 12 in. × 6 in. section in each make, the English has a weight of 57 lbs. per ft., and carrying strength of 45·79 tons ; the critical number is = 45·79 ÷ 57 = 0·803 ; the foreign section has a weight 44 lbs. and safe load 22·1 tons ; the critical number is therefore = 22·1 ÷ 44 = 0·502 ; therefore for this particular section, if the foreign material is worth £5 per ton, the English is worth £8. I will take next the 9 in. × 7 in. section, a very useful one where headway is short, but one that is more expensive than others on account of the difficulties experienced in rolling girders with flanges so wide in proportion to the depth. The English section has a weight of 58 lbs. per foot ; load 34·53 tons ; critical number = 34·53 ÷ 58 = 0·595. The foreign section has, weight 52 lbs. per foot ; load 18·8 tons ; critical number 18·8 ÷ 52 = 0·361. If the foreign section is worth £5 a ton, the English is worth £8 5s.

Now there is nothing approaching this difference in prices, and it therefore appears that although foreign steel can be delivered in our own manufacturing centres at a lower price than that at which we can produce our own material, yet the quality of the English steel joists is so much better that it is cheaper to use them.

These are heavy sections, so I will now compare some very light ones, 4¾ in. × 1¾ in. English steel joists have, weight 9·25 lbs. per ft. ; load 2·42 tons ; critical number = 2·42 ÷ 2·25 = 0·261. The foreign section has, weight 6½ lbs. per ft. ; load 1·24 tons ; critical number = 1·24 ÷ 6½ = 0·180 ; so in this case, taking the foreign steel at £5 per ton, the English would be worth about £7 5s.

These strengths, although published, are not guaranteed in either case except under special specifications ; in fact it would be quite out of the question for any merchant to guarantee the strengths of stock-joists, which may come from different rollings ; joists that are to be guaranteed must, as a rule, be specially rolled to order.

There would be no use in carrying this investigation farther, as sufficient examples have been taken to show in which direction the advantage lies, so I will now proceed to consider the arrangement of floors in regard to the use of main girders for carrying the floor-joists.

For this purpose the floor-joists will be assumed to be fixed two feet apart from centre to centre, and to carry a total load of two hundredweight per superficial foot of floor area.

Now assuming that the spans of the floors are sufficiently great to necessitate or to justify the use of main girders to carry the joists, some general principles may be laid down— to be subsequently proved—which will help in determining the direction of such main girders.

If the floor load has ultimately to come upon the walls, it should be brought there in the most direct manner possible, that is, it should not pass through girders un-necessarily ; and even if it is ultimately to be taken by girders, it should be brought directly to them and not be passed to one girder which rests upon another, before reaching its ultimate support, if this can be avoided.

This leads to the axiom that all the ends of floor-joists

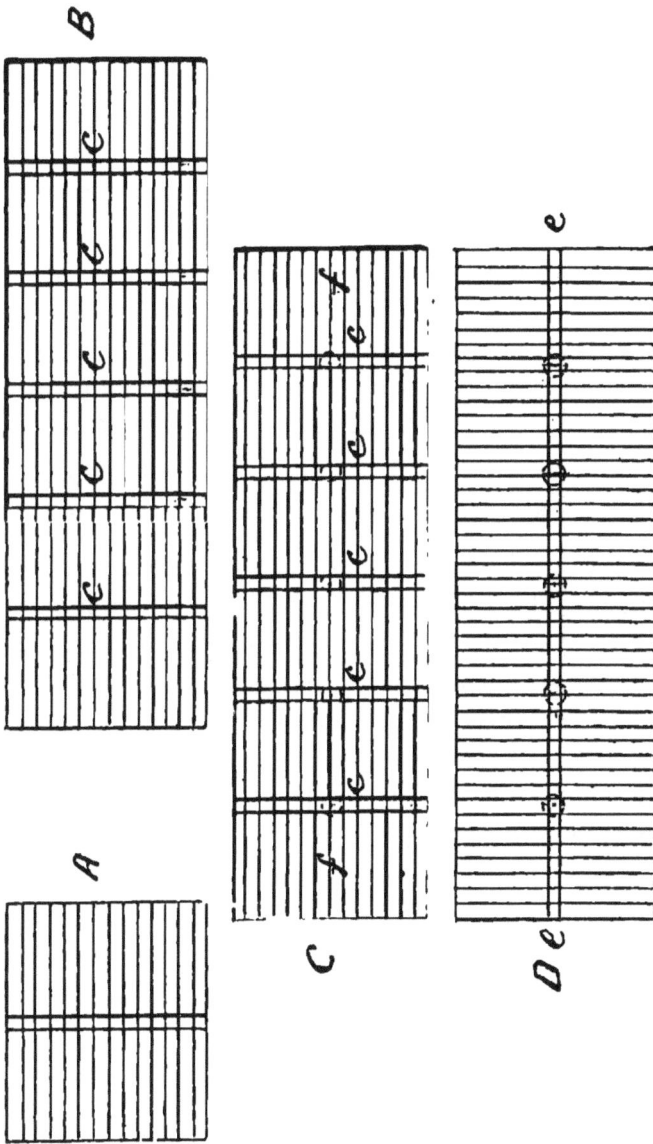

Fig. 42.

which can be arranged to rest directly upon the bearing walls should do so ; and this is much better for the wall as well, because the floor load is then distributed equally along it instead of being concentrated at the ends of main girders, which in many cases necessitates the building of strong piers, or the use of cast-iron stanchions in the walls, thereby materially increasing the cost of the structure.

In Fig. 42, some plans of floors are shown. That at *A* represents a floor 30 feet square ; the point to be decided is whether it is cheaper to use steel joists over the whole span, or to carry on one or more intermediate main girders. Taking the whole span, 30 feet, the load upon each floor-joist will be

$$= 30 \text{ ft.} \times 2 \text{ ft.} \times 2 \text{ cwt.} = 120 \text{ cwt.} = 6 \text{ tons.}$$

I shall work all these calculations for English steel joists and girders, and use one-third as the factor of safety. It is needless to say that I have not space at my disposal to work out the strength of every element, and I shall therefore use a table of strengths which I have verified, assuming the steel to have an ultimate tensile strength of 32 tons per sectional square inch. This would require 8 in. × 5 in. joists weighing 34 lbs. per foot run. The width of the room being 30 feet, there will be fifteen bays and therefore fourteen joists : the bearing to be allowed at each end for this span is 9 inches, therefore the total length of each joist will be 31 feet 6 inches, and the length of joists for the whole floor will be

$$31 \cdot 5 \times 14 = 441 \text{ lineal feet,}$$

and the total weight will be 14,994 lbs. It is convenient to keep these weights in pounds for comparison.

If one main girder is used across the centre of the room, both span and load on each joist will be halved ; the load

will be 3 tons and the span 15 feet.   The nett span is less
by half the width of the main girder, but it is not advisable
to rely on the edge of the girder flange as the limit of
span.

For the floor-joists, we have to find a section to suit 3
tons over 15 feet, and this will be satisfied by a 5 in. $\times$
3 in. joist weighing 13·5 lbs. per foot run. The total length
of joists will be the same as before, and therefore their total
weight,

$$= 441 \text{ ft.} \times 13\cdot5 \text{ lbs.} = 5953\cdot5 \text{ lbs.}$$

The load to be carried by the girders is evidently one-
half of the floor load, which

$$= 30 \times 15 \times 2 \text{ cwt.} = 900 \text{ cwt.} = 45 \text{ tons.}$$

This load would require two rolled girders 16 in. $\times$ 6 in.
placed side by side and bolted together; the weight of the
two is 136 lbs. per foot run, and the length, allowing a foot
at each end for bearing, 32 feet; the weight will therefore
be

$$= 32 \times 136 \text{ lbs.} = 4352 \text{ lbs.}$$

The total weight of the steel-work will therefore be, in this
arrangement,

Floor joists	5953·5 lbs.
Main girder	4352·0 ,,
	10,305·5 ,,

against 14,994 lbs. for joists taking the whole span. This
shows a saving of metal equal to 4689 lbs., or 31·27 per
cent. We will now see if any further saving can be made
by using two main girders, which would divide the floor
into three equal parts. The effective spans of the floor-
joists would then be reduced to ten feet, and the load on

each to 2 tons, which would require floor joists 4 in. ×
1¾ in., weighing 8·5 lbs. per foot. The total weight of floor
joists would then be

$$= 441 \text{ ft.} \times 8\text{·}5 \text{ lbs.} = 3748\text{·}5 \text{ lbs.}$$

Each main girder will carry one-third of the floor, there
fore the load upon it will be

$$= 30 \times 10 \times 2 \text{ cwt.} = 30 \text{ tons.}$$

An 18 in. × 7 in. rolled girder weighing 78 lbs. per foot
will carry this with a little margin; the weight of the two
main girders will therefore be

$$= 2 \times 32 \times 78 \text{ lbs.} = 4992 \text{ lbs.},$$

and the total weight of steel work,

Floor-joists	3748·5 lbs.
Main girders	4992·0 ,,
	8740·5 ,,

Which shows a further saving, that over the plain floor
amounting to 6253·5 lbs. or 41·7 per cent. (nearly).

The smallest joist on the list is 3 in. × 1¼ in., and this
would carry 1·37 tons over a six-foot span; the floor load
on a joist of this span will be—

$$= 6 \times 2 \times 2 \text{ cwt.} = 34 \text{ cwt.} = 1\text{·}2 \text{ tons.}$$

So if this joist is used the floor may be divided into five by
the main girders, of which four would then be required, the
load on each main girder being

$$= 30 \times 6 \times 2 \text{ cwt.} = 360 \text{ cwt.} = 18 \text{ tons.}$$

A 15 in. × 5 in. girder weighing 60 lbs. per foot would do
this work, and the weight of the four main girders would be

$$= 4 \times 32 \times 60 \text{ lbs.} = 7680 \text{ lbs.}$$

The 3 in. × 1¼ in. joists weigh 5 lbs. per foot, making the joist weight

$$= 441 \text{ ft.} \times 5 \text{ lbs.} = 2205 \text{ lbs.}$$

and the total weight of steel work in the floor—

Floor-joists	2205 lbs.
Main girder	7680 ,,
	9885 ,,

which is in excess of that found when using 4 in. × 1¾ in. joists, to which we should therefore keep. In working out these trials it has been assumed that the most convenient depth of main girder can be used without restriction, and that the floor joists can run over them, but if there is a limited headway the case will be altered, because shallower joists must be used or the floor joists must be supported on angle-irons riveted to their webs; if the first, the weight of the main girders will be greater. Thus, instead of one 18 in. × 7 in. girder, we might have to use two 12 in. × 6 in. rolled girders side by side for each main girder, which would then have a weight of 114 lbs., making the weight of the two main girders

$$= 2 \times 32 \times 114 \text{ lbs.} = 7296 \text{ lbs.}$$

and the total weight of steel in the floor,

Floor-joists	3748·5 lbs.
Main girder	7296·0 ,,
	11044·5 ,,

So in this case the plan with one main girder would be cheaper than that with two.

It is evident that so much depends upon the conditions surrounding each case that no general rule can be laid down, and there is no means of avoiding the drudgery of trying one scheme after another until the most economical one is determined.

To prevent the projection of the whole depth of the main girders below the floor-joists, the latter, instead of running over them, may have their ends resting upon angle-irons riveted to the webs of the main girders at such a level that the top of the main girder is flush with the underside of the floor finish, and with the top of the concrete, where a concrete floor is used, or with the top of the floor joists where this material is not adopted. The addition of the angle-irons will, of course, add both in weight of material and cost of riveting to the price of the main girders, and if the main girders are flush with the floor-joists on the top surface there will also be the expense of notching off the top flanges of the latter to allow their ends to butt up against the webs of the main girders. There is also another expense incurred in placing the floor-joists between, instead of upon, the main girders, and that arises from having to cut them to dead lengths instead of using them with the mill margin of one inch long or short.

Where flat ceilings are indispensable, but the floor thickness is not limited, the joists may rest upon the bottom flanges of the main girders, and concrete ridges or sleeper-joists, according to the material used, be provided to carry the floor finish clear of the tops of the main girders. It will not be necessary to joggle the bottom flanges of the joists, unless those of the main girders are thicker than an inch, as this can be made up in the ceiling material, or by concrete, if that is used for the bottom of the floor.

A floor of a different form is shown at B, Fig. 42. This is assumed to be 28 feet wide between the walls, and 90 feet in length; and that there is sufficient headway below to allow the introduction of main girders with the floor-joists running over them.

For a workshop or warehouse floor I should not care to use any joist of a section less than $4\frac{3}{4}$ in. $\times$ $1\frac{3}{4}$ in., on account of the liability to impact from falling weights, and the turning over and over of heavy cases to get them along the floor.   A packing-room floor is particularly subject to concussions of this sort if the goods handled are heavy. Under localized loads of heavy machines or large stoves, special joists of greater strength must be fixed.

At a load of 2 cwt. per foot super of floor, and joists 2 feet apart, the $4\frac{3}{4}$ in. $\times$ $1\frac{3}{4}$ in. joist will carry over a span of 15 feet 6 inches, or it would support a central load of $15\frac{1}{4}$ cwt.   If the room is divided into six equal bays, the joist-span from centre to centre of girders will be 15 feet, and this will also be the distance from the centres of the end girders to the insides of the end walls.

The load upon each of the five main girders c will be

$$= 28 \times 15 \times 2 \text{ cwt.} = 840 \text{ cwt.} = 42 \text{ tons.}$$

A rolled girder 20 in. $\times$ 8 in. weighing 90 lbs. per foot will carry this load with a slight margin.   Now arises a question whether any economy will result from putting columns under the main girders, as shown at C.   By so doing both load and span are halved, and the strength of the main girders is to be determined for a span of 14 feet and a load of 21 tons.   A girder 10 in. $\times$ 6 in. weighing 43 lbs. per foot will meet the case, this would save in girder-weight 47 lbs. per foot, which, taking the girders as 30 feet long, will save on each girder,

$$47 \times 30 = 1,410 \text{ lbs.}$$

Suppose the height of column required to support the girder

to be 14 feet and its diameter 6 ins., then by the formula already given its breaking resistance per sectional square inch will be, as

$$r = \frac{14}{0\cdot 5} = 28 = \frac{36}{1 + \frac{(28)^2}{400}} = 12\cdot 16 \text{ tons.}$$

and taking six as a factor of safety, the working stress

$$s = \frac{12\cdot 16}{6} = 2\cdot 02 \text{ tons per square inch.}$$

The load on each column will be 21 tons—one-half of each load on the spans on either side—therefore the sectional area of the columns will be

$$\frac{21}{2} = 10\cdot 5 \text{ square inches (nearly).}$$

This would only require the thickness of $\frac{5}{8}$ inch of metal, but in such a position I should not make it less than $\frac{3}{4}$ inch thick; this will make the sectional area 12·37 square inches, and as a bar of cast iron one foot long weighs 3·2 lbs. the weight of the column will be,

$$= 12\cdot 37 \times 3\cdot 2 = 39\cdot 58 \text{ lbs. per foot of length.}$$

For a rough estimate three feet may be added to the length to account for the cap and base, and the total weight of the column will be, by this method,

$$17 \text{ ft.} \times 39\cdot 58 \text{ lbs.} = 672\cdot 86 \text{ lbs.}$$

There is not very much difference between the cost of columns per cwt. and that of rolled girders, so in this case there would be an evident saving by using the columns. It may be convenient to carry the girders C right across in one length, as by so doing they form ties to steady the columns, and their own deflection is very materially reduced, but there is no gain in strength, the maximum moment of stress being the same as it would be if the spans were separate and free, but instead of occurring at the centre of the span it accrues immediately over the supporting central column.   If the girder is not continuous, its two parts should be joined over the column by cleats riveted on to the ends of each—one on each side of the web—and then bolted or, preferably, riveted together by their free limbs.

Care is to be taken to avoid all labours that are not absolutely needed, for the cost of executing these in builder's ironwork is very heavy; this is due, in a great measure, to the discontinuity of the work. If holes have to be drilled at considerable distances apart, and only a few of them in each piece of work, each hole must necessarily cost a great deal more than if a great number of holes are to be drilled in close succession; the expenditure in labour is proportionately excessive, for the lifting and lowering of a girder to drill one hole in it costs as much as if twenty were drilled.

Drilling on the site of the structure is yet more expensive, because the drilling tackle has to be shifted for each hole, and the positions in which the work has to be done are often very awkward.

If it is decided to use columns, which in this case show a saving in the main girders c, we can effect a further saving by running the main girders longitudinally from e to e as shown at D. In the plan C, only one-sixth of the floor load passes directly on to the walls, and that sixth is represented by the load passing from the end bays f f, on to the end walls.

Now, if the longitudinal girders are adopted, and the

floor-joists run across the building, as shown at D, one-half of the floor load passes directly on to the side walls, and the other half is taken by the longitudinal main girders on to the columns.

Each span of the longitudinal girder will be 15 feet, reckoning from centre to centre of columns, and the floor-load upon each span will therefore be,

$$15 \times 14 \times 2 \text{ cwt.} = 420 \text{ cwt.} = 21 \text{ tons.}$$

This is the same load as that on each span of the main girders c in plan C, but it has to be carried over a larger span—15 feet instead of 14 feet; this will require a somewhat heavier section, but it is supplied by the maximum weight rolled of 10 in. × 6 in. girders, which is 48 lbs. per foot run.

I will now compare the weights of the transverse main girders with those of the longitudinal girders, allowing one foot bearing on the walls in each case.

The weight of the transverse girders would be 43 lbs. per foot, therefore five, each 30 feet long, would weigh

$$5 \times 30 \times 43 \text{ lbs.} = 6450 \text{ lbs.}$$

The longitudinal girders will in all reach a length of 92 feet, so their weight will be,

$$92 \times 48 \text{ lbs.} = 4416 \text{ lbs.}$$

As the span of the floor-joists is reduced the same section can be used as in the former case, that is $4\frac{3}{4}$ in. × $1\frac{3}{4}$ in. There is therefore a clear saving in weight of main girders amounting to

$$6450 - 4416 = 2034 \text{ lbs.,}$$

something over $31\frac{1}{2}$ per cent. of the total weight of the main girders.

While dealing with the matter of using appropriate sections for different loads, I must not omit to point out that these are not always immediately procurable; the medium or normal sections are kept in stock by the manufacturers or their agents, but if the maximum or minimum weights are required they must be rolled for the purpose. If the order for one such section is large, say twenty tons, the manufacturers may put the rolls in specially; if not, the purchaser must wait until the rolls go in to fill a number of orders sufficient to cover the expense of changing the rolls. Time is generally an object in these matters, and therefore it is very frequently necessary to use such joists or girders as can be bought out of stock; in regard to minimum sections, this matters only to the extent of the extra weight, and therefore cost, of the stock sections; but if a maximum weight of any particular section is required and cannot be obtained, we must use the next normal section above it. For instance, if the 10 in. × 6 in. girder referred to above, at its maximum weight of 48 lbs. per foot run, cannot be obtained, a deeper one must be used.

In all structures of small extent, especially where there are many different spans of floors, the architect should design his floors to suit such "stock" sections of joists as are easily obtainable; this he can do without difficulty, as manufacturers and merchants are always willing to furnish architects and builders with their section sheets and books of strengths; for alterations of sections after the plans and elevations are completed will almost always cause a disturbance of floor levels and lead to a good deal of trouble, and perhaps serious error.

If columns or stanchions have been cast to measurements suited to the use of girders ten inches deep, and it is found afterwards that the nearest section procurable is twelve inches deep, either the columns must go to waste or the building must rise two inches more at each storey where the girders occur, than was originally intended.

If neither columns or stanchions are used the inconvenience is not so serious, especially if the necessity of the alteration is discovered before the stone templates are fixed to receive the main girders, as then it will only mean that the main girders must project two inches lower beneath the ceiling than was originally intended, unless it is thought desirable to give the extra height to the building required to retain the clear headway first decided on.

It is obviously much more economical to use deep rolled girders than shallower compound ones, but as circumstances frequently require the adoption of the latter, a simpler way of finding what plates are necessary to make up a certain amount of strength is now shown.

Suppose a girder is required 30 feet span to carry an equally distributed load of 60 tons, and let two 16 in. × 6 in. rolled steel joists be used with flange plates riveted on the top and bottom.

The safe load on the 16 in. × 6 in. joists will be, at one-third the breaking strength = 43·2 tons, this will leave to be taken up by the flange plates 60 − 43·2 = 16·8 tons. The safe resistance of the steel flange plates may be taken at 10 tons per sectional square inch, then, proceeding in exactly the same way as with any other built-up girder, the sectional area of each flange at the centre of the span must not be less than—

$$\frac{W \times l}{8 \times d \times s} = \frac{16 \cdot 8 \times 30}{8 \times 1\frac{1}{3} \times 10} = 4 \cdot 72 \text{ square inches.}$$

This is the nett sectional area required, and the loss by rivet holes in the flanges of the rolled joists must be made up in the plates. The rivet holes will be $\frac{3}{4}$ inch in diameter, and the thickness of flange of the joist where the rivet holes will occur is 0.82 inch, the loss from two rivet holes will be $= 0\cdot75 \times 2 \times 0\cdot82 = 1\cdot23$ square inches, so the net sectional area of flange plate must be $= 4\cdot72 + 1\cdot23 = 5\cdot95$ square inches. A plate 12 inches wide by $\frac{1}{2}$-inch thick would give this, but to make up for the rivet holes in the plate, another $1\frac{1}{2}$ inches must be added to the width, making it $13\frac{1}{2}$ inches. The width in practice would be 14 inches, and the thickness $\frac{1}{2}$ inch. These flange plates need not necessarily be carried the whole length of the girder, for where the moment of stress drops to that which the rolled girders can carry unaided, the flange plates may terminate. The safe load for the girders without the plates being 43·2 tons distributed, the maximum moment corresponding is—

$$= -\frac{W \times l}{8} = -\frac{43\cdot2 \times 30}{8} = -162 \text{ foot-tons.}$$

There being a load of 60 tons upon a girder of 30 feet span, the load per lineal foot is 2 tons $= w$; then, by the formula previously given the moment of stress M at any point distant x feet from either point of support is—

$$M = \frac{w \cdot x^2}{2} - \frac{w \cdot l \cdot x}{2} = x^2 - l\,x.$$

The values of M and l are known, being 162 and 30 respectively, so the equation becomes

$$-162 = x^2 - 30\,x.$$

This is an incomplete quadratic equation, to be treated in the usual way, thus—

$$x^2 - 30\,x + (15)^2 = -162 + (15)^2 = -162 \times 225 = 63.$$

Taking the square root of each side of the equation, we get

$$x - 15 = + \sqrt{63} = \pm 7\cdot937$$

$$x = 15 \pm 7\cdot937 = 22\cdot937, \text{ or } 7\cdot063 \text{ feet.}$$

To prove this it is as well to check the result by the formula for M—

$$M = x^2 - lx = (7\cdot063)^2 - 7\cdot063 \times 30$$
$$= - 162\cdot004 \text{ foot-tons.}$$

The excess ·004 is due to dropped decimals. Working with the second value of x—

$$M = x^2 - lx = (22\cdot937)^2 - 22\cdot937 \times 30$$
$$= - 162 \text{ feet-tons.}$$

The length of flange plates required will therefore be equal to the total span of the girder, less twice the lower value of x,

$$= 30 - 2 \times 7\cdot063 = 15\cdot874 \text{ feet};$$

the plates would be made 16 feet long.

In this matter of cutting off the plates at the points where they are not required, another trouble sometimes occurs. If the floor-joists are carried on the top of the girder, they must have a level bed throughout, and therefore the spaces between the ends of the flange plates and the ends of the span must be packed to take the floor joists; and in most cases it will be more economical to carry the plate on the top flange through the whole length of the girder.

The rivet heads throughout the top flange should be countersunk flush with the top of the flange plates, so that they may not foul the ends of the floor joists.

Where machinery in motion is carried upon the floors,

special precautions are to be taken to prevent the vibration of the machine from putting a racking stress upon the steel-work of the floor.

So far as the wrenching stresses are concerned, these will be taken up by a solid plate or framing, upon which the machine is fixed, so that no stress is put upon the steel-work of the floor except that arising from the weight of the machine, and the general vibration due to the motion.

The introduction of special joists or girders to carry concentrated local loads leads up to the circumstances under which combinations of girders may be used, arranged so that one may afford relief to another.

The simplest case is shown by a girder having a normal span of considerable length, but which has under its centre another girder at right angles to it and partly supporting it, in common language, taking part of the load off it. Suppose, for instance, there is a space 40 feet long and 12 feet wide, with a heavily loaded girder running longitudinally along the centre, and that no supporting columns can be put under it, but a transverse girder under the centre is admissible. There are two ways of using this, making the longitudinal girder in two lengths, so as to throw half the total load upon the centre of the transverse girder; or, carrying the longitudinal girder across in one length, and so making use of the strength of its section over the central auxiliary girder. In the first arrangement the transverse girder becomes a bearing girder, not an auxiliary one, and it is treated in the usual way as a girder loaded at the centre. The primary girder in the second case appears under conditions approaching those which obtain with a continuous girder of two spans, but these are not reached because the central support is elastic, and therefore allows a central deflection of the primary girder.

The nicety of calculation in this combination turns upon the point that it is not only the absolute strengths of the primary and auxiliary girders which must be proportioned to the respective shares of load to be carried by each, but their sections must be so proportioned, the material is to be so disposed on each side of the neutral axis, that under these relative loads the deflection of both girders at the point of connection shall be normally equal, otherwise the stresses will not be properly apportioned, and we shall not know how much load is apportioned to each element.

If, for instance, it is desired that the relieving girder shall pick up one-third of the total load, its section must be so proportioned that with that central load its deflection shall be equal to that of the primary girder under two-thirds of the total load distributed. Now, although we may not be able to ascertain the exact modulus of elasticity of the metal, yet we may fairly assume that if both the girders are made from the same working of iron or steel, that the modulus will be the same for both of them, and this being so, the laws of deflection will become practically applicable ; these laws, as demonstrated in a previous chapter, show that :—

The deflection varies directly as the load and as the cube of the span of the girder, or length of the cantilever, and inversely as m, where $m = b\,d^3 - b'\,d'^3 - \&c. - b^n.\,d^{n3}$. It also varies inversely as the modulus of elasticity, which is included in the constants given for each class of work. If a girder carrying a uniformly-distributed load is partly supported at the centre, the deflection at that point will be the total deflection due to the uniformly-distributed load, less a deflection (upwards) due to the sustaining force at the centre. This I have proved by experiments. If, for instance, there is a girder 30 feet span, loaded with 20 tons, and the value of m is 44,000, the central deflection under this load will be—

$$D = \frac{W\,l^3}{44 \cdot 8\,m} = \frac{20 \times 30^3}{44 \cdot 8 \times 44,000} = 0 \cdot 274 \text{ inch.}$$

If there is a supporting force in the centre of the beam equal to 6 tons, the deflection equivalent to this would be—

$$D = \frac{W\,l^3}{28\,m} = \frac{6 \times 30^3}{28 \times 44,000} = 0 \cdot 131 \text{ inch.}$$

Hence the actual central deflection of the beam will be—

$$0 \cdot 274 - 0.131 = 0 \cdot 143 \text{ inch.}$$

The point of maximum stress will not be at the centre of the span, the curve of stress being, as it were, caught up there.

Let l = span of girder in feet.

Let w = load in tons per lineal foot.

Let R = reaction at one point of support.

Let P = upward sustaining force at centre of girder.

Let M = moment of stress in foot-tons at any point distant x feet from the nearest end support.

On each end support one-half of the total load, less one-half of the central force P, will act; therefore,

$$R = \frac{w.l}{2} - \frac{P}{2}$$

and,

$$M = \frac{w\,x^2}{2} - R\,x = \frac{w\,x^2}{2} - \frac{w\,l\,x}{2} + \frac{P\,x}{2}$$

We will now find the points of maximum stress. When this is reached and the moment about to diminish we may

imagine an indefinitely small increase of x during which
the moment of stress remains constant; here, then, the
increase of the positive quantity must be equal to that of
the negative quantity; let, then, x become $x + a$, a being
indefinitely small; then—

$$M = \frac{w}{2}(x + a)^2 + \frac{P}{2}(x + a) - \frac{w.l}{2}(x + a)$$

$$= \frac{w}{2}(x^2 + 2ax + a^2) + \frac{P}{2}\cdot(x + a) - \frac{w.l}{2}(x + a).$$

The quantity a being originally very small compared
with x, the quantity a^2 is so much smaller that it may be
neglected; then the increase of the positive quantity is—

$$= wax + \frac{Pa}{2};$$

and that of the negative quantity,

$$= \frac{w.l.a}{2};$$

equating these, we have,

$$w.a.x + \frac{Pa}{2} = \frac{wl.a}{2};$$

therefore,

$$x = \frac{w.l.a}{2w.a} - \frac{Pa}{2w.a} = \frac{l}{2} - \frac{P}{2w},$$

which is the value of x corresponding to the maximum
moment of stress.

Suppose, now, a uniformly-loaded girder, to be assisted
by another girder of similar section placed immediately
beneath it, and supporting it at the centre, the deflections
would be equal; what would be the relations of the maxi-

mum stresses upon them ? The maximum stress upon the
second girder will be at the centre of the span, where the
load comes upon it, and the moment of stress there will be,

$$= -\frac{P\,l}{4}$$

The deflection of this girder will be—

$$D = \frac{P\,l^3}{28\,m}.$$

Let $W = w.l$, then the deflection at the centre of the first
girder will be—

$$D = \frac{w.l^4}{44\cdot8\,m} - \frac{P\,l^3}{28.\,m}.$$

But these two deflections must be equal, therefore ;

$$\frac{w.l^4}{44\cdot8\,m} - \frac{P\,l^3}{28\cdot m} = \frac{P\,l^3}{28\cdot m}$$

and,

$$\frac{4\,w.l^4}{44\cdot8\,m} = \frac{P.\,l}{14\,m} ;$$

whence,

$$P = 0\cdot3125\ w.\,l.$$

The maximum stress upon the uniformly-loaded girder
will be—

$$M = \frac{w}{2}\left(\frac{l}{2} - \frac{P}{2\,w}\right)^2 - \frac{w.l}{2} \times \left(\frac{l}{2} - \frac{P}{2\,w}\right)$$
$$+ \frac{P}{2}\left(\frac{l}{2} - \frac{P}{2\,w}\right) = \tfrac{1}{4}\left(P.l - \frac{w.l^2}{2} - \frac{P^2}{2\,w}\right)$$

the maximum moment on the second girder is—

$$M' = -\frac{P.l}{4}.$$

If $r =$ the ratio between these moments, then —

$$r = \frac{M}{M'} = -1 + \frac{w.l}{2\,P} + \frac{P}{2\,w\,l}$$

If it is desired to divide the load in any particular manner there are two ways in which it may be done—one, by interposing some element between the girders so that their deflections are not equal; and the other, by making the girders of different sections so that the value of m is not the same for both.

Let it be required to keep the maximum stress per sectional square inch the same for both girders.

In the first instance, let some intermediate girder be used, so that the deflections of the two girders are not necessarily equal; then, if the sections are kept alike, the maximum moments must be equal.

$$\frac{P\,l}{4} - \frac{W.l}{8} - \frac{P^2\,l}{8\,W} = \frac{P\,l}{4}$$

therefore,

$$P = 0{\cdot}268\ W.$$

The interposed element must, under the load P, have a deflection equal to the difference of the deflections of the two principal girders.

For the second case I will assume, for simplicity of treatment, that solid rectangular beams are used, and that their deflections are to be equal. It is evident that the force P must produce half the deflection, as a central load, that the load W would cause equally distributed.

Let $D^1 =$ the actual deflection of the two girders, then—

$$D' = \frac{1}{2} \times \frac{W\, l^3}{44 \cdot 8\, m'} \, ;$$

also,

$$D' = \frac{P\, l^3}{28\, m} \, ;$$

equating these we have,

$$\frac{W\, l^3}{89 \cdot 6\cdot\, m'} = \frac{P\, l^3}{28\, m} \, ;$$

where m' refers to the uniformly-loaded girder, and m to the auxiliary centrally-loaded girder, then—

$$28\, W m = 89.6\, P m'.$$

Some ratio between W and P must be decided upon or else one between m and m'.

Let $P = \frac{W}{2}$; then

$$28\, W.\, m = 89 \cdot 6\, \frac{W}{2}\, m' \, ;$$

therefore,

$$m = \frac{89 \cdot 6\, m'}{56} = 1 \cdot 6\, m'.$$

But the same stress per sectional square inch is also to be kept. The maximum moment of stress on the first girder is,

$$M' = \frac{P\, l}{4} - \frac{W\, l}{8} - \frac{P^2\, l}{8\, W}$$

$$= \frac{W\, l}{8} - \frac{W\, l}{8} - \frac{W\, l}{32} = -\frac{W\, l}{32}$$

The maximum moment of stress upon the second, or auxiliary girder, is—

$$M = -\frac{P\,l}{4} = -\frac{W\,l^2}{8}$$

The moment of resistance of the auxiliary girder must therefore be four times that of the first girder, or—

$$\frac{s\,.\,b\,.\,d^2}{6} = \frac{4\,.\,s\,.\,b'\,d^2}{6}\,;$$

and s, the stress per sectional square inch, must be the same for both, therefore,

$$\frac{b\,d^2}{6} = \frac{2\,.\,b'\,.\,d^2}{3}$$

and,

$$b\,d^2 = 4\,b'\,d'^2$$

It has been shown above that to satisfy the conditions stated,

$$m = 1\cdot6\,m'\,;$$

therefore,

$$b\,d^3 = 1\cdot6\,b'\,d'^3.$$

Hence, as equals divided by equals are equal,

$$\frac{b\,d^3}{b\,d^2} = \frac{1\cdot6\,b'\,d'^3}{4\,b'\,d'^2}\,;$$

whence

$$d = 0\cdot4\,d.'$$

I will take an example to illustrate this. Let the span of the first beam be 6 feet, and the distributed load $1\cdot2$ tons. Then $P = 0\cdot6$ tons.

The maximum moment of stress upon the first girder—

$$= -\frac{Wl}{32} = -\frac{1{\cdot}2 \times 6 \times 12}{32} = 2{\cdot}7 \text{ inch-tons.}$$

The multiplier 12 is used to reduce the span, which is given in feet, to inches. On the second girder,

$$M = -\frac{W.l}{8} = -\frac{1{\cdot}2 \times 6 \times 12}{8} = 10{\cdot}8 \text{ inch-tons.}$$

The moments of resistance must, of course, be equal to these moments of stress; therefore, if $s = 4$ tons, and $b' = 1$ inch,

$$\frac{s.\,b'\,d'^2}{6} = \frac{4\,b'\,d'^2}{6} = \frac{2\,.\,d'^2}{3} = 2{\cdot}7 \text{ inch-tons ;}$$

therefore,

$$d'^2 = \frac{8{\cdot}1}{2} = 4{\cdot}05 \text{ inches,}$$

and $d' = \sqrt{4{\cdot}05} = 2{\cdot}012$ in. ; $d = 0{\cdot}4\,d' = 0{\cdot}805$ in. ; $b\,d^2 = 4\,b'\,d'^2$; therefore,

$$b = \frac{4\,.\,b'\,d'^2}{d^2} = \frac{4 \times 1 \times 4{\cdot}05}{0{\cdot}648} = 25 \text{ inches.}$$

If these deductions are correct, the calculated deflections of the beams should be equal ; in the first beam, $m' = 8{\cdot}145$, and

$$D' = \frac{W.\,l^3}{89{\cdot}6\,m'} = \frac{1{\cdot}2 \times 6^3}{89{\cdot}6 \times 8{\cdot}145} = 0{\cdot}355 \text{ inch.}$$

For the second beam, $m = 13{\cdot}041$; and,

$$D = \frac{W\,l^3}{56\,m} = \frac{1{\cdot}2 \times 6^3}{56 \times 13{\cdot}041} = 0{\cdot}355 \text{ inch.}$$

This proves the accuracy of the formula deduced from the previous investigation of the question.

This inquiry shows that, under certain circumstances, auxiliary girders may be used with advantage practically, but cannot often be used in builder's ironwork to give the economy which they would indicate.

The matter of deflection is one of primary importance in buildings on account of the nature of the load carried. Deflection in a railway bridge is of no consequence—assuming that it is within the limits of safety—a train passes over and causes deflection during its passage, after which the bridge resumes its normal form; but deflection to any extent in a domestic or a public building will cause a ceiling to crack, and therefore become unsightly, or if it occurs in a girder carrying a wall, the bond of the brickwork will become broken, and the stability of the wall endangered.

The object, then, of the auxiliary girder is to stiffen the primary one, and it may with advantage be practically regarded as halving the span of the latter. In point of fact the maximum stress on a continuous girder of two equal spans is equal to that on a free girder of half the length, but it occurs over the central support instead of in the centre of the span, and the load on the central support is greater than half the total load on the two spans.

The points of contra-flexure on each side of the central support are, under a full load, distant one-quarter of the span on each side, and this accords with a pressure on the central supports equal to five-eighths of the total load on the two spans, and for this the auxiliary girder should be calculated.

Although no weight is saved by making the girder continuous instead of in two separate spans, there is a gain in stiffness, the deflection in the former case being about two-fifths of that in the latter.

In the case referred to of a girder 40 feet long, between

supports, being relieved by a secondary girder 12-feet span, under its centre the maximum central load upon the latter will be five-eighths of the total load on the former. Let the load upon the primary girder be 1·5 tons per lineal foot, then the central load, which cannot be exceeded, on the secondary girder will be—

$$= 40 \times 1\cdot5 \times \frac{5}{8} = 37\cdot5 \text{ tons.}$$

A rolled girder 12 in. × 6 in., weighing 57 lbs. per lineal foot, gives a working strength—at one-third the breaking weight—of 38·1 tons distributed over the length; therefore two of these, side by side, would carry the same central load; they should be bolted together at the centre, as there is almost sure to be a tendency to spread apart laterally.

The maximum stress upon the primary girder cannot exceed that which would be brought upon a girder 20 feet span by an equally distributed load of 30 tons.

A rolled steel girder 15 in. by 6 in., weighing 61 lbs. per lineal foot, will carry—at one-third the breaking weight—30·41 tons, so this will serve the purpose almost exactly.

The different ways in which girders may be arranged to act together are innumerable, but it is unnecessary to multiply examples because, when the general principles upon which such combinations are based are understood, their application to any particular case becomes obvious.

In arranging the iron and steel work for a building, the circumstances under which it will be erected must be kept constantly in mind, and also the nature of the locality and the means of access to it. If the roads leading to the site are bad or very steep this will impose a limit upon the size of the pieces to be carried, and in one case we may be compelled to make a main girder in separate spans, whereas in another, where the carriage is easier, the girder may be made in a continuous length. Where the work is to be executed in a town of which the streets are narrow, there is also the question of turning the girders round to get them into the building to be considered.

The re-building, or altering the internal arrangements, of old premises often presents a problem of complicated aspect, which becomes further involved when restrictions are put upon the contractor to prevent any cessation of business or inconvenience to customers during the alterations.

The very heavy loads which are concentrated upon columns and stanchions must in all cases be spread beneath them to cover sufficient surface to reduce the pressure per foot super to that which the subsoil is capable of supporting. This is not so simple and easily settled a matter as it looks at first sight; the varieties of foundations are unending, and their proper adaptation to the duties required of them is vital to the permanence of the superstructure. The stratagems resorted to to obtain firm beds are as numerous as the qualities of the materials encountered; a shingly river-bed has formed a good foundation for a bridge when enclosed on each side by piling so as to form a coffer-dam, and loose sand has been made into a solid mass of concrete by injecting cement into it. On some of the foundations in Westminster and near the river side there is a load of over four tons per superficial foot.

The strength of a foundation depends upon the strength of the material used and upon the way it is bedded; the fact of a certain stone possessing a given resistance to compression does not take into account stresses due to hollow bedding, from which slovenly work I have known very thick and strong sandstone to split across, causing a drop in the superincumbent floors.

The first cracking of the material shows the limit of its resistance so far as practical use is concerned, and although in many experiments a general ratio of one to two has been found for the cracking compared with the crushing load, this is by no means universal. From some special experiments made a few years back the following results are found for the average resistance of some varieties of bricks—they are given in tons per square foot :—

	Cracked.	Crushed.
Recessed red brick—Moraud's machine . .	86·8	161·6
Wire-cut ,, ,, Porter's ,, . .	104·5	228·1
Recessed top and bottom red brick— Scholefield's machine	120·9	192·4
Wire-cut red brick—Johnson's machine .	125·8	233·0
Red brick ground in mortar mill . . .	143·9	200·3
,, ,, pugged in pug mill . . .	92·7	163·2
,, ,, Bulwer's machine	109·6	220·4
Recessed reddish brick—Notts Brick Company	152·1	204·9
Smith's red brick	46·5	64·2
Cowley's stock brick	43·4	56·6
Burham gault brick, wire-cut	75·6	114·9
Blue Staffordshire brick	136·3	254·7
Oldham red brick	71·0	144·0

The average crushing resistance of Portland cement at different ages has been ascertained to be : at three months old, 240 tons per square foot ; at six months, 339 tons ; and at nine months, 443 tons. How these figures are used must depend upon the length of time which will be allowed to elapse between the laying of the cement and using the building in the construction of which it is employed.

The strength of the brickwork will not necessarily be equal to that of the bricks of which it is composed, but there is no reason why it should not be if the joints are true and are perfectly filled, especially if cement is the material used in making them. On this subject, as in so many others connected with the building trade, we have unfortunately little or no information of a scientific character to guide us.

If a pier is built of blue Staffordshire bricks truly made and with the joints accurately filled up with Portland cement, there is no logical reason why its cracking strength should fall short of that of the brick itself, and any tendency of the pier to split by the separating of the bricks is resisted by the tenacity of the cement between them. Then if six be taken as a factor of safety, and working from the cracking—not the ultimate—resistance, such a pier should safely support 22·7 tons per square foot of horizontal area. To obtain such results extra cost must be incurred, for the best workmanship must be secured, and for this the closest and most faithful supervision will be required, so for ordinarily good work we should not risk putting more than half this load upon it. It may, however, be observed that as a smaller pier properly built will be more reliable than a larger one carelessly thrown together, what is expended in workmanship may be saved in material.

Very careful supervision should be exercised over all the materials brought upon a job, and in nothing must this be more unceasing than in regard to mortars, cements, and concretes, which afford great facilities for the introduction of rubbish to make up bulk. Coarse mortar may consist of one part lime to four of coarse gravelly sand, or crushed ballast. Fine mortar : one part lime to two or three of sharp river sand. Lime concrete : one part lime to four of gravel and two of sand. Hydraulic mortar : one part of blue lias lime to two-and-a-half of burnt clay ground together ; or, one of blue lias lime to six of sharp sand, one of puzzolana, and one of calcined ironstone,

Cement concrete : one of cement to six parts of burnt ballast, broken stone, or gravel. For many purposes concrete made of cement and furnace slag, or mill cinders, is largely used, but I should not use it for foundations; and in all cases the materials mixed with the cement must be free from loam, mud, fine sand, clay, or any kind of dirt. All bricks, stones, and the like, to be used with Portland cement, should first be thoroughly soaked in water, and the materials, both for mortars and concretes, should be thoroughly mixed dry before the water for combining them is added.

The crushing resistances of stones commonly used in buildings, in tons per square foot, are as under :

Aberdeen grey granite	. 700·0	Compact limestone .	. 495·0
Cornish granite	. 900·0	Purbeck . . .	. 602·2
Mount Sorrell . .	. 822·4	Anglesea. . .	. 487·2
Arbroath sandstone	. 506·8	Portland oolite stone	. 210·0
Bramley Fall . .	. 388·9	Craigleith sandstone	. 339·8
Caithness . .	. 417·2	Derby grit . .	. 200·0
Red (Cheshire) .	. 140·4	Yorkshire paving .	. 367·3

The bases of columns and stanchions should be always truly faced and the bed-stones worked to a truly horizontal surface; even if it is a little rough the roughness may be filled up by a layer of cement. Felt should never be placed between the metal and the stone, it will dry in time and become dust, and so to the extent of its thickness let the column or stanchion down ; for the same reason it should not be put under the ends of main girders, or between the caps and bases of columns running up in tiers. If the faces of the latter are properly faced or planed they will make a true and solid joint without any substance being interposed between them.

So far the columns and stanchions which come directly

upon the masonry foundations have been regarded as
simply resting upon them, which in the majority of cases is
quite sufficient, as the superincumbent load will prevent
any chance of displacement. If, however, the load carried
is light, and there is any possibility of the column being run
against by heavy vans, its position may be made more
secure by casting a small projection, C (Fig. 43), on the base,
which projection fits into a cavity in the bedstone, A. B
is the lower part of the shaft of the column, and D the top
of the footings upon which the bed-stone is set. If the base
of the column is to be faced up in a lathe it will be most
convenient to make the projection cylindrical.

When the floors carried by the columns or stanchions are
subject to violent vibrations, as from the action of machinery, some additional steadiness may be obtained by the
use of holding-down bolts arranged as shown in Fig. 44.
A is the bed-stone resting upon the footings D, and upon
it is the base of a stanchion B, held down by bolts $C C$.
These bolts are screwed at their upper ends to receive the
nuts $F F$, under which the base is strengthened by bosses
cast upon it. The lower ends of the bolts are tapered as
shown, and tapered recesses are sunk in the bed-stone A,
just wide enough at the top to admit the widest ends of
the bolts. When the bolts are put in position the spaces
$E E$ are filled in with cement, or secured by running in
melted sulphur. It used to be common practice formerly
to run these in with lead, but this should not be done, as
the neck of the bolt will rapidly corrode where it is in
contact with the lead under the influence of moisture.

Instead of having tapered bolts to hold the stanchion
down, "rag" bolts are frequently used. These are of the
form shown at G, and have each side roughed to afford a
hold on the filling material. The amount of steadiness

Fig. 43.

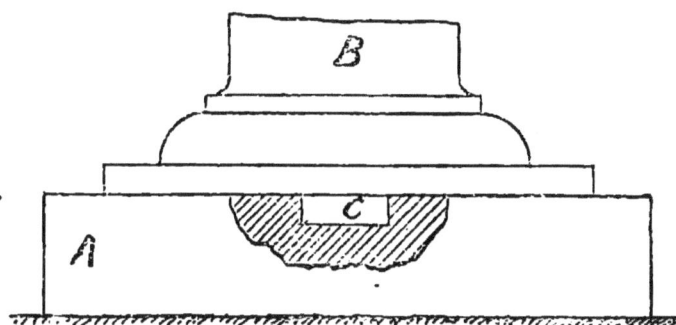

Fig. 44.

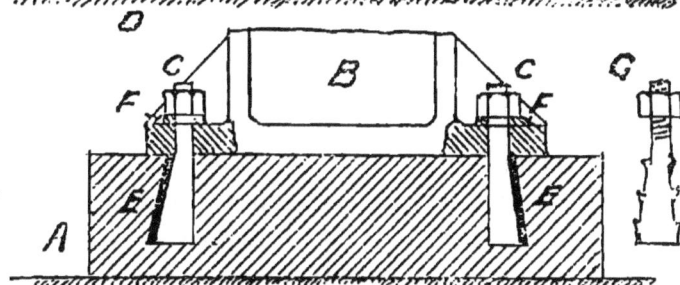

Fig. 45.

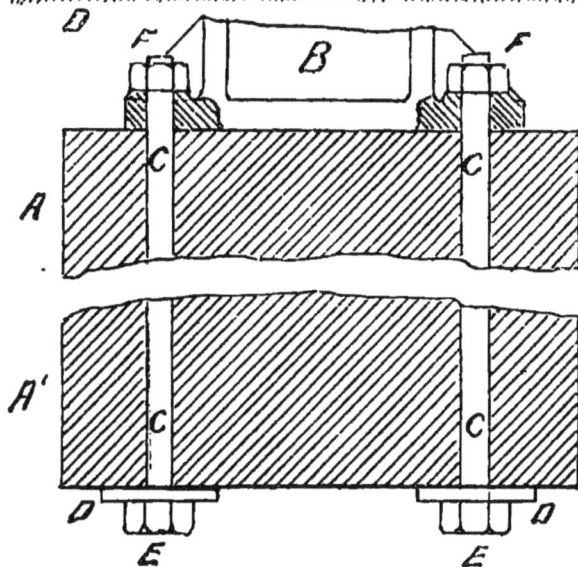

thus secured depends upon the weight of the bed-stone and the strength of its connection with the footings below.

If a greater weight must be secured for steadying the column, this can be effected by taking the holding-down bolts through the footings as well as through the bed-stone as shown at Fig. 45. A is the upper and A' the lower part of the foundation shown, broken across to save space. Before putting in the foundation, bolts $C\,C$, long enough to reach through the foundation and the base of the stanchion B, are put in position, and have resting upon their heads $E\,E$, stout washers D, of cast or wrought iron, above which the foundation is built, sufficient play being allowed around the bolts $C\,C$ to admit of their exact adjustment at the top to the holes in the base of the stanchion, which is secured by nuts $F\,F$. Light cranes are frequently attached to columns and stanchions, and if in such a case there is no great permanent load on the top some such means of obtaining stability must be resorted to. If the foundation is made with a few courses of brick in cement at the bottom, and then carried up in concrete, the bolts must be surrounded by wooden tubes to prevent the concrete from rigidly enclosing them ; when the foundation is carried up to its proper level the holding-down bolts may be secured in their exact positions, and the spaces round them filled in by running in cement before putting the stanchions in place.

The matters connected with the foundation being satisfactorily and finally settled, we turn our attention to the superstructure. If a start is made with the building before the plans and sections are absolutely completed, trouble will commonly arise in connection with the staircases, or their landings, or both. When the skeleton of a building is

mainly of iron or steel, the stairs should be kept in a sort
of well running from the bottom to the top floor ; by taking
this course jibs and cantilever trimming joists are avoided,
and in most cases the staircase can be constructed independently of any iron or steel work ; but if this is not practicable, the landings may be so arranged as to be carried by
girders running across the well and taking bearings in the
walls on each side, and if carriages are required to support
the ends of the steps, these can run from one landing girder
to the next above or below, as the circumstances require,
but the joints between the carriages and landing girders
should be rigorously vertical or horizontal, to prevent the
possibility of a thrust being brought upon the latter.   The
connection of the carriages to the girders is often a troublesome and unsatisfactory matter when the former are of rolled
iron or steel, but this difficulty may be avoided by using
cast-iron carriages, which may be made with ends suited to
any kind of connection that may be necessary or expedient.

In Fig. 46 is shown a staircase trimming, in plan, in
which a jib must be used.   This jib, $D\,B$, may be strong
enough if its wall-hold at the end D is sufficient to counter-
act the loads brought upon it by the joist F and by the
trimmer $C\,B$.   If not, the trimmer $C\,B$ must be strongly
fixed to the joist F, as shown by the section $C'\,C'$.   The
trimmer $C\,B$ is taken as a rolled joist 6 in. $\times$ 5 in. double
cleated into the joist F.   There will be two bolts C, one on
each side of the trimmer $C\,B$; these bolts will show an insufficient strength to resist the stresses put upon them by $C\,B$,
therefore a tie-plate $G\,G$ must be used to connect the top flange
of $C\,B$ with the joist F, and this joist must also be 6 in.
$\times$ 5 in. section to afford room for the bolts connecting the tie-
plate $G\,G$ with its top flange.   No connecting-plate is required
on the bottom flanges as there will only be a thrust there.

The effect of the jib load will be to exert torsion on the joist F. The end of the trimmer $B\,C$ is notched on the top and bottom flanges and double cleated to the jib $D\,B$.

If the jib can be made the continuation of a girder passing through another room its stability will be insured, and then an ordinary joist, or at most a 6 in. × 3 in. joist, single cleated at each end, may be used in place of the 6 in. × 5 in. cantilever $B\,C$, and for the span shown a 6 in. × 3 in. joist will be sufficient for F, and $4\frac{3}{4}$ in. × $1\frac{1}{4}$ in. for the joists $A\,A$.

Fig. 47 shows a horizontal section of a wall $A\,B$, with the flues $C\,C\,C$ in it ; these may be flues from fire-places ; or fresh or foul-air flues required for purposes of ventilation ; but, unless the walls or piers in which they are formed are of unusual thickness, the brickwork in the line of $G\,G$ in front of the flues will not afford sufficient length of bearing for the floor joists $F\,F$ if the width of the room is considerable, and as such flues mostly occur in the walls of hospitals, technical schools, and other large public buildings, the rooms will often require joists of 20 to 24 feet span. The best course to pursue is to trim round the flues as shown. The joists $F\,F$ are kept short and their ends run into a trimmer $E\,E$, carried by joists $D\,D$, made stronger than the ordinary floor-joists to carry the extra load brought upon them by $E\,E$. This load will not, however, augment the stress upon the joists $D\,D$ to a very great extent, on account of the close proximity of the bearings of the trimmer upon the joists to those of the latter at $H\,H$ on the wall $A\,B$. To arrive at some idea of the difference of strengths required between the ordinary joists F, for we should not lessen the section for the slight reduction of span where they run into the trimmer $E\,E$, and the trimmer joists $D\,D$, an example may be taken. Let the

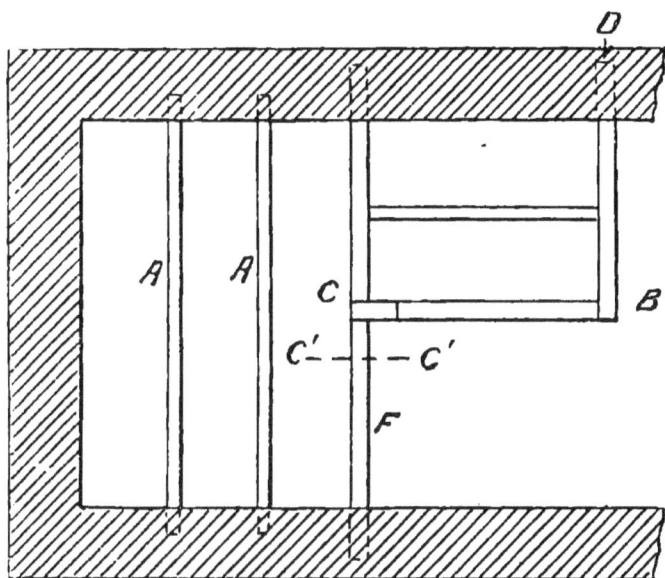

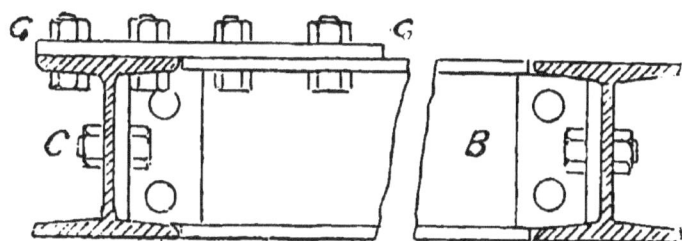

Section on C'C'

Fig. 46.

width of the room be 24 feet, and the distance apart of the joists 2 feet from centre to centre. The floor is assumed to be of solid concrete or some other cognate form of fire-resisting floor, weighing $\frac{1}{2}$ cwt. per superficial foot, and the live load taken as $1\frac{1}{2}$ cwt. per superficial foot, making alto-gether 2 cwt. per superficial foot. The area of floor carried by one joist will be $24 \times 2 = 48$ square feet, and the load at 2 cwt. per square feet $= 4\cdot8$ tons equally distributed. An 8 in. $\times$ 4 in. English steel joist, with a breaking tensile resistance of 28 tons per sectional square inch and weighing 28 lbs. per foot, gives, at one-third the breaking weight, a resistance equal to $5\cdot1$ tons equally distributed, so that would be the section chosen for the ordinary joists F.

Assuming the projection of the brickwork containing the flues to be 9 inches, the distance of the centre of the trimmer E E from the face of the wall will be one foot, therefore the span of the two ordinary joists coming upon it will be 23 feet only; for this difference we should not alter the size of the section, the load upon each will be $23 \times 2 \times 2$ cwt. $= 4\cdot6$ tons, and of this one-half goes upon the trimmer at two feet from its nearest support D, and therefore the maximum moment of stress upon E E $= 2\cdot3 \times 2 = 4\cdot6$ foot-tons, and the trimmer being 6 feet long; the equally distributed load to produce an equal maxi-mum moment of stress is thus found—

$$\frac{W\,l}{8} = 4\cdot6 \text{ ft.-tons} = \frac{W \times 6}{8};$$

therefore

$$W = \frac{8 \times 4\cdot6}{6} = 6\cdot13 \text{ tons.}$$

A joist $5\frac{1}{2}$ in. $\times$ 2 in. weighing 12 lbs. per foot run, gives sufficient strength, but the bearing on its lower flange would be less than one inch, and it would be necessary in

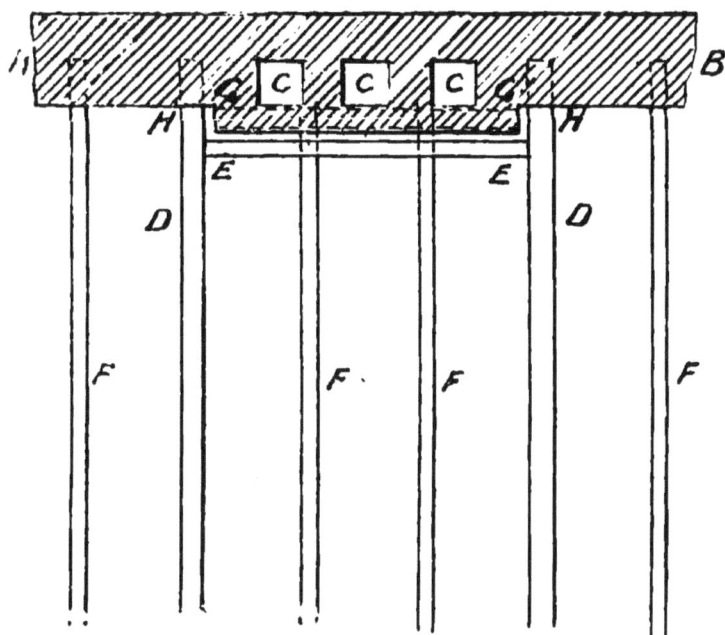

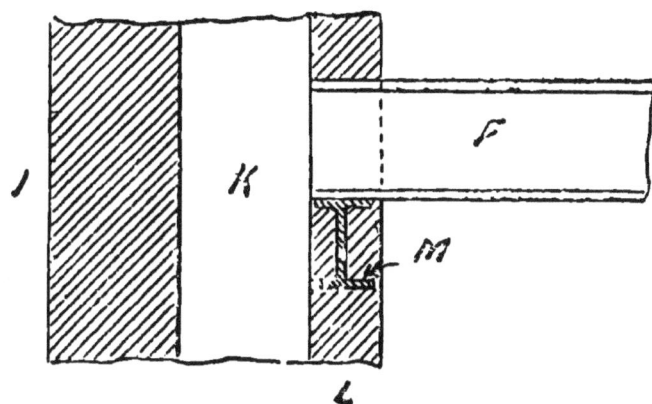

Fig. 47.

all cases to cleat in the ends of the joists $F F$, but if an 8 in. × 4 in. trimmer were used there would be nearly two inches bearing, and cleating could be dispensed with in floors in which a permanent centering is used, such as the "Fawcett" floor, but if plain concrete is used the ends referred to must be cleated.

The labours put on joists are so expensive that it would be cheaper to use the heavier joist as a trimmer if the cleating can be saved; in fact, as prices are at the time of writing, the cleating would cost nearly twice as much as the extra weight in an 8 in. × 4 in. over that in a 5½ in. × 2 in. joist. Of course, the joists cannot move laterally when fire-clay or terra-cotta lintels or permanent centres are placed between them and so determine their spacing.

We now come to the trimmer joists $D D$; each of these has to carry its own normal load of 4·8 tons equally distributed and the concentrated load brought upon it by the trimmer $E E$, which is one-half of the total load on the latter, or 2·3 tons at one foot from the bearing H. The amount of load thrown upon the bearing at the other end of joist D by the concentrated load will be

$$\frac{2\cdot 3 \times 1}{24} = 0\cdot096 \text{ ton.}$$

The load per lineal foot on the joist $= \dfrac{4\cdot8}{24} = 0\cdot2$ ton.

Let x equal the distance of any given point in the joist d from the bearing on the wall opposite $A B$, and M the moment of stress at that point, then,

$$M = \frac{w \cdot x^2}{2} - \frac{w \cdot l \cdot x}{2} - 0\cdot096\, x$$

$$= \frac{0\cdot2 \times x^2}{2} - \frac{0\cdot2 \times 24 \times x}{2} - 0\cdot096\, x$$

$$= 0\cdot1 \times x^2 - 2\cdot496\, x$$

The value of x which gives the maximum moment of stress must next be found. From the point of maximum stress it will commence decreasing in each direction, but we may imagine two points taken infinitesimally close together, so that no appreciable difference occurs between them ; let a be an indefinitely small increment of x; then

$$M = 0\cdot1\ (x + a)^2 - 2\cdot496\ (x + a)$$
$$= 0\cdot1(x^2 + 2\ x\ a + a^2) - 2\cdot496\ (x + a).$$

As a is taken indefinitely small, a^2, which will be much smaller, may be disregarded, then

$$0\cdot1\ (x^2 + 2\ x\ a) - 2\cdot496\ (x + a) = 0\cdot1\ x^2 - 2\cdot496\,x;$$

subtracting $0\cdot1\,x^2$ and $-\ 2\cdot496\ x$ from each side of the equation, we have

$$0\cdot2\ \times x\,.\,a = 2\cdot496\,a\ ;$$
whence
$$x = 12\cdot48\ \text{feet.}$$

Replacing x in the general equation by this value the maximum moment of stress is found,

$$M = 0\cdot1\ \times\ \overline{12\cdot48}^2 - 2\cdot496\ \times\ 12\cdot48$$
$$= -\ 15\cdot575\ \text{foot-tons.}$$

To find the equivalent distributed load this is equated with the general formula for the maximum load due to the latter, thus

$$-\ 15\cdot575 = -\ \frac{W\,l}{8} = -\ \frac{W \times 24}{8}\ ;$$
therefore,
$$W = 5\cdot19\ \text{tons.}$$

The section adopted for the ordinary joists, 8 in. × 4 in., has a sufficient margin of strength to suffice for the joists $D D$ also, but had the trimmer $E E$ been a little farther from the wall $A B$ this would not have been the case ; so, although the calculation seems unnecessary in this example, yet it should always be made, nothing in connection with constructional work of any description should ever be guessed at, and it is not often that two exactly similar cases occur.

The expense of the trimming may be saved by building a joist into the brickwork at $G G'$ as shown in section at I. K is the flue and L the brickwork in front of it, and into this is built a joist M, to receive the ends of the joists F and spread their loads along the brickwork. The objection to this is that it interrupts the vertical continuity of the wall, for if the part L is sufficiently thin to require it, it will occupy all the thickness, and it would really be better to put a strong stone sill from G to G in the position which the joist M would occupy.

The grouping together of a number of air-flues or chimney-flues is generally awkward where the floors are carried by iron or steel joists—unless their courses have been considered and absolutely settled before the floors are set out, and this is another strong argument against letting contracts on uncompleted plans. If any difficulty arises, neither the builder or the contractor for the steel and iron-work will be prepared to stand the loss ensuing, and the architect will be slow to certify for extra payment, although the variation may have been due to the inadequacy of his contract plans, because his client will be annoyed at the growing cost of the structure, and the result is delay and an antagonistic disposition between the different trades— and why ?—the answer is simple, because the whole work was not in the first instance designed under the direction of a man who understood all the branches of his business.

It would be practically impossible for one man to plan the whole of a large and complicated structure personally, and also to prepare the details of each part; but he should have the mental capacity to grasp the whole matter when the details have been prepared by assistants, and to notice at once any incongruity which may have arisen from the very necessity of having different individuals to detail different parts of the work; and without this "hitches" of greater or less consequence must occur during the execution of the contracts.

CHAPTER VIII.

IRON AND STEEL DOORS AND TRAPS.

THERE is a good deal of iron and steel work used in buildings which, although it is not, strictly speaking, constructional work, must be considered in the general arrangement of the structure ; it cannot be left without danger of inconvenience for after-consideration.

Iron doors for separating different parts of a warehouse in case of fire are of the greatest importance, as on their construction and arrangement the fate of a large block of buildings will often depend. An ideal fireproof building is one in which the contents of one room having become ignited, that room can be so closed in that those contents will burn away without transmitting through the walls or floors sufficient heat to ignite the materials in adjoining rooms, or will smoulder out through lack of air to maintain combustion. If all access of air is cut off, the fire, in the case of ordinary merchandise, will soon be brought under control ; in warehouses where drugs and chemicals are stored it is a different matter, and a great deal also depends upon the heat-generating capacity of the substances consumed.

Next to explosives—which cut the matter short by blowing the building to pieces—inflammable liquids are the most difficult to provide against, and the more volatile the greater the danger.

If we knew the particular kind of conflagration which would be likely to occur in any given building, especial precautions might be taken, but, as matters stand, this can hardly be ascertained, and therefore the best we can do is to guard against every probable danger.

Although there has been a general idea among the ignorant that fire must work upward, yet a study or even common observation of fires in buildings shows that it will spread in any direction; it will commence in the basement of one warehouse and if not checked burn up to the roof, and from thence burn downward through the next warehouse, and where the contents are liquid, or easily liquefied, the rapidity of progress is such that the whole structure is enveloped in flame in a marvellously short time.

The isolating doors are of two sorts, those in the walls and those in the floors, generally termed trap-doors.

Doors in walls will first be considered: these are of two classes, sliding doors and hinged doors. Whichever class of door is used the requirements are the same; the door must resist a very high temperature without buckling or shrinking away from its surfaces of contact with the walls and floor; it must admit of opening and closing with freedom, and so fit that it will not allow a draught of air to pass under or around it into a room in which a fire has started, nor let a flame pass outwards from such a room to the next, nor must the heat pass through it in such intensity as to extend a conflagration.

Very little consideration is required to show that to secure the last efficiency the doors must be fitted in pairs with some space between them, for when a door has become red-hot its radiant heat will speedily inflame some kinds of merchandise, but if this heat has to cross three or four feet of air it will take a long time to heat the second door sufficiently to become dangerous. Heat-resisting paints and compositions are useful in this connection, for although it may be cracked off the door next to the fire by the expansion of the metal, or even burnt off, that on the face of the opposite door will protect it from becoming hot for some time.

For air and flame-tight properties the sliding door possesses the advantage that when closed all its edges are surrounded—assuming that it is properly designed—without any adventitious arrangements, but the hinged door must have some mechanical appliance to close the space between its lower edge and the floor, as if this were done by a ledge for the bottom of the door to press against, such ledge would be a source of inconvenience, and perhaps danger to customers passing from one department of a large business house to another, and such considerations are of the first importance among a trading community.

For these doors iron is to be preferred to steel, because its melting-point is much higher, and therefore it will hold out longer. In fact I have never heard of a case where wrought iron has been liquefied in a conflagration, although I have seen some in which the cast-iron columns have yielded to the heat in that way.

These warehouse doors, as usually made, present in front elevation the appearance shown at A, Fig. 48. D is a vertical section in the line $F F$. The front elevation might pass for that of a wooden door, but the construction is widely different, and not in favour of the iron one. The stiles and rails $B B$, $B' B'$, $C C C$, are flush riveted to a plate E, which forms the whole back of the door; this plate, for ordinary sizes, is one-quarter of an inch thick and the stiles and rails three-eighths of an inch thick. The striking difference between this door and a wooden one is

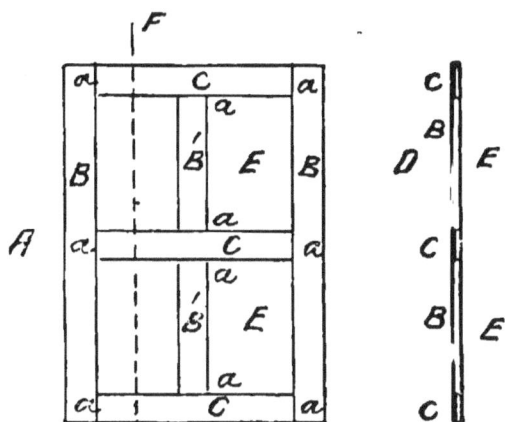

Fig. 48.

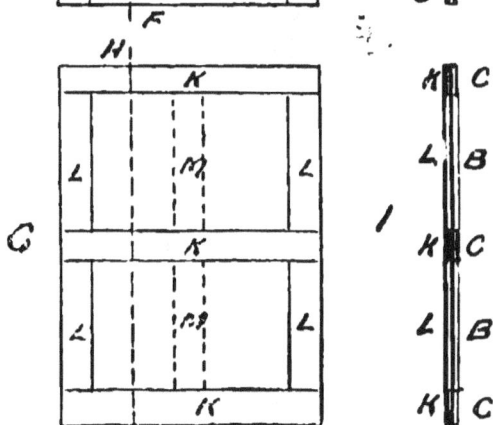

Fig. 49.

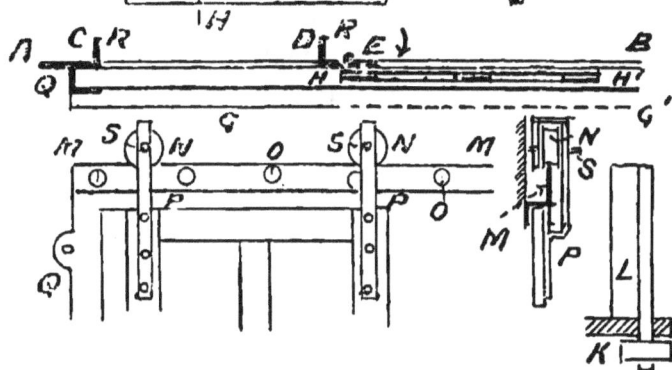

Fig. 50.

that there is no connection of the stiles and rails to each other at the places of contact a; if the plate-back tends to warp, the pseudo-joints a open, and in fact the frame only acts as local stiffening. For this reason a door running between grooves is safer than a hinged door shutting into a rebate, because the grooved channels at the top and bottom of the door will prevent its buckling at those edges.

An iron door may, however, be made with a continuous frame, so to speak, by riveting rails and stiles on behind, as shown at G, Fig. 49. The rails $K\,K\,K$ stretch right across the back of the door, and so tie the stiles $B\,B$ in front; the back stiles $L\,L$ make the back framing flush. The dotted lines $M\,M$ show the positions of the front stiles $B'\,B'$.

At I is shown a vertical section of the door on the line $H\,H$, the same letters being retained for the front-framing as are used in Fig. 48. It may be objected that the weight is materially increased by the addition of the back-framing, and the point to be decided is whether the additional security is worth the extra cost. The weight need not be so very largely augmented; it is not necessary to make the rails and stiles so thick when they are fixed on both back and front of the plate E, as when they are on one side only. There will be nearly 50 per cent. more holes to punch or drill, but the number of rivets will not be increased, so there will be no extra cost for riveting. The construction of the door itself need not be affected by the manner in which it is to be hung, but the fittings will be entirely different for a sliding door from those for one that is hinged, as will also be the grooves or rebates into or against which they are designed to close. In all cases the edges of the doors should be planed true.

In Fig. 50 are shown the fittings required with sliding doors. $A\,B$ is a horizontal section which shows in plan the grooved iron in which the bottom of the door slides, and in section the jambs C and D and the closing flap E. An extension G forms a sill to the doorway, and is let in flush with the cement finish of the floor.

The object of the flap E is to close any slight opening between, which may occur between the back of the door $H H'$ and the jamb D. This flap is carried by a bottom pin I, shown in the detail L, and it is steadied by a similar pin which takes into a hole in the top guide-iron of the door. Around the lower pin is a spiral spring K, of which one end is attached to the pin and the other to a casing, and tension is put upon it in fixing, so that it will tend to revolve the flap E in the direction of the arrow M. While the door $H H$ is in the position shown, that is, open, the flap is held back by it, but when the door is closed and its front edge run into the recess F in the jamb C, the flap E will close over the back edge H', and prevent the passage of flame between it and the jamb D. The spring K will prevent the flap from being forced open by a draught induced by fire in front of the door. The groove in which the door runs will seldom exceed one inch in width, even with doors with back and front frames, but it is advisable to have a bar to drop into this groove when the door is opened for the day, not only to prevent the chance of any one tripping over it, but also to exclude anything which, by falling in and lodging there, might cause delay in closing the door when sudden emergency calls for it. In case of fire time is the first consideration, and although the properly fitted bar can be lifted out in a fraction of a second, a substance accidentally lodged and perhaps jammed into the groove might cause the loss of several minutes during its removal, thus allowing a free current of air through the doorway in either direction.

Doors of this description are carried upon small wheels or runners attached either to the top or bottom of the door. If top suspension is used, a flat bar $M M$ is fixed to the wall over the door by bolts or studs, and distanced from the wall by pipe packings on the studs to allow room for grooved wheels $N N$ to run upon the edge of the bar; the studs are shown at $O O O$, &c.

The wheels $N N$ run upon bolts S, which pass through, and are held by, straps P. The upper parts of these straps are bent over, as shown, to afford holds for the bolts on each side of the wheels which run upon them as dead centres. The lower ends of the straps P are securely riveted to the stiles of the door, as shown.

If the door is to be carried upon bottom runners, the sill G must be extended to G', and runners $T T$, Fig 50A, fitted to dead centres $U U$. The runners will be plain on the edges and the sill flat. One dead centre is shown in detail at V. W is a section of the lower part of the stile and plate of a single-framed door, and into this the stud U is flush riveted as shown, the riveted end being turned or swaged down from the solid, so as to give a firm hold to the body of the stud. The other end of the stud is turned to form a bearing for the runner T, and a shoulder is formed at the bottom of the threaded part, on which the nut Y is screwed to prevent the nut from jamming the runner. These studs should be carefully made, for if the runners do not move freely, the door may "stick" at a critical moment with very serious results.

The top and bottom grooves, and the jambs C and D, require to be of neat and true sections; they are best made of rolled iron or steel, but if such sections cannot be obtained, these parts must be cast, and on several occasions I have seen the work so executed in a very satisfactory

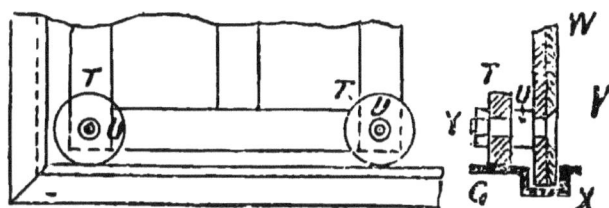

Fig. 50A.

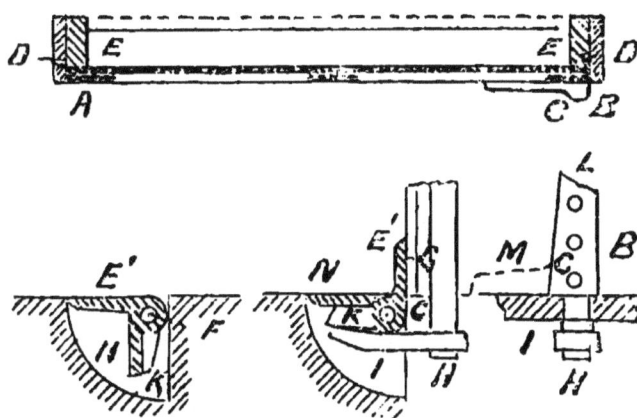

Fig. 51.

manner.   Of course, a very superior quality of cast iron is required in such cases.   The attachment to the wall must be perfect, for the jambs must be on the face to allow the door to slide back when opened.   Ribs R R may be built into the wall, or they may serve as return sides to protect the edges of the door opening, and the jambs secured by rag-bolts built into the walls with screwed ends projecting to pass through lugs, of which one is shown at Q. Nuts on these ends secure the jamb C, but for the jamb D there must be cast on it some thickened bosses to allow room for countersunk nuts.

In Fig. 51 are shown some fittings required for hinged doors. A B is a horizontal section of the door, which shuts into rebates D D, formed by riveting together two bars of different widths, and securing them in the usual way to the sides of the doorway.   The door is supported by a pintle C, which passes through a floor-plate, and has an extension H, the use of which will be subsequently described.   At the top of the door, and in the same axial line with the pintle C, is another one passing into a hole in a top plate to keep the door in its proper position.   The pintles are forged on to tangs L, which are riveted or bolted to the back stile of the door.   If additional strength is required, another tang, shown by dotted lines at M, can be forged on the pintle and riveted or bolted to the bottom rail of the door.   As the bottom edge of the door must clear the floor to allow of its closing easily, there will be a small space to be closed by some kind of a flap; this is supplied, in the example shown, by a revolving threshold E E, which has a chequered top surface to afford a foothold. E' is a transverse section of the revolving threshold, shown in the position it occupies when the door is open, and E'' another section showing it as it appears after the door is shut. At each end of the threshold is a pin F, which is supported in a bearing in the floor-plate, and upon these pins it can turn freely. There is a recess in the floor in which this piece works, and a rib N closes it when the part E' is again at the door. The mode of operating the threshold automatically is as follows :—

At the end of the threshold nearest the hinge of the door is a tail-piece K, and on the extension H of the bottom pintle of the door a cam-piece I, firmly fixed on at right angles to the surface of the door; when the door is being closed this cam-piece presses against the tail-piece K, and so causes the threshold to revolve upon its bearings through a quarter of a circle, and close the space between the bottom of the door and the floor, as shown at E''. The cam-piece I may have a spring upon it to keep the threshold pressed close against the door.

If a hinged door has a considerable width of opening there will be a tendency to rack and buckle at the free edge, or even to take a twist, and this calls for a different arrangement of framing from that shown in Figs. 48 and 49. Instead of having central vertical stiles, a strut $A\,B$, Fig. 52, is used, placed diagonally, so that the framing acts as a bracket. The strut $A\,B$ holds up the front stile C through its connection with the plate D.

If the door is high as well as wide, it will be advisable to have two struts on the face, as shown by $E\,E$. Folding doors should not be used where single ones can be introduced, as they are liable under great heat to warp apart at the meeting line. This can only be avoided by having an ample supply of locking bars, and this occupies a good deal of time in fastening up and unfastening.

In Fig. 53 A is the plan of the top of a trap-door; B the plan of the underside; and $C\,D$ is a vertical section taken along the line $g\ g$. The critical difference between vertical doors and traps is that the latter, if they occur in floors, have to carry a working or live load in addition to their own dead weight.

The top of the trap is formed of a chequered plate to give a foothold, and to its underside is riveted an angle-iron frame $e\ e\ e\ e$, which is made by notching out the angle-iron at each corner, and then bending it to an exactly rectangular form and thoroughly welding up the corners, so that the frame shall be quite solid throughout. If the load upon the trap is light, a margin $e'\ e'$ is left round the angle-iron to form a lip, which rests upon the bearing when the trap is shut; but if heavy loads are to be anticipated, the frame is to be made with the flat limb of the angle-iron outwards, as shown in section at K, in which the angle-frame e itself rests upon the bearing angle-iron i. In the section $C\ D$ it is the edge e' of the plate that brings the load on to the bearing-frame $i\ i$. The construction of the bearing-frame is the same in both cases.

To support the transverse stress one or more tee-irons f —according to the size of the trap and its shape—are riveted on the underside, all rivet-heads on the top being of course kept flush with the surface.

The bearing-frame is made of two angle-irons riveted together, one being rather shallower than the other, in order to form a rebate for the trap to drop in. The outer and deeper angle-iron $h\ h$ is fitted to the floor, being bolted to joists or girders or let in cement or concrete according to the mode of construction adopted in building. The construction of the trap will not be altered if it is hinged on to the opening, but in that case the hinges must be so placed that their rivets pass through the horizontal limbs of the frame $e\ e\ e\ e$.

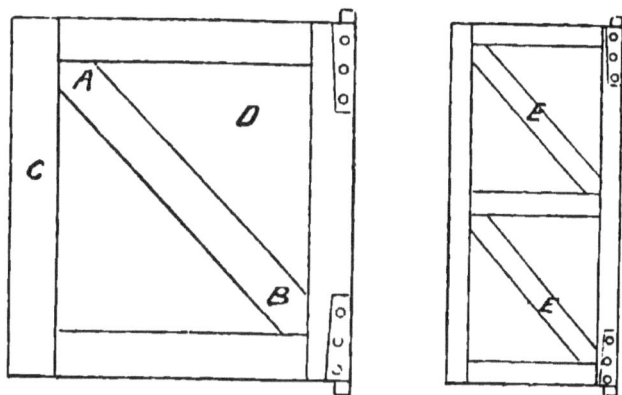

Fig. 52.

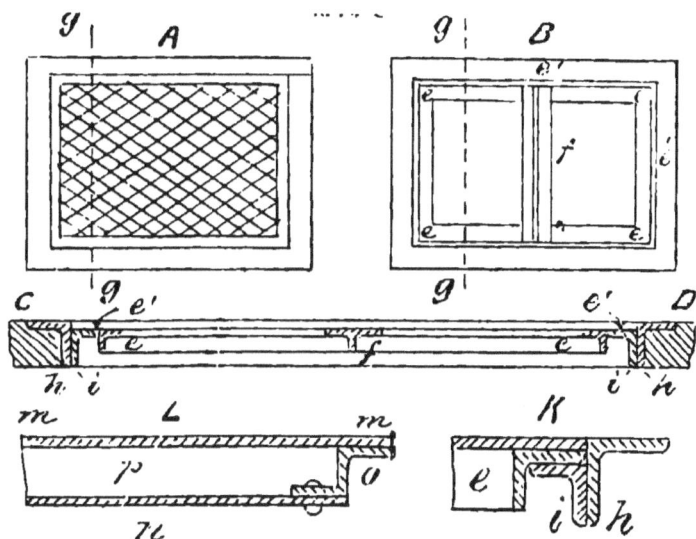

Fig. 53.

If the trap-door is intended to be fire-resisting it should be made double, as shown partly in section at L. The top plate m, and bottom plate n, are both riveted to a frame o, made of Zed iron. If there are any cross-bearers, such as f, these must be riveted on to the top plate m before it is riveted on to the Zed iron, which having the plate n previously riveted on to it will cover in the whole. The space p between the plates may be filled in with slag-wool, or other non-conducting material.

For the purpose of getting hold of the traps to raise them, sunk handles or bars for hooks may be fitted, and to form the recesses for them holes will have to be cut in the top plates and recessed castings bolted on to the under-side, and in the recesses the rings rest, or the bars are formed.

It is sometimes found necessary to have hook holes in vertical iron doors, but wherever this is the case they must be furnished with dropping shields to cover the holes automatically directly the hook is removed. A hinged door can, of course, have a handle on each side of any form that may be fancied, but any ordinary handle on the back of a sliding door would foul the jamb in opening, and if the door slides into a wall it cannot have a projecting handle on either side.

To insure keeping the edges of the door and trap-plates true they should be—as they are usually only quarter inch thick—pretty closely riveted to the framing, say three or four inches pitch, the rivets being half-inch to five-eighths in diameter, flush riveted both sides.

CHAPTER IX.

SPECIFICATIONS AND QUANTITIES.

In connection with every work of importance the architect should prepare a specification and a schedule of quantities for the prices to be filled in by persons invited or permitted to tender for its execution.

The specification is a most important document, and if one is not supplied the contractors are left entirely in the dark as to the strength and quality of the materials to be supplied, and each can use his own discretion in these matters. This is not fair to any one, and it certainly does not tend towards obtaining the best class of work. The principle of letting competing contractors design the constructional metal-work themselves is a very fallacious one, as, in the keen quest of business, some may be tempted to cut down the strength to a point which leaves the durability of the work very doubtful, and the architect moreover ought certainly not to have to depend upon the manufacturers for the execution of the professional part of the work—that connected with the designs.

In Chapter I. I have given the qualities of various materials obtainable, but it appears to be advisable to point out the strengths generally suitable for the class of work under consideration, with the view of embodying it in a specification which shall also deal with the quality of the workmanship. The following clauses may be of assistance in framing different specifications, although they may not all be required in any one.

CAST IRON.—All cast iron to be of such quality as to show a grey fracture and not to break with a less tensile stress than 7 tons per sectional square inch. A bar of the cast iron, 3 feet 6 inches long, 2 inches deep, and 1 inch wide, is not to break with a less central load than 21 cwt. when resting upon supports 3 feet apart. Two test bars are to be made with every cast for transverse testing and two bars for tensile testing. The latter shall be planed down to exact dimensions, and all of them tested as the architect shall direct. All columns and stanchions of cast iron to be faced at both ends truly parallel, and the rough sand surfaces rubbed smooth with coke, or otherwise, after which each casting shall receive one coat of red oxide paint before leaving the works, another coat of paint being applied after the column or stanchion is fixed, the colour of the second coat to be decided by the architect. The faced ends of the castings are to be smeared with a mixture of white lead and tallow to keep them from rusting before the work is fixed. Holes for connection bolts to be drilled out truly and in exact position where so indicated on the drawings ; other bolt holes may be cast in. All cores in columns to be so stiffened by core-bars and studs as to insure a uniform thickness of metal all round. All faces for the connection of brackets, cast-iron bracing bars, or other cast-iron work, to be truly planed, and also the faces which come in contact with them. All castings of every description to be cleanly made to the dimensions shown on the drawings, and to be free from blowholes, sponginess, cinder, scales, cracks, or other defects. Should any part of any casting break off, a new part is not, under any circumstances, to be " burnt " on   if such part cannot be satisfactorily attached with bolts or screws a new casting must be made to replace the damaged one. No wrought-iron sheets or plates are to be cast in the thickness of any column.

WROUGHT-IRON plates, flat bars, angle, tee, and channel-iron to have an ultimate tensile resistance of not less than 21 tons per sectional square inch with the grain and 18 tons across the grain, with elongations before fracture of 12 per cent. and 9 per cent. respectively in a length of eight inches. A plate or bar $\frac{1}{4}$ inch thick is to admit of being bent double cold without breaking. When a bar or plate is broken it shall show a fibrous fracture free from sparkling places and from cinder or scoria. Rivet-iron to show an ultimate tensile strength of not less than 24 tons per sectional square inch, and a shearing resistance of not less than 20 tons per sectional square inch, and an extension of about 10 per cent. in eight inches.

WROUGHT-IRON BUILT GIRDERS.—All the plates, bars, covers, and joint-plates to be made of the sizes marked upon the drawings, and with the number, size, and position of the rivets in each part exactly as shown. The edges of the web plates are to be truly planed (not chipped or filed) after they have been straightened and any winding taken out, in order that they may meet each other and the inside surfaces of the flanges accurately and make close joints. The flange-plates are to be carefully straightened (or if for a curved girder bent to the required curve) in proper straightening rolls, and if the flanges are more than 12 inches wide, or consist of more than one layer of plates each, then the edges of the plates are to be planed to bring them to the same width to fit together properly. The ends of the plates that meet to form a joint are to be accurately planed at right angles to the length of the plate to insure uniform bearing through out the surfaces of contact. This particularly refers to the upper or compression flange, and will also apply to the bottom flanges of continuous girders.

All joggles in angle, tee, and other bars are to be accu rately formed—preferably by hydraulic pressure—to fi over the parts of the work on which they lie, and all end to be fitted to the parts against which they abut.

All packing pieces are to be fitted to the spaces they ar intended to fill, and where necessary the edges are to b planed to prevent their appearing beyond the edges of th parts between which they are interposed. All rivet-hole are to be drilled out of the solid, or punched ⅛ of an incl less in diameter than the finished size, and broached out t the finished size in a broaching machine. Where automati drilling or punching machines are not used, the rivet-hole must be set out with centre punches from templates accu rately made, and the holes made with "nipple" punches After the holes have been broached out to full size, th sharp "arrises" on the edges must be removed by an obtus hand-drill. Where flush or countersunk headed rivets ar necessary the countersink is not to be deeper than one third of the thickness of the plate, and the angle of counter sink is to be 45 degrees to the surface of the plate.

The rivets used are to have cup heads of a height not les than three-quarters of the diameter of the body of th rivet, and the diameter of the finished head is to be nearl twice the diameter of the rivet.

The rivets are to be thoroughly closed by hydrauli riveters, and sufficient length of point must be allowed t fill the rivet-snap, and prevent it from indenting the plate if any collar forms around the rivet-head from excess of metal, this collar is not to be cut off.

In such riveting as must be done by hand, through th

position precluding the use of the hydraulic riveter, the rivets are to be made etxra hot and riveted up without loss of time, to insure the filling of the rivet-holes and proper contraction of the rivets after the heads are formed.

BOLTS AND NUTS.—All bolts used for connecting parts having drilled or broached holes, are to have turned bodies; those that are used for castings having holes cast in them may be left black. The heads of the bolts to be not less than three-quarters of the diameter of the bodies in thickness, and two diameters in width across the angles, the nuts to be equal in thickness to the diameters of the bolts, and to two diameters across the angles. The threads on the bolts and in the nuts are to be cleanly cut—not squeezed up in dies, and by taps—so as to give the full strength of the metal, and as a test of this any bolt may be taken from a parcel and drawn asunder in a machine which grasps the head and the nut ; should the head pull off, or the nut strip the thread off before the bolt tears asunder, that parcel of bolts may be rejected at the option of the architect.

WROUGHT STEEL plates, flat bars, angles, tees, and channels, to have an ultimate tensile resistance not less than 27 or more than 30 tons per sectional square inch, with an elongation before fracture of 18 per cent. in a length of 10 inches ; plates and flat bars up to $\frac{3}{4}$ inch thick to bear bending double cold without breaking. Fracture of plates or bars when broken to show homogeneous texture.

WROUGHT STEEL BUILT GIRDERS.—The clauses for wrought-iron built girders apply also to those made of steel.

GENERAL SUGGESTIONS.—In the above clauses I have given a scanty outline of a specification for good work, but each undertaking will require special treatment, to cover every condition by which it is affected. We cannot close our eyes to the fact that, if money is limited, and the best work cannot be afforded, the specification must be relaxed to meet this contingency, and the most expedient course to follow under such circumstances is to save where possible in labour, not in the quality or strength of material. Constructional iron and steel work in buildings is usually encased in concrete or plaster, or built in brickwork, or concealed by cornices, and therefore rough edges do not meet the eye after the building is finished, and so where girders of the built type are to be covered in, the edges may be left unplaned—but the ends of the plates meeting at joints must be true for the sake of strength. The lengths of bolts for connecting ought, in every case, to be carefully ascertained, and the proper lengths applied in the proper places, otherwise a rickety joint is the result, but there is no saving in money, and there is risk of accident after. It has often happened that when a labourer has been left to do a mechanic's job, and he has picked up a bolt too long for the connection he has to make, that instead of taking it out and finding one of a suitable length, he will put several washers upon the bolt, to allow the nut to get a hold on it, and these washers by no means fit the bolts. Now the plainest common sense will show that a bolt so fixed has no hold upon the work at all, one washer may slip one way, and another in some other direction, like a broken puzzle, hidden perhaps in concrete, but leaving a joint unmade. Therefore it may be well to introduce into the specification a clause to the effect that "no washers shall be used under either the heads or nuts of bolts connecting different parts of the work together, and the bolts shall be screwed sufficiently far up to render the use of washers unnecessary."

Cast-iron columns and stanchions should, wherever it is possible, be cast vertically, but it is obviously useless to specify this if there is no foundry in the neighbourhood in which it can be carried out. As a matter of fact, the great bulk of such castings for builder's work is cast in a horizontal position, and then the only care that can be taken is to have a good head, in which the spongy part may be taken up; and under such a system of casting, columns should never be used when stanchions will serve the purpose as well.

When the floor is to be of concrete, and the girders and columns or stanchions are to be encased, a clause should be inserted in the specification to the effect that no concreting or encasing is to be done until the architect, or his representative, has inspected the joints of the stanchions and the main girders, and certified that they are in accordance with the general specification.

This inspection rests as a rule with the clerk of the works, for it is not possible for the architect to exercise personal supervision over all the works he has designed; it is therefore of the first importance that the clerk of the works should be a man who has a general knowledge of constructional iron work, as well as of all the other trades employed upon the building. Such a man is independent of the contractors, and is worth a salary that will put him beyond the temptations of presents or their equivalents.

There is commonly specified a time for maintenance of work after its completion, and this may lead to some complications if the clause dealing with it is not clearly defined, for the question of sub-letting of contracts enters into it.

The foundations, brickwork, masonry, and carpentry, together with the finishers' work, may be let to a general contractor, and the constructional ironwork independently to another contractor, or the whole work may be let to the general contractor, who sublets the constructional ironwork, and any special floors which form part of the design.

The way in which the contract is let will very materially affect the cost of the work; for instance, if the general contractor sublets the ironwork he must have his profit on it, but if the ironwork is made a separate contract, the general contractor's profit is saved, as money is always to be saved by avoiding the middle-man.

There may be an idea that trouble is saved by putting the whole responsibility of the execution of the work in a satisfactory manner upon one man or one firm, but this any competent architect should ignore, because the responsibility really rests upon him as the designer of the work, and he can certainly do better for his clients by dividing the contracts in regard to builders' work and engineers' work, than by letting the latter be a sub-contract under the former.

ESTIMATION OF WEIGHTS.—Simple and accurate methods of calculating the weights of the materials used are indispensable to the constructional engineer and architect. The data for arriving at the required results are very copiously supplied by books of tables which give the weights per foot run for some classes of material and per foot super for others, but it does not always happen that such tables are at hand, and then it becomes necessary to calculate the weights without them. It happens very fortunately that the only data to be committed to memory are remarkably simple; a bar of wrought-iron 1 in. square and one yard long weighs ten pounds, which is also the weight of a square foot of wrought iron ¼ in. thick. Mild steel weighs two per cent. heavier than wrought iron, and cast iron is about five per cent. lighter—if it is good grey iron—but this will vary with the different qualities, and may be less or more. In rough castings the rounding in the internal corners adds to the weight, and it is practically accurate enough to calculate them as if they were wrought iron. Small brackets, stiffeners, and other such appendages of castings may be taken out in cubic inches; one cubic inch weighs 0·262 lbs.

If, then, there is a run of uniform section in wrought or cast iron, the weight per lineal foot will be found by multiplying the sectional area in square inches by ten and dividing by three; for example, the weight per lineal foot of a wrought-iron bar, 9 in. wide by $\frac{3}{4}$ in. thick, is required : the sectional area is $9 \times 0·75 = 6·75$ square inches, and

$$\frac{6·75 \times 10}{3} = 22·5 \text{ lbs. per lineal foot.}$$

A few examples will most clearly show how to make these calculations in practice. Let the weight be required of a CAST-IRON STANCHION, as shown in Fig. 54, where A is a horizontal section, and B a front elevation, C and D plans of the cap and base. The stanchion is H section, 15 ft. high and 9 in. by 9 in. in the shaft; the cap is 15 in. square and the base 18 in.; these are both $1\frac{1}{4}$ in. thick, and the rest of the work is 1 in. thick. Stiffeners F occur every 3 ft. of the height, on both sides, to strengthen the stanchions against twisting or lateral yielding of the flanges $H H$. For the support of the cap and base brackets E and G are provided.

The sectional area of the shaft is

$$= 2 \{9 \times 1\} + 7 \times 1 = 25 \text{ square inches.}$$

The 7 in. × 1 in. refers to the central web. The weight

per foot $= 25 \times \dfrac{10}{3} = 83\dot{\cdot}3$ lbs.; and the weight of the whole shaft $= 83\dot{\cdot}3 \times 15$ ft. $= 1250$ lbs. A plate 1 foot square and 1 inch thick is taken as weighing 40 lbs., therefore the weight of the base $= 1\cdot5 \times 1\cdot5 \times 1\frac{1}{4} \times 40 = 112\cdot5$ lbs.; and the cap $1\cdot25 \times 1\cdot25 \times 1\frac{1}{4} \times 40 = 78\cdot1$ lbs. There are eight stiffeners F, each measuring 7 in. by 4 in. by 1 in. thick, so their total weight will be $8 \times 7 \times 4 \times 1 \times 0\cdot262 = 58\cdot688$ lbs. The overhang of the cap is 3 inches each way, so the six brackets will be each 3 in. by 3 in. by 1 in.; two of them will make up a 3 in. square, so their weight will be $= 3$ squares; 3 in. $\times$ 3 in. $\times$ 1 in. $\times$ 0\cdot262 $= 7\cdot074$ lbs. The projection of the base is $4\frac{1}{2}$ in. each way, so the weight of base brackets will be

$$= 3 \times 4\cdot5 \times 4\cdot5 \times 1 \times 0\cdot262 = 15\cdot916 \text{ lbs.}$$

Adding all these weights together, the total weight of the stanchion is found thus :—

		In.	In.	In.	Ft.	In.	Lbs.
No. 1 Shaft	. . .	9 ×	9 × 1 × 15	0	=		1250·000
,, 1 Base	. . .	18 × 18 × 1¼			=		112·500
,, 1 Cap	. . .	15 × 15 × 1¼			=		78·100
,, 8 Stiffeners	. .	7 × 4 × 1			=		56·688
,, 6 Base brackets	.	4½ × 4½ × 1			=		15·916
,, 6 Cap ,,	.	3 × 3 × 1			=		7·074
							1520·278

which is equal to 13 cwts. 2 qrs. 8 lbs.

The weight of a cast-iron column is made up by the shaft, cap, and base. The sectional area of the shaft is equal to the difference in area of two circles corresponding respectively to the inside and outside diameters of the column; the area of a circle is found by multiplying the square of the diameter by 0·785, therefore the weight per foot—if d

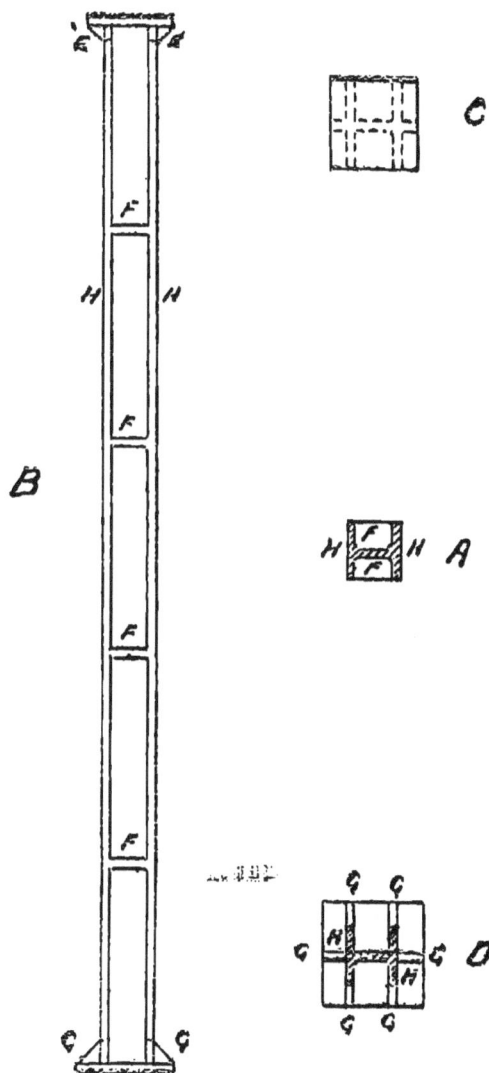

Fig. 54.

and d' = the external and internal diameters respectively,
will be

$$= 0.785 \left\{ d^2 - d'^2 \right\} \times \frac{10}{3}$$

Let the external diameter be 12 in. and the metal $1\frac{1}{4}$
in. thick, then the internal diameter will be $12 - 2 \times 1\frac{1}{4}$
$= 9.5$ in., and the weight per foot

$$= 0.785 \left\{ \overline{12}^2 - \overline{9.5}^2 \right\} \times \frac{10}{3} = 140.64 \text{ lbs.}$$

If the column is taper the mean diameters may be taken.
Let the cap and base be plain—any ornamental work being
screwed to it afterwards—then the area of the inside circle
must be deducted from each; let there be four brackets to
the base and four to the cap running to the corners and
having heights equal to their projection from the shaft, the
same as in the stanchion just dealt with. Let the base be
2 ft. square by $1\frac{1}{2}$ in. thick, and the cap 18 in. by $1\frac{1}{2}$ in.
thick, the brackets being the same thickness as the shaft.
The column is taken as of the same diameter all the way
up and 18 ft. in height between the base and cap. The
weight of the shaft will be 140.64 lbs. × 18 ft. = 2531.52
lbs. The weight of base = {(24 in. × 24 in.) − (0.785
× 9.5)} × 1.5 × 0.262 = 190.651 lbs. The weight of
cap will be = {(18 in. × 18 in.) − (0.785 × 9.5)} ×
1.5 × 0.262 = 99.475 lbs.

The width of the base across the angles is very nearly
34 in. and this will give a bracket projection $= \dfrac{34 - 12}{2}$
$= 11$ in.; the cap measures a little over 25 in. across the
angles which gives a bracket projection $\dfrac{25 - 12}{2} = 6.5$ in.
These are to the extreme corners, but the brackets will be

stopped where their sides meet the edges of the base and cap, and the extreme corner cut off; this will make their lengths respectively, 10·5 and 6 in., and their weights, taking two brackets to form a square, will be : base brackets, 2 × 10·5 in. × 10·5 in. × 1¼ × 0.262 = 72·213 lbs.; cap brackets, 2 × 6 in. × 6 in. × 1¼ × 0·262 = 23·58 lbs.

Summing up these several weights the total weight of the column is found.

		Ft. In.		Ft. In.		Ft. In.		Lbs.
No. 1 Shaft	.	. 1 0 dia.	×	1¼	× 18 0	=	2531·520
„ 1 Base	.	. 2 0 „	× 2 0	×	1½	=	190·651
„ 1 Cap	.	. 1 6 „	× 1 6	×	1½	×	99·475
„ 4 Base brackets		10½ „	×	10½	×	1¼	=	72·213
„ 4 Cap	.	. 6 „	×	6	×	1¼	=	23·580
								2917,439

which is equal to 1 ton 6 cwts. 0 qrs. 5 lbs.

Octagonal and hexagonal columns are sometimes used, and occasionally grouped columns or clustered; in these cases the sectional area is found by measuring along the centre of the thickness and multiplying the length so found by the thickness.

The next example I shall take is a WROUGHT-IRON GIRDER built up of plates, angle-iron, tee-iron and flat bars riveted together with rivets ¾-in. diameter and 4-in. pitch.   In Fig. 55, A shows a side elevation, B a transverse section at the centre of the girder, and C a transverse section near the end where there is only one plate in each flange.   The length of the girder is 40 ft., its depth between flanges, 4 ft. ; all the flange plates and the end plates are 18 in. wide by ½ in. thick.   The inner tier of flange plates is 40 ft. long with a joint near the centre, the second tier is 30 ft. long, also with a joint near the centre—these two joints being made by a cover plate 11 ft. long; the outer tier consists of one flange plate 20 ft. long. The angle-irons are $3\frac{1}{2}$ in. by $3\frac{1}{2}$ in. by $\frac{1}{2}$ in. thick, with one joint in each covered by an angle-bar 3 in. by 3 in. by $\frac{5}{8}$ in. thick, round backed to fit into the main angle-iron. The web consists of plates 4 ft. wide by $\frac{3}{8}$ in. thick and 8 ft. long, the joints being brought under some of the tee-iron stiffeners, which are 6 in. by 3 in. by $\frac{1}{2}$-in. thick and 3 ft. 11 in. long, riveted on with each end upon the main angle-iron, and having a $\frac{1}{2}$ in. packing strip underneath the central part; these strips will be 6 in. wide and 3 ft. 5 in. long to fit between the vertical limbs of the main angle-irons. There will also be packing strips under the end angle-irons which connect the end plates with the web. The weights per foot run will be as follows :—

$$\text{Web plates} \quad . \quad \overset{\text{In.}}{48} \times \overset{\text{In.}}{\frac{3}{8}} \times \frac{10}{3} = 60 \text{ lbs. per foot run.}$$

$$\text{Flange} \quad . \quad . \quad 18 \times \frac{1}{2} \times \frac{10}{3} = 30 \quad ,, \qquad ,,$$

$$\text{Packing bars} \quad . \quad 6 \times \frac{1}{2} \times \frac{10}{3} = 10 \quad ,, \qquad ,,$$

$$,, \qquad ,, \quad . \quad 3\frac{1}{2} \times \frac{1}{2} \times \frac{10}{3} = 5\cdot83 \quad ,, \qquad ,,$$

$$\text{Angle-iron} \quad . \quad (3\tfrac{1}{2} + 3\tfrac{1}{2} - \tfrac{1}{2}) \times \tfrac{1}{2} \times \frac{10}{3} = 10\cdot8\dot{3} \text{ lbs. per foot run.}$$

$$,, \qquad ,, \quad . \quad (3 + 3 - \tfrac{5}{8}) \times \tfrac{3}{8} \times \frac{10}{3} = 11\cdot19 \quad ,, \qquad ,,$$

$$\text{Tee-iron} \quad . \quad . \quad (6 + 3 - \tfrac{1}{2}) \times \tfrac{1}{2} \times \frac{10}{3} = 14\cdot16 \quad ,, \qquad ,,$$

It is usual to allow 5 per cent. on the weight of the plates and bars for that of the rivet heads. It is always advisable to keep a copy of the details of weights, in case of any

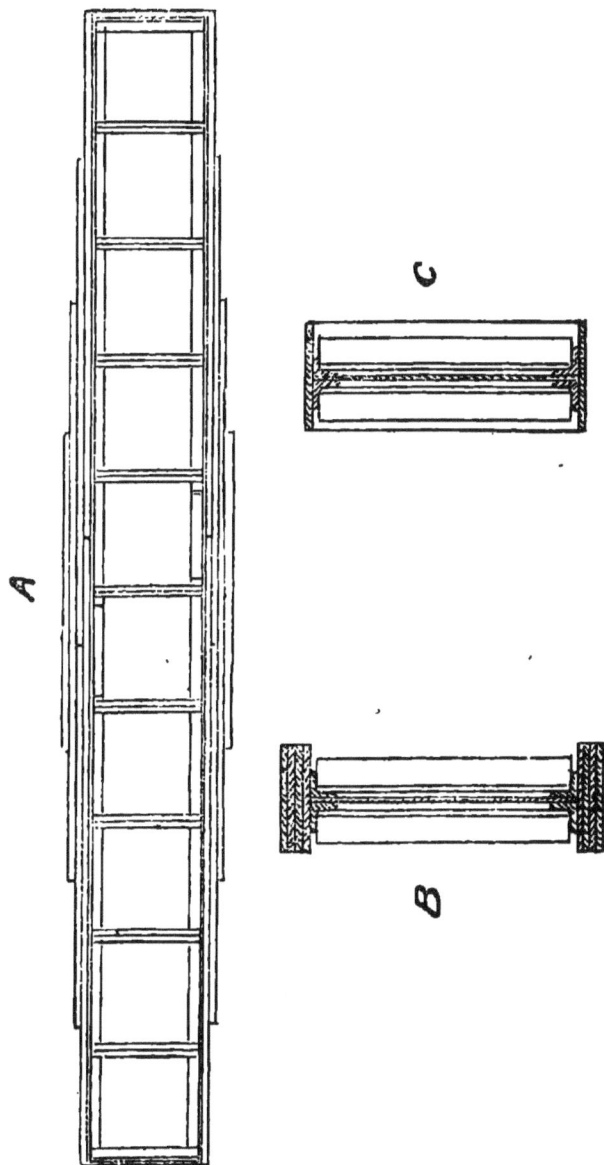

Fig. 55.

question as to accuracy arising, so that they are self-explanatory. I have found the following arrangement of schedule very convenient for this purpose.

Weight of No. 1 Girder 40 ft. long, 4 ft. deep, and 1 ft. 6 in. wide.

	No.	Ft. In.	Ft. In.	In.		Lbs.			Lbs.
Web plates	1..	39 11	× 4 0	× $\frac{3}{8}$	=	39·91	@ 60	=	2394·6
Flange ,,	2..	40 0	× 1 6	× $\frac{1}{2}$	=	80·00	,, 30	=	2400·0
,, ,,	2..	30 0	× 1 6	× $\frac{1}{2}$	=	60·00	,, 30	=	1800·0
,, ,,	2..	20 0	× 1 6	× $\frac{1}{2}$	=	40·00	,, 30	=	1200·0
Cover ,,	2..	11 0	× 1 6	× $\frac{1}{2}$	=	22·00	,, 30	=	660·0
End ,,	2..	4 0	× 1 6	× $\frac{1}{2}$	=	8·00	,, 30	=	240·0
Main angle-irons	4..	39 11	× 3½ × 3½	× ½	=	159·66	,, 10·83	=	1729·2
End ,, ,,	4..	3 11	× 3½ × 3½	× ½	=	15·66	,, 10·83	=	169·6
Cover ,, ,,	4..	3 0	× 3 × 3	× $\frac{3}{8}$	=	12·00	,, 11·19	=	134·3
Tee-irons	18..	3 11	× 6 × 3	× ½	=	70·50	,, 14·16	=	998·3
Packing strips	18..	3 5	× 6	× ½	=	61·50	,, 10·00	=	615·0
,, ,,	4..	3 5	× 3½	× ½	=	13·66	,, 5·83	=	79·7

$$12420·7$$

Rivet heads . . . . 5 per cent. = 621·0

$$13041·7$$

which is equal to 5 tons 16 cwts. 1 qr. 21 lbs.

In calculating the weights of bolts and nuts it is practically accurate enough to add 7 diameters to the length of the bolt measured from under head to point, and treat it as round iron. For instance, the length to be added to a $\frac{3}{4}$-in. bolt will be 0·75 × 7 = 5·25 inches.

If the girder in the example were steel the weight would be greater by 2 per cent., that is 260 lbs. = 2 cwt. 1 qr. 8 lbs.

Next let the weight be required of a compound girder, composed of two rolled steel girders 12 in. × 6 in., having a web thickness of 0·61 in., and a mean thickness of flange equal to 0·87 in., with 2 steel plates 12 in. wide by ½ in. thick, riveted on the top and bottom flanges, the length of girder over all being 20 feet.

The sectional area of the flanges of one rolled girder is $2 \times 6 \times 0·87 = 10·44$ sq. in., and that of the web, $\{12 - 2 \times 0·87\} \times 0·61 = 6·258$ sq. in., making the whole sectional area $= 16·698$ sq. in. The sectional area of each plate is 12 in. $\times$ ½ in. $= 6$ sq. in. So the total sectional area of the compound girder will be

No. 2 rolled steel girders, 12 in. x 6 in. = 33·396 sq. in.
,,   4 flange plates, 12 in. x ½ in.   = 24·000   ,,
                                           57·396 sq. in.

Remembering that steel is 2 per cent. heavier than iron, and that rivet heads amount to another 5 per cent., the weight of the compound girder will be

$$1·07 \times 57·396 \times \frac{10}{3} = 204·71 \text{ lbs. per foot run,}$$

making the weight of the entire girder $= 204·71 \times 20 = 4094·2$ lbs. $= 1$ ton 16 cwt. 2 qrs. 6 lbs.

Although it is not convenient to carry about a bulky pocket-book of tables of weights, a few multipliers which may be written on a card will be convenient as an aid to memory, until use impresses them upon the mind. They are derived from the data already given, and the method of finding them is shown below, all multipliers being given for pounds per lineal foot of the materials to which they refer. As a bar of wrought iron 1 in. sq. and 1 yard long weighs 10 lbs., a piece of the same section 1 foot long weighs 3⅓ or 3·3 lbs. The sectional area of a round bar is equal to the square of its diameter multiplied by 0·7854.

The required multipliers will therefore be as follows, the scantlings being in inches.

Flat iron bars and plates $\left\{\begin{array}{l}\text{Breadth} \times \text{thickness} \times 3\cdot333 = \text{weight of} \\ \text{1 ft. length in lbs.}\end{array}\right.$

Round iron bars . . $\left\{\begin{array}{l}\text{Diameter squared} \times 2\cdot618 = \text{weight of 1 ft.} \\ \text{length in lbs.}\end{array}\right.$

Angle-irons . . . $\left\{\begin{array}{l}\text{Sum of the sides, less the thickness} \times \text{thick-} \\ \text{ness} \times 3\cdot333 = \text{weight of 1 ft. length} \\ \text{in lbs.}\end{array}\right.$

Tee-irons . . . $\left\{\begin{array}{l}\text{Height over web, plus width, less the} \\ \text{thickness} \times \text{thickness} \times 3\cdot333 = \text{weight of} \\ \text{1 ft. length in lbs.}\end{array}\right.$

Channel irons . . $\left\{\begin{array}{l}\text{Sum of the width of back and the two} \\ \text{sides, less twice the thickness} \times \text{thickness} \\ \times 3\cdot333 = \text{weight of 1 ft. length in lbs.}\end{array}\right.$

When the material is steel replace the multiplier 3·333 by 3·4, and 2·618 by 2·67. A superficial foot of wrought iron weighs 5 lbs. for every eighth of an inch of thickness, and steel 5·1 lbs. per eighth of an inch of thickness. A circular plate of wrought iron weighs the square of its diameter in feet multiplied by 3·927 lbs. for every eighth of an inch of thickness; for steel plates the multiplier is 4 lbs.

For rough purposes the same mutipliers may be used for cast as for wrought iron.

It will be observed that the limbs of angle-irons, tee-irons, channels, and joist sections are not parallel, or of equal thickness across the breadth, as some of them must have a taper to allow of their leaving the rolls easily, and with all it makes the rolling rather easier. It is not always a simple matter to say where the mean thickness is, but the sectional area can be calculated from the weight by multiplying by three and dividing by ten; an angle-iron which weighs 12

lbs. per foot run has a sectional area $= 12 \times \dfrac{3}{10} = 3\cdot6$ sq. inches.

In those girders that are built up of angles, and tees, or channels in connection with plates, it is advisable to specify the weight per foot instead of the thickness, especially if the work is to be let at a lump sum contract, where the tendency of the maker is naturally not to put more than the estimated weight into the job. There must also be a limit to excess when the work is charged by weight, in this case it is usual to allow what is called a rolling margin of $2\frac{1}{2}$ per cent. on the running weight of the sections.

In regard to rolled girders and joists it is most economical to select minimum weights where possible, because in these the webs have the least weight in proportion to the flanges ; for it is obvious that as the rolls are moved wider apart the web is thickened exactly as much as the flange is widened, and therefore the bulk of the additional metal goes where it is of the least use, and the percentage of added strength is frequently not half of that of the additional weight of metal.

In order to be able to compare questions of weight and strength with a view to adopting that section which is most economical, it is indispensable to have a method of readily calculating weights, and that is one reason why I enter so fully into this part of the subject, which, being less interesting than other parts is more likely to be overlooked by the student.

In order to obtain some practical knowledge of constructional iron and steel work, it is highly desirable that the student should pass a year or two in a foundry and girder yard where good work is executed, such as that required for bridges and large roofs, so that he will be able to appreciate the quality of work when he sees it ; if then in after-practice he chooses from motives of economy to accept

T

rough work, it will not be because he does not know the difference.

In inspecting work the conditions under which it has been let should always be remembered ; if a strict specification has been supplied and agreed to by the contractor, we can claim execution of the work in accordance with its clauses ; but, if the only stipulation has been for a certain strength, or for a contractor's guarantee that the work shall be so executed as to be sufficiently strong for the purpose it is required to serve, we cannot demur for rough or even slovenly workmanship, so long as the dimensions are accurate and the different parts admit of being fixed together in accordance with the plans.

INDEX.

THE B. SIMCHES MEMORIAL LIBRARY MANAYUNK

CROSBY LOCKWOOD & SON'S PUBLICATIONS.

PRACTICAL BUILDING CONSTRUCTION: a Handbook for Students, and a Book of Reference for Persons engaged in Building. By J. P. ALLEN. 8vo, 450 pages, with 1,000 Illustrations, 12s. 6d. cloth.

SCIENCE OF BUILDING: an Elementary Treatise on the Principles of Construction. By E. WYNDHAM TARN, M.A. Third Edition, Enlarged. Fcap. 8vo, with 59 Engravings, 4s. cloth.

MECHANICS OF ARCHITECTURE: A Treatise on Applied Mechanics, especially adapted to the Use of Architects. By E. WYNDHAM TARN, M.A. Second Edition, Enlarged. Crown 8vo, with 125 Illustrations, 7s. 6d. cloth.

GRAPHIC AND ANALYTIC STATICS, in their Practical Application to the Treatment of Stresses in Roofs, Solid Girders, Bridges, Arches, Piers, and other Frameworks. By R. H. GRAHAM, C.E. Second Edition, Enlarged. 8vo, 16s. cloth.

HANDYBOOK FOR THE CALCULATION OF STRAINS in Girders and Similar Structures, and their Strength. By W. HUMBER. Fifth Edition, Crown 8vo, with 100 Illustrations and 3 Plates, 7s. 6d. cloth.

GRAPHIC TABLE for Facilitating the Computation of the Weights of Wrought Iron and Steel Girders, &c. By J. H. WATSON BUCK. On a Sheet, 2s. 6d.

TRUSSES OF WOOD AND IRON: Practical Applications of Science in Determining the Stresses, Breaking Weights, Safe Loads, Scantlings, and Details of Construction. With Complete Working Drawings. By W. GRIFFITHS. Oblong 8vo, 4s. 6d. cloth.

EXPANSION OF STRUCTURES BY HEAT. By JOHN KEILY, C.E. Crown 8vo, 3s. 6d. cloth.

IRON BRIDGES OF MODERATE SPAN: their Construction and Erection. By H. W. PENDRED. Second Edition, 12mo, with 40 Illustrations, 2s. cloth.

ENGINEERING CHEMISTRY: a Practical Treatise for the Use of Analytical Chemists, Engineers, Ironmasters, &c. By H. J PHILLIPS, F.I.C. Second Edition, Enlarged. Crown 8vo, 10s. 6d. cloth.

CONCRETE: ITS NATURE AND USES. A Book for Architects, Builders, Contractors, and Clerks of Works. By G. L. SUTCLIFFE, A.R.I.B.A. Crown 8vo, 350 pages, with Illustrations, 7s. 6d. cloth.

LOCKWOOD'S BUILDER'S PRICE BOOK FOR 1896. A Comprehensive Handbook of the Latest Prices and Data for Builders, Architects, Engineers, and Contractors. Re-constructed, Re-written, and Greatly Enlarged by FRANCIS J. W. MILLER. Crown 8vo. 800 closely-printed pages, 4s. cloth.

London: CROSBY LOCKWOOD & SON, 7, Stationers' Hall Court, E.C.

CROSBY LOCKWOOD & SON'S PUBLICATIONS.

MECHANICAL ENGINEER'S POCKET-BOOK of Tables, Formulæ, Rules, and Data. By D. KINNEAR CLARK, M.INST.C.E. Third Edition, Revised. Fcap. 8vo, 6s., leather.

ENGINEERING ESTIMATES, COSTS, AND ACCOUNTS. By A GENERAL MANAGER. Demy 8vo, 12s., cloth.

WORKS' MANAGER'S HANDBOOK of Modern Rules, Tables, and Data for Engineers, Millwrights, Machinists, &c. By W. S. HUTTON, C.E. Fifth Edition, Revised, with Additions. 8vo, 15s., cloth.

PRACTICAL ENGINEER'S HANDBOOK: a Treatise on Engines, Boilers, and other Engineering Work. By W. S. HUTTON, C.E. Fifth Edition, Revised. 8vo, 18s., cloth.

STEAM BOILER CONSTRUCTION: a Handbook for Engineers, Boiler-makers, and Steam Users. By W. S. HUTTON, C.E. 8vo, 18s., cloth.

HANDBOOK ON THE STEAM ENGINE: with especial reference to Small and Medium-sized Engines. By H. HAEDER, C.E. Translated from the German, with Additions and Alterations, by H. H. P. POWLES, A.M.Inst.C.E. Second Edition, Revised. Crown 8vo, with nearly 1,100 Illustrations, 9s. cloth.

TEMPLETON'S PRACTICAL MECHANIC'S WORKSHOP COMPANION. Seventeenth Edition, Modernised and Enlarged by W. S. HUTTON, C.E. Fcap. 8vo, 6s. leather.

ELECTRIC LIGHT: its Production and Use. By J. W. URQUHART, C.E. Fifth Edition, carefully Revised, with Additions. Crown 8vo, with 145 Illustrations, 7s. 6d. cloth.

ELECTRIC LIGHT FITTING: a Handbook for Working Electrical Engineers. By J. W. URQUHART, C.E. Second Edition, Revised, with Additions. Crown 8vo, 5s. cloth.

ELECTRICAL ENGINEER'S POCKET-BOOK of Modern Rules, Formulæ, Tables, and Data. By H. R. KEMPE, A.M.Inst.C.E. Second Edition, Revised and Enlarged. Royal 32mo, 5s. leather.

LOCKWOOD'S DICTIONARY OF TERMS USED IN THE PRACTICE OF MECHANICAL ENGINEERING. Comprising upwards of 6,000 Definitions. Second Edition, with Additions. Crown 8vo, 7s. 6d.

TABLES AND MEMORANDA FOR MECHANICS, Engineers, &c. By FRANCIS SMITH. Sixth Edition, Revised and Enlarged. Waistcoat pocket size, 1s. 6d. leather.

London: CROSBY LOCKWOOD & SON, 7, Stationers' Hall Court, E.C.

WEALE'S SERIES

OF

SCIENTIFIC AND TECHNICAL

WORKS.

" It is not too much to say that no books have ever proved more popular with or more useful to young engineers and others than the excellent treatises comprised in WEALE'S SERIES."—**Engineer.**

𝔄 𝔑𝔢𝔴 𝔆𝔩𝔞𝔰𝔰𝔦𝔣𝔦𝔢𝔡 𝔏𝔦𝔰𝔱.

Capio Lumen

CROSBY LOCKWOOD AND SON,

7, STATIONERS' HALL COURT, LONDON, E.C.

1904.

CIVIL ENGINEERING & SURVEYING.

Civil Engineering.
By HENRY LAW, M.Inst.C.E. Including a Treatise on HYDRAULIC
ENGINEERING by G. R. BURNELL, M.I.C.E. Seventh Edition, revised,
with LARGE ADDITIONS by D. K. CLARK, M.I.C.E. . . . **6/6**

Pioneer Engineering:
A Treatise on the Engineering Operations connected with the Settlement of
Waste Lands in New Countries. By EDWARD DOBSON, M.INST.C.E.
With numerous Plates. Second Edition **4/6**

Iron Bridges of Moderate Span:
Their Construction and Erection. By HAMILTON W. PENDRED. With 40
Illustrations **2/0**

Iron and Steel Bridges and Viaducts.
A Practical Treatise upon their Construction for the use of Engineers,
Draughtsmen, and Students. By FRANCIS CAMPIN, C.E. With Illus. **3/6**

Constructional Iron and Steel Work,
As applied to Public, Private, and Domestic Buildings. By FRANCIS
CAMPIN, C.E. **3/6**

Tubular and other Iron Girder Bridges.
Describing the Britannia and Conway Tubular Bridges. By G. DRYSDALE
DEMPSEY, C.E. Fourth Edition **2/0**

Materials and Construction:
A Theoretical and Practical Treatise on the Strains, Designing, and Erection of Works of Construction. By FRANCIS CAMPIN, C.E. . . **3/0**

Sanitary Work in the Smaller Towns and in Villages.
By CHARLES SLAGG, Assoc. M.Inst.C.E. Third Edition . . **3/0**

Construction of Roads and Streets.
By H. LAW, C.E., and D. K. CLARK, C.E. Sixth Edition, revised, with
Additional Chapters by A. J. WALLIS-TAYLER, A.M. Inst. C.E. . **6/0**

Gas Works,
Their Construction and Arrangement and the Manufacture and Distribu-
tion of Coal Gas. Originally written by S. HUGHES, C.E. Ninth Edition.
Revised, with Notices of Recent Improvements, by HENRY O'CONNER,
A.M. Inst. C.E., Author of "The Gas Engineers' Pocket Book."
[*Just Published.* **6/0**

Water Works
For the Supply of Cities and Towns. With a Description of the Principal
Geological Formations of England as influencing Supplies of Water. By
SAMUEL HUGHES, F.G.S., C.E. Enlarged Edition **4/0**

The Power of Water,
As applied to drive Flour Mills, and to give motion to Turbines and other
Hydrostatic Engines. By JOSEPH GLYNN, F.R.S. New Edition . **2/0**

Wells and Well-Sinking.
By JOHN GEO. SWINDELL, A.R.I.B.A., and G. R. BURNELL, C.E. Revised
Edition. With a New Appendix on the Qualities of Water. Illustrated **2/0**

The Drainage of Lands, Towns, and Buildings.
By G. D. DEMPSEY, C.E. Revised, with large Additions on Recent
Practice, by D. K. CLARK, M.I.C.E. Third Edition . . . **4/6**

The Blasting and Quarrying of Stone,
For Building and other Purposes. With Remarks on the Blowing up of
Bridges. By Gen. Sir J. BURGOYNE, K.C.B. **1/6**

Foundations and Concrete Works.
With Practical Remarks on Footings, Planking, Sand, Concrete Béton,
Pile-driving, Caissons, and Cofferdams. By E. DOBSON. Ninth Ed. **1/6**

Pneumatics,
Including Acoustics and the Phenomena of Wind Currents, for the Use of Beginners. By CHARLES TOMLINSON, F.R.S. Fourth Edition . **1/6**

Land and Engineering Surveying.
For Students and Practical Use. By T. BAKER, C.E. Nineteenth Edition, Revised and Extended by F. E. DIXON, A.M. Inst. C.E., Professional Associate of the Institution of Surveyors. With numerous Illustrations and two Lithographic Plates **2/0**

Mensuration and Measuring.
For Students and Practical Use. With the Mensuration and Levelling of Land for the purposes of Modern Engineering. By T. BAKER, C.E. New Edition by E. NUGENT, C.E. **1/6**

MINING AND METALLURGY.

Mining Calculations,
For the use of Students Preparing for the Examinations for Colliery Managers' Certificates, comprising numerous Rules and Examples in Arithmetic, Algebra, and Mensuration. By T. A. O'DONAHUE, M.E., First-Class Certificated Colliery Manager. **3/6**

Mineralogy,
Rudiments of. By A. RAMSAY, F.G.S. Fourth Edition, revised and enlarged. Woodcuts and Plates **3/6**

Coal and Coal Mining,
A Rudimentary Treatise on. By the late Sir WARINGTON W. SMYTH, F.R.S. Eighth Edition, revised by T. FORSTER BROWN . . . **3/6**

Metallurgy of Iron.
Containing Methods of Assay, Analyses of Iron Ores, Processes of Manufacture of Iron and Steel, &c. By H. BAUERMAN, F.G.S. With numerous Illustrations. Sixth Edition, revised and enlarged **5/0**

The Mineral Surveyor and Valuer's Complete Guide.
By W. LINTERN. Fourth Edition, with an Appendix on Magnetic and Angular Surveying **3/6**

Slate and Slate Quarrying:
Scientific, Practical, and Commercial. By D. C. DAVIES, F.G.S. With numerous Illustrations and Folding Plates. Fourth Edition . . **3/0**

A First Book of Mining and Quarrying,
With the Sciences connected therewith, for Primary Schools and Self-Instruction. By J. H. COLLINS, F.G.S. Second Edition . . . **1/6**

Subterraneous Surveying,
With and without the Magnetic Needle. By T. FENWICK and T. BAKER, C.E. Illustrated **2/6**

Mining Tools.
Manual of. By WILLIAM MORGANS, Lecturer on Practical Mining at the Bristol School of Mines **2/6**

Mining Tools, Atlas
Of Engravings to Illustrate the above, containing 235 Illustrations of Mining Tools, drawn to Scale. 4to **4/6**

Physical Geology,
Partly based on Major-General PORTLOCK's "Rudiments of Geology." By RALPH TATE, A.L.S., &c. Woodcuts **2/0**

Historical Geology,
Partly based on Major-General PORTLOCK's "Rudiments." By RALPH TATE, A.L.S., &c. Woodcuts **2/6**

Geology, Physical and Historical.
Consisting of "Physical Geology," which sets forth the Leading Principles of the Science; and "Historical Geology," which treats of the Mineral and Organic Conditions of the Earth at each successive epoch. By RALPH TATE, F.G.S. **4/6**

MECHANICAL ENGINEERING.

The Workman's Manual of Engineering Drawing.
By JOHN MAXTON, Instructor in Engineering Drawing, Royal Naval
College, Greenwich. Eighth Edition. 300 Plates and Diagrams . **3/6**

Fuels: Solid, Liquid, and Gaseous.
Their Analysis and Valuation. For the Use of Chemists and Engineers.
By H. J. PHILLIPS, F.C.S., formerly Analytical and Consulting Chemist
to the Great Eastern Railway. Fourth Edition. .   .   .   **2/0**

Fuel, Its Combustion and Economy.
Consisting of an Abridgment of "A Treatise on the Combustion of Coal and
the Prevention of Smoke." By C. W. WILLIAMS, A.I.C.E. With Extensive Additions by D. K. CLARK, M.Inst.C.E. Fourth Edition . **3/6**

The Boilermaker's Assistant
In Drawing, Templating, and Calculating Boiler Work, &c. By J. COURT-
NEY, Practical Boilermaker. Edited by D. K. CLARK, C.E. . **2/0**

The Boiler-Maker's Ready Reckoner,
With Examples of Practical Geometry and Templating for the Use of
Platers, Smiths, and Riveters. By JOHN COURTNEY. Edited by D. K.
CLARK, M.I.C.E. Fifth Edition **4/0**
₊ The last two Works in One Volume, half-bound, entitled "THE BOILER-
MAKER'S READY-RECKONER AND ASSISTANT." By J. COURTNEY and
D. K. CLARK. Price **7/0.**

Steam Boilers:
Their Construction and Management. By R. ARMSTRONG, C.E. Illustrated
1/6

Steam and Machinery Management.
A Guide to the Arrangement and Economical Management of Machinery.
By M. POWIS BALE, M.Inst.M.E. **2/6**

Steam and the Steam Engine,
Stationary and Portable. Being an Extension of the Treatise on the Steam
Engine of Mr. J. SEWELL. By D. K. CLARK, C.E. Fourth Edition **3/6**

The Steam Engine,
A Treatise on the Mathematical Theory of, with Rules and Examples for
Practical Men. By T. BAKER, C.E. **1/6**

The Steam Engine.
By Dr. LARDNER. Illustrated **1/6**

Locomotive Engines.
By G. D. DEMPSEY, C.E. With large Additions treating of the Modern
Locomotive, by D. K. CLARK, M.Inst.C.E. **3/0**

Locomotive Engine-Driving.
A Practical Manual for Engineers in charge of Locomotive Engines. By
MICHAEL REYNOLDS. Eleventh Edition. 3s. 6d. ; cloth boards . **4/6**

Stationary Engine-Driving.
A Practical Manual for Engineers in charge of Stationary Engines. By
MICHAEL REYNOLDS. Seventh Edition. 3s. 6d. ; cloth boards . **4/6**

The Smithy and Forge.
Including the Farrier's Art and Coach Smithing. By W. J. E. CRANE.
Fourth Edition **2/6**

Modern Workshop Practice,
As applied to Marine, Land, and Locomotive Engines, Floating Docks,
Dredging Machines, Bridges, Ship-building, &c. By J. G. WINTON.
Fourth Edition, Illustrated **3/6**

Mechanical Engineering.
Comprising Metallurgy, Moulding, Casting, Forging, Tools, Workshop
Machinery, Mechanical Manipulation, Manufacture of the Steam Engine,
&c. By FRANCIS CAMPIN, C.E. Third Edition .   .   .   **2/6**

Details of Machinery.
Comprising Instructions for the Execution of various Works in Iron in the
Fitting-Shop, Foundry, and Boiler-Yard. By FRANCIS CAMPIN, C.E. **3/0**

Elementary Engineering:
A Manual for Young Marine Engineers and Apprentices. In the Form of Questions and Answers on Metals, Alloys, Strength of Materials, &c. By J. S. BREWER. Fifth Edition **1/6**

Power in Motion:
Horse-power Motion, Toothed-Wheel Gearing, Long and Short Driving Bands, Angular Forces, &c. By JAMES ARMOUR, C.E. Third Edition **2/0**

Iron and Heat,
Exhibiting the Principles concerned in the Construction of Iron Beams, Pillars, and Girders. By J. ARMOUR, C.E. **2/6**

Practical Mechanism,
And Machine Tools. By T. BAKER, C.E. With Remarks on Tools and Machinery, by J. NASMYTH, C.E. **2/6**

Mechanics:
Being a concise Exposition of the General Principles of Mechanical Science, and their Applications. By CHARLES TOMLINSON, F.R.S. . . **1/6**

Cranes (The Construction of),
And other Machinery for Raising Heavy Bodies for the Erection of Build-ings, &c. By JOSEPH GLYNN, F.R.S. **1/6**

NAVIGATION, SHIPBUILDING, ETC.

The Sailor's Sea Book:
A Rudimentary Treatise on Navigation. By JAMES GREENWOOD, B.A. With numerous Woodcuts and Coloured Plates. New and enlarged Edition. By W. H. ROSSER **2/6**

Practical Navigation.
Consisting of THE SAILOR'S SEA-BOOK, by JAMES GREENWOOD and W. H. ROSSER; together with Mathematical and Nautical Tables for the Working of the Problems, by HENRY LAW, C.E., and Prof. J. R. YOUNG . **7/0**

Navigation and Nautical Astronomy,
In Theory and Practice. By Prof. J. R. YOUNG. New Edition. **2/6**

Mathematical Tables,
For Trigonometrical, Astronomical, and Nautical Calculations; to which is prefixed a Treatise on Logarithms. By H. LAW, C.E. Together with a Series of Tables for Navigation and Nautical Astronomy. By Professor J. R. YOUNG. New Edition **4/0**

Masting, Mast-Making, and Rigging of Ships.
Also Tables of Spars, Rigging, Blocks; Chain, Wire, and Hemp Ropes, &c., relative to every class of vessels. By ROBERT KIPPING, N.A. . **2/0**

Sails and Sail-Making.
With Draughting, and the Centre of Effort of the Sails. By ROBERT KIPPING, N.A. **2/6**

Marine Engines and Steam Vessels.
By R. MURRAY, C.E. Eighth Edition, thoroughly revised, with Addi-tions by the Author and by GEORGE CARLISLE, C.E. . . . **4/6**

Naval Architecture:
An Exposition of Elementary Principles. By JAMES PEAKE . . **3/6**

Ships for Ocean and River Service,
Principles of the Construction of. By HAKON A. SOMMERFELDT . **1/6**

Atlas of Engravings
To Illustrate the above. Twelve large folding Plates. Royal 4to, cloth **7/6**

The Forms of Ships and Boats.
By W. BLAND. Tenth Edition, with numerous Illustrations and Models **1/6**

ARCHITECTURE AND THE
BUILDING ARTS.

Constructional Iron and Steel Work,
As applied to Public, Private, and Domestic Buildings. By FRANCIS
CAMPIN, C.E. **3/6**

Building Estates:
A Treatise on the Development, Sale, Purchase, and Management of Building Land. By F. MAITLAND. Third Edition **2/0**

The Science of Building:
An Elementary Treatise on the Principles of Construction. By E. WYND-
HAM TARN, M.A. Lond. Fourth Edition **3/6**

The Art of Building:
General Principles of Construction, Strength, and Use of Materials, Working
Drawings, Specifications, &c. By EDWARD DOBSON, M.R.I.B.A. .   **2/0**

A Book on Building,
Civil and Ecclesiastical. By Sir EDMUND BECKETT, Q.C. (Lord GRIM-
THORPE). Second Edition **4/6**

Dwelling-Houses (The Erection of),
Illustrated by a Perspective View, Plans, and Sections of a Pair of Villas, with
Specification, Quantities, and Estimates. By S. H. BROOKS, Architect   **2/6**

Cottage Building.
By C. BRUCE ALLEN. Twelfth Edition, with Chapter on Economic Cot-
tages for Allotments, by E. E. ALLEN, C.E. **2/0**

Acoustics in Relation to Architecture and Building:
The Laws of Sound as applied to the Arrangement of Buildings. By Professor T. ROGER SMITH, F.R.I.B.A. New Edition, Revised .   .   **1/6**

The Rudiments of Practical Bricklaying.
General Principles of Bricklaying; Arch Drawing, Cutting, and Setting;
Pointing; Paving, Tiling, &c. By ADAM HAMMOND. With 68 Woodcuts
1/6

The Art of Practical Brick Cutting and Setting.
By ADAM HAMMOND. With 90 Engravings **1/6**

Brickwork :
A Practical Treatise, embodying the General and Higher Principles of
Bricklaying, Cutting and Setting; with the Application of Geometry to Roof
Tiling, &c. By F. WALKER **1/6**

Bricks and Tiles,
Rudimentary Treatise on the Manufacture of; containing an Outline of the
Principles of Brickmaking. By E. DOBSON, M.R.I.B.A. Additions by
C. TOMLINSON, F.R.S. Illustrated **3/0**

The Practical Brick and Tile Book.
Comprising: BRICK AND TILE MAKING, by E. DOBSON, M.INST.C.E.;
Practical BRICKLAYING, by A. HAMMOND; BRICK-CUTTING AND SETTING,
by A. HAMMOND. 550 pp. with 270 Illustrations, half-bound .   .   **6/0**

Carpentry and Joinery—
THE ELEMENTARY PRINCIPLES OF CARPENTRY. Chiefly composed from the
Standard Work of THOMAS TREDGOLD, C.E. With Additions, and TREATISE
ON JOINERY, by E. W. TARN, M.A. Eighth Edition .   .   .   **3/6**

Carpentry and Joinery—Atlas
Of 35 Plates to accompany and Illustrate the foregoing book. With
Descriptive Letterpress. 4to **6/0**

A Practical Treatise on Handrailing;
Showing New and Simple Methods. By GEO. COLLINGS. Third Edition, including a TREATISE ON STAIRBUILDING. With Plates . . . **2/6**

Circular Work in Carpentry and Joinery.
A Practical Treatise on Circular Work of Single and Double Curvature. By GEORGE COLLINGS. Fourth Edition **2/6**

Roof Carpentry:
Practical Lessons in the Framing of Wood Roofs. For the Use of Working Carpenters. By GEO. COLLINGS **2/0**

The Construction of Roofs of Wood and Iron;
Deduced chiefly from the Works of Robison, Tredgold, and Humber. By E. WYNDHAM TARN, M.A., Architect. Fourth Edition . . . **1/6**

The Joints Made and Used by Builders.
By WYVILL J. CHRISTY, Architect. With 160 Woodcuts . . **3/0**

Shoring
And its Application: A Handbook for the Use of Students. By GEORGE H. BLAGROVE. With 31 Illustrations **1/6**

The Timber Importer's, Timber Merchant's, and Builder's Standard Guide.
By R. E. GRANDY **2/0**

Plumbing:
A Text-Book to the Practice of the Art or Craft of the Plumber. With Chapters upon House Drainage and Ventilation. By WM. PATON BUCHAN. Ninth Edition, with 512 Illustrations **3/6**

Ventilation:
A Text Book to the Practice of the Art of Ventilating Buildings. By W. P. BUCHAN, R.P., Author of "Plumbing," &c. With 170 Illustrations **3/6**

The Practical Plasterer:
A Compendium of Plain and Ornamental Plaster Work. By W. KEMP **2/0**

House Painting, Graining, Marbling, & Sign Writing.
With a Course of Elementary Drawing, and a Collection of Useful Receipts. By ELLIS A. DAVIDSON. Eighth Edition. Coloured Plates . **5/0**

** *The above, in cloth boards, strongly bound,* **6/0**

A Grammar of Colouring,
Applied to Decorative Painting and the Arts. By GEORGE FIELD. New Edition, enlarged, by ELLIS A. DAVIDSON. With Coloured Plates . **3/0**

Elementary Decoration
As applied to Dwelling Houses, &c. By JAMES W. FACEY. Illustrated **2/0**

Practical House Decoration.
A Guide to the Art of Ornamental Painting, the Arrangement of Colours in Apartments, and the Principles of Decorative Design. By JAMES W. FACEY **2/6**

** *The last two Works in One handsome Vol., half-bound, entitled* "HOUSE DECORATION, ELEMENTARY AND PRACTICAL," *price* **5/0**.

Portland Cement for Users.
By HENRY FAIJA, A.M.Inst.C.E. Third Edition, Corrected . . **2/0**

Limes, Cements, Mortars, Concretes, Mastics, Plastering, &c.
By G. R. BURNELL, C.E. Fifteenth Edition **1/6**

Masonry and Stone-Cutting.
The Principles of Masonic Projection and their application to Construction. By EDWARD DOBSON, M.R.I.B.A. **2/6**

Arches, Piers, Buttresses, &c.:
Experimental Essays on the Principles of Construction.   By W. BLAND. **1/6**

Quantities and Measurements,
In Bricklayers', Masons', Plasterers', Plumbers', Painters', Paperhangers', Gilders', Smiths', Carpenters' and Joiners' Work. By A. C. BEATON. **1/6**

The Complete Measurer:
Setting forth the Measurement of Boards, Glass, Timber and Stone. By R. HORTON. Sixth Edition **4/0**

Guide to Superficial Measurement:
Tables calculated from 1 to 200 inches in length, by 1 to 108 inches in breadth. For the use of Architects, Surveyors, Engineers, Timber Merchants, Builders, &c. By JAMES HAWKINGS. Fifth Edition . . **3/6**

Light:
An Introduction to the Science of Optics. For the Use of Students of Architecture, Engineering, and other Applied Sciences. By E. W. TARN, M.A. **1/6**

Hints to Young Architects.
By GEORGE WIGHTWICK, Architect. Sixth Edition, revised and enlarged by G. HUSKISSON GUILLAUME, Architect **3/6**

Architecture—Orders:
The Orders and their Æsthetic Principles. By W. H. LEEDS. Illustrated. **1/6**

Architecture—Styles:
The History and Description of the Styles of Architecture of Various Countries, from the Earliest Period. By T. TALBOT BURY . . **2/0**
₄ ORDERS AND STYLES OF ARCHITECTURE, *in One Vol.,* **3/6.**

Architecture—Design:
The Principles of Design in Architecture, as deducible from Nature and exemplified in the Works of the Greek and Gothic Architects. By EDW. LACY GARBETT, Architect. Illustrated **2/6**
₄ *The three preceding Works in One handsome Vol., half-bound, entitled* "MODERN ARCHITECTURE," *price* **6/0.**

Perspective for Beginners.
Adapted to Young Students and Amateurs in Architecture, Painting, &c. By GEORGE PYNE **2/0**

Architectural Modelling in Paper.
By T. A. RICHARDSON. With Illustrations, engraved by O. JEWITT **1/6**

Glass Staining, and the Art of Painting on Glass.
From the German of Dr. GESSERT and EMANUEL OTTO FROMBERG. With an Appendix on THE ART OF ENAMELLING **2/6**

Vitruvius—The Architecture of.
In Ten Books. Translated from the Latin by JOSEPH GWILT, F.S.A., F.R.A.S. With 23 Plates **5/0**
N.B.—This is the only Edition of VITRUVIUS *procurable at a moderate price.*

Grecian Architecture,
An Inquiry into the Principles of Beauty in. With an Historical View of the Rise and Progress of the Art in Greece. By the EARL OF ABERDEEN. **1/0**
₄ *The two preceding Works in One handsome Vol., half-bound, entitled* "ANCIENT ARCHITECTURE," *price* **6/0.**

INDUSTRIAL AND USEFUL ARTS.

Cements, Pastes, Glues, and Gums.
A Guide to the Manufacture and Application of Agglutinants. With 900 Recipes and Formulæ. By H. C. STANDAGE **2/0**

Clocks, Watches, and Bells for Public Purposes.
A Rudimentary Treatise. By EDMUND BECKETT, LORD GRIMTHORPE, LL.D., K.C., F.R.A.S. Eighth Edition, with new List of Great Bells and an Appendix on Weathercocks. [*Just published.* **4/6**
*** *The above, handsomely bound, cloth boards,* **5/6**.

Electro-Metallurgy,
Practically Treated. By ALEXANDER WATT. Tenth Edition . **3/6**

The Goldsmith's Handbook.
Containing full Instructions in the Art of Alloying, Melting, Reducing, Colouring, Collecting and Refining, Recovery of Waste, Solders, Enamels, &c., &c. By GEORGE E. GEE. Sixth Edition **3/0**

The Silversmith's Handbook,
On the same plan as the GOLDSMITH'S HANDBOOK. By G. E. GEE. **3/0**
*** *The last two Works, in One handsome Vol., half-bound,* **7/0**.

The Hall-Marking of Jewellery.
Comprising an account of all the different Assay Towns of the United Kingdom; with the Stamps and Laws relating to the Standards and Hall Marks at the various Assay Offices. By GEORGE E. GEE . . **3/0**

French Polishing and Enamelling.
Numerous Recipes for making Polishes, Varnishes, &c. By R. BITMEAD. **1/6**

Practical Organ Building.
By W. E. DICKSON, M.A. Second Edition, Revised, with Additions **2/6**

Coach-Building:
A Practical Treatise. By JAMES W. BURGESS. With 57 Illustrations **2/6**

The Cabinet-Maker's Guide
To the Entire Construction of Cabinet-Work. By R. BITMEAD . **2/6**

The Brass Founder's Manual:
Instructions for Modelling, Pattern Making, &c. By W. GRAHAM . **2/0**

The Sheet-Metal Worker's Guide.
For Tinsmiths, Coppersmiths, Zincworkers, &c. By W. J. E. CRANE. **1/6**

Sewing Machinery:
Its Construction, History, &c. By J. W. URQUHART, C.E. . . **2/0**

Gas Fitting:
A Practical Handbook. By JOHN BLACK. New Edition . . **2/6**

Construction of Door Locks.
From the Papers of A. C. HOBBS. Edited by C. TOMLINSON, F.R.S. **2/6**

The Model Locomotive Engineer, Fireman, and Engine-Boy.
By MICHAEL REYNOLDS **3/6**

The Art of Letter Painting made Easy.
By J. G. BADENOCH. With 12 full-page Engravings of Examples . **1/6**

The Art of Boot and Shoemaking.
Measurement, Last-fitting, Cutting-out, Closing, &c. By J. B. LENO. **2/0**

Mechanical Dentistry:
By CHARLES HUNTER. Fourth Edition **3/0**

Wood Engraving:
A Practical and Easy Introduction to the Art. By W. N. BROWN . **1/6**

Laundry Management.
A Handbook for Use in Private and Public Laundries . . . **2/0**

AGRICULTURE, GARDENING, ETC.

Draining and Embanking:
A Practical Treatise. By Prof. JOHN SCOTT. With 68 Illustrations **1/6**

Irrigation and Water Supply:
A Practical Treatise on Water Meadows, Sewage Irrigation, Warping, &c.; on the Construction of Wells, Ponds, Reservoirs, &c. By Prof. JOHN SCOTT. With 34 Illustrations **1/6**

Farm Roads, Fences, and Gates:
A Practical Treatise on the Roads, Tramways, and Waterways of the Farm; the Principles of Enclosures; and the different kinds of Fences, Gates, and Stiles. By Prof. JOHN SCOTT. With 75 Illustrations . **1/6**

Farm Buildings:
A Practical Treatise on the Buildings necessary for various kinds of Farms, their Arrangement and Construction, with Plans and Estimates. By Prof. JOHN SCOTT. With 105 Illustrations **2/0**

Barn Implements and Machines:
Treating of the Application of Power and Machines used in the Threshing-barn, Stockyard, Dairy, &c. By Prof. J. SCOTT. With 123 Illustrations.
2/0

Field Implements and Machines:
With Principles and Details of Construction and Points of Excellence, their Management, &c. By Prof. JOHN SCOTT. With 138 Illustrations . **2/0**

Agricultural Surveying:
A Treatise on Land Surveying, Levelling, and Setting-out; with Directions for Valuing Estates. By Prof. J. SCOTT. With 62 Illustrations . **1/6**

Farm Engineering.
By Professor JOHN SCOTT. Comprising the above Seven Volumes in One, 1,150 pages, and over 600 Illustrations. Half-bound . . . **12/0**

Outlines of Farm Management.
Treating of the General Work of the Farm; Stock; Contract Work; Labour, &c. By R. SCOTT BURN **2/6**

Outlines of Landed Estates Management.
Treating of the Varieties of Lands, Methods of Farming, Setting-out of Farms, Roads, Fences, Gates, Drainage, &c. By R. SCOTT BURN . **2/6**

Soils, Manures, and Crops.
(Vol. I. OUTLINES OF MODERN FARMING.) By R. SCOTT BURN . **2/0**

Farming and Farming Economy.
(Vol. II. OUTLINES OF MODERN FARMING.) By R. SCOTT BURN **3/0**

Stock: Cattle, Sheep, and Horses.
(Vol. III. OUTLINES OF MODERN FARMING.) By R. SCOTT BURN **2/6**

Dairy, Pigs, and Poultry.
(Vol. IV. OUTLINES OF MODERN FARMING.) By R. SCOTT BURN **2/0**

Utilization of Sewage, Irrigation, and Reclamation of Waste Land.
(Vol. V. OUTLINES OF MODERN FARMING.) By R. SCOTT BURN . **2/6**

Outlines of Modern Farming.
By R. SCOTT BURN. Consisting of the above Five Volumes in One, 1,250 pp., profusely Illustrated, half-bound **12/0**

Book-keeping for Farmers and Estate Owners.

A Practical Treatise, presenting, in Three Plans, a system adapted for all classes of Farms. By J. M. WOODMAN. Fourth Edition . . **2/6**

Ready Reckoner for the Admeasurement of Land.

By A. ARMAN. Revised and extended by C. NORRIS. Fifth Edition **2/0**

Miller's, Corn Merchant's, and Farmer's Ready Reckoner.

Second Edition, revised, with a Price List of Modern Flour Mill Machinery, by W. S. HUTTON, C.E. **2/0**

The Hay and Straw Measurer.

New Tables for the Use of Auctioneers, Valuers, Farmers, Hay and Straw Dealers, &c. By JOHN STEELE **2/0**

Meat Production.

A Manual for Producers, Distributors, and Consumers of Butchers' Meat. By JOHN EWART **2/6**

Sheep:

The History, Structure, Economy, and Diseases of. By W. C. SPOONER, M.R.V.S. Fifth Edition, with fine Engravings **3/6**

Market and Kitchen Gardening.

By C. W. SHAW, late Editor of "Gardening Illustrated" . . **3/6**

Kitchen Gardening Made Easy.

Showing the best means of Cultivating every known Vegetable and Herb, &c., with directions for management all the year round. By GEORGE M. F. GLENNY. Illustrated **1/6**

Cottage Gardening:

Or Flowers, Fruits, and Vegetables for Small Gardens. By E. HOBDAY.
1/6

Garden Receipts.

Edited by CHARLES W. QUIN **1/6**

Fruit Trees,

The Scientific and Profitable Culture of. From the French of M. DU BREUIL. Fifth Edition, carefully Revised by GEORGE GLENNY. With 187 Woodcuts **3/6**

The Tree Planter and Plant Propagator:

With numerous Illustrations of Grafting, Layering, Budding, Implements, Houses, Pits, &c. By SAMUEL WOOD **2/0**

The Tree Pruner:

A Practical Manual on the Pruning of Fruit Trees, Shrubs, Climbers, and Flowering Plants. With numerous Illustrations. By SAMUEL WOOD **1/6**

⁎ *The above Two Vols. in One, handsomely half-bound, price* **3/6**.

The Art of Grafting and Budding.

By CHARLES BALTET. With Illustrations **2/6**

MATHEMATICS, ARITHMETIC, ETC.

Descriptive Geometry,
An Elementary Treatise on; with a Theory of Shadows and of Perspective, extracted from the French of G. MONGE. To which is added a Description of the Principles and Practice of Isometrical Projection. By J. F. HEATHER, M.A. With 14 Plates **2/0**

Practical Plane Geometry:
Giving the Simplest Modes of Constructing Figures contained in one Plane and Geometrical Construction of the Ground. By J. F. HEATHER, M.A. With 215 Woodcuts **2/0**

Analytical Geometry and Conic Sections,
A Rudimentary Treatise on. By JAMES HANN. A New Edition, re-written and enlarged by Professor J. R. YOUNG **2/0**

Euclid (The Elements of).
With many Additional Propositions and Explanatory Notes; to which is prefixed an Introductory Essay on Logic. By HENRY LAW, C.E. . **2/6**

* *Sold also separately, viz:—*

Euclid. The First Three Books. By HENRY LAW, C.E. . . . **1/6**

Euclid. Books 4, 5, 6, 11, 12. By HENRY LAW, C.E. . . . **1/6**

Plane Trigonometry,
The Elements of. By JAMES HANN. **1/6**

Spherical Trigonometry,
The Elements of. By JAMES HANN. Revised by CHARLES H. DOW-LING, C.E. **1/0**

* *Or with " The Elements of Plane Trigonometry," in One Volume,* **2/6**

Differential Calculus,
Elements of the. By W. S. B. WOOLHOUSE, F.R.A.S., &c. . . **1/6**

Integral Calculus.
By HOMERSHAM COX, B.A. **1/6**

Algebra,
The Elements of. By JAMES HADDON, M.A. With Appendix, containing Miscellaneous Investigations, and a Collection of Problems . . **2/0**

A Key and Companion to the Above.
An extensive Repository of Solved Examples and Problems in Algebra. By J. R. YOUNG **1/6**

Commercial Book-keeping.
With Commercial Phrases and Forms in English, French, Italian, and German. By JAMES HADDON, M.A. **1/6**

Arithmetic,
A Rudimentary Treatise on. With full Explanations of its Theoretical Principles, and numerous Examples for Practice. For the Use of Schools and for Self-Instruction. By J. R. YOUNG, late Professor of Mathematics in Belfast College. Thirteenth Edition **1/6**

A Key to the Above.
By J. R. YOUNG. **1/6**

Equational Arithmetic,
Applied to Questions of Interest, Annuities, Life Assurance, and General Commerce; with various Tables by which all Calculations may be greatly facilitated. By W. HIPSLEY **1/6**

Arithmetic,
Rudimentary, for the Use of Schools and Self-Instruction. By JAMES HADDON, M.A. Revised by ABRAHAM ARMAN . . . **1/6**

A Key to the Above.
By A. ARMAN **1/6**

Mathematical Instruments:

Their Construction, Adjustment, Testing, and Use concisely Explained. By J. F. HEATHER, M.A., of the Royal Military Academy, Woolwich. Fifteenth Edition, Revised, with Additions, by A. T. WALMISLEY, M.I.C.E. Original Edition, in 1 vol., Illustrated **2/0**

‧ *In ordering the above, be careful to say "Original Edition," or give the number in the Series* (32), *to distinguish it from the Enlarged Edition in* 3 *vols.* (*as follows*)—

Drawing and Measuring Instruments.

Including—I. Instruments employed in Geometrical and Mechanical Drawing, and in the Construction, Copying, and Measurement of Maps and Plans. II. Instruments used for the purposes of Accurate Measurement, and for Arithmetical Computations. By J. F. HEATHER, M.A. . **1/6**

Optical Instruments.

Including (more especially) Telescopes, Microscopes, and Apparatus for producing copies of Maps and Plans by Photography. By J. F. HEATHER, M.A. Illustrated **1/6**

Surveying and Astronomical Instruments.

Including—I. Instruments used for Determining the Geometrical Features of a portion of Ground. II. Instruments employed in Astronomical Observations. By J. F. HEATHER, M.A. Illustrated. . . . **1/6**

‧ *The above three volumes form an enlargement of the Author's original work,* "*Mathematical Instruments,*" *price* **2/0**. (*Described at top of page.*)

Mathematical Instruments:

Their Construction, Adjustment, Testing and Use. Comprising Drawing, Measuring, Optical, Surveying, and Astronomical Instruments. By J. F. HEATHER, M.A. Enlarged Edition, for the most part entirely re-written. The Three Parts as above, in One thick Volume. **4/6**

The Slide Rule, and How to Use It.

.Containing full, easy, and simple Instructions to perform all Business Calculations with unexampled rapidity and accuracy. By CHARLES HOARE, C.E. With a Slide Rule, in tuck of cover. Eighth Edition . . **2/6**

Logarithms.

With Mathematical Tables for Trigonometrical, Astronomical, and Nautical Calculations. By HENRY LAW, C.E. Revised Edition . . **3/0**

Compound Interest and Annuities (Theory of).

With Tables of Logarithms for the more Difficult Computations of Interest, Discount, Annuities, &c., in all their Applications and Uses for Mercantile and State Purposes. By FEDOR THOMAN, Paris. Fourth Edition . **4/0**

Mathematical Tables,

For Trigonometrical, Astronomical, and Nautical Calculations ; to which is prefixed a Treatise on Logarithms. By H. LAW, C.E. Together with a Series of Tables for Navigation and Nautical Astronomy. By Professor J. R. YOUNG. New Edition **4/0**

Mathematics,

As applied to the Constructive Arts. By FRANCIS CAMPIN, C.E., &c. Third Edition **3/0**

Astronomy.

By the late Rev. ROBERT MAIN, F.R.S. Third Edition, revised and corrected to the Present Time. By W. T. LYNN, F.R.A.S. . . . **2/0**

Statics and Dynamics,

The Principles and Practice of. Embracing also a clear development of Hydrostatics, Hydrodynamics, and Central Forces. By T. BAKER, C.E. Fourth Edition **1/6**

BOOKS OF REFERENCE AND MISCELLANEOUS VOLUMES.

A Dictionary of Painters, and Handbook for Picture Amateurs.
Being a Guide for Visitors to Public and Private Picture Galleries, and for Art-Students, including Glossary of Terms, Sketch of Principal Schools of Painting, &c. By PHILIPPE DARYL, B.A. **2/6**

Painting Popularly Explained.
By T. J. GULLICK, Painter, and JOHN TIMBS, F.S.A. Including Fresco, Oil, Mosaic, Water Colour, Water-Glass, Tempera Encaustic, Miniature, Painting on Ivory, Vellum, Pottery, Enamel, Glass, &c. Sixth Edition **5/0**

A Dictionary of Terms used in Architecture, Building, Engineering, Mining, Metallurgy, Archæology, the Fine Arts, &c.
By JOHN WEALE. Sixth Edition. Edited by R. HUNT, F.R.S. . **5/0**

Music:
A Rudimentary and Practical Treatise. With numerous Examples. By CHARLES CHILD SPENCER **2/6**

Pianoforte,
The Art of Playing the. With numerous Exercises and Lessons. By CHARLES CHILD SPENCER **1/6**

The House Manager.
A Guide to Housekeeping, Cookery, Pickling and Preserving, Household Work, Dairy Management, Cellarage of Wines, Home-brewing and Wine-making, Gardening, &c. By AN OLD HOUSEKEEPER . . **3/6**

Manual of Domestic Medicine.
By R. GOODING, M.D. Intended as a Family Guide in all cases of Accident and Emergency. Third Edition, carefully revised . . **2/0**

Management of Health.
A Manual of Home and Personal Hygiene. By Rev. JAMES BAIRD **1/0**

Natural Philosophy,
For the Use of Beginners. By CHARLES TOMLINSON, F.R.S. . . **1/6**

The Elementary Principles of Electric Lighting.
By ALAN A. CAMPBELL SWINTON, M.INST.C.E., M.I.E.E. Fifth Edition **1/6**

The Electric Telegraph,
Its History and Progress. By R. SABINE, C.E., F.S.A., &c. . . **3/0**

Handbook of Field Fortification.
By Major W. W. KNOLLYS, F.R.G.S. With 163 Woodcuts . . **3/0**

Logic,
Pure and Applied. By S. H. EMMENS **1/6**

Locke on the Human Understanding,
Selections from. With Notes by S. H. EMMENS . . . **1/6**

The Compendious Calculator
(*Intuitive Calculations*). Or Easy and Concise Methods of Performing the various Arithmetical Operations required in Commercial and Business Transactions; together with Useful Tables, &c. By DANIEL O'GORMAN. Twenty-eighth Edition, carefully revised by C. NORRIS . . . **2/6**

Measures, Weights, and Moneys of all Nations.
With an Analysis of the Christian, Hebrew, and Mahometan Calendars.
By W. S. B. WOOLHOUSE, F.R.A.S., F.S.S. Seventh Edition **2/6**

Grammar of the English Tongue,
Spoken and Written. With an Introduction to the Study of Comparative
Philology. By HYDE CLARKE, D.C.L. Fifth Edition. **1/6**

Dictionary of the English Language.
As Spoken and Written. Containing above 100,000 Words. By HYDE
CLARKE, D.C.L. **3/6**

Composition and Punctuation,
Familiarly Explained for those who have neglected the Study of Grammar.
By JUSTIN BRENAN. Nineteenth Edition. **1/6**

French Grammar.
With Complete and Concise Rules on the Genders of French Nouns. By
G. L. STRAUSS, Ph.D. **1/6**

English-French Dictionary.
Comprising a large number of Terms used in Engineering, Mining, &c.
By ALFRED ELWES **2/0**

French Dictionary.
In two Parts—I. French-English. II. English-French, complete in
One Vol. **3/0**

French and English Phrase Book.
Containing Introductory Lessons, with Translations, Vocabularies of Words,
Collection of Phrases, and Easy Familiar Dialogues **1/6**

German Grammar.
Adapted for English Students, from Heyse's Theoretical and Practical
Grammar, by Dr. G. L. STRAUSS **1/6**

German Triglot Dictionary.
By N. E. S. A. HAMILTON. Part I. German-French-English. Part II.
English-German-French. Part III. French-German-English. **3/0**

German Triglot Dictionary.
(As above). Together with German Grammar, in One Volume **5/0**

Italian Grammar.
Arranged in Twenty Lessons, with Exercises. By ALFRED ELWES. **1/6**

Italian Triglot Dictionary,
Wherein the Genders of all the Italian and French Nouns are carefully
noted down. By ALFRED ELWES. Vol. I. Italian-English-French. **2/6**

Italian Triglot Dictionary.
By ALFRED ELWES. Vol. II. English-French-Italian **2/6**

Italian Triglot Dictionary.
By ALFRED ELWES. Vol. III. French-Italian-English **2/6**

Italian Triglot Dictionary.
(As above). In One Vol. **7/6**

Spanish Grammar.
In a Simple and Practical Form. With Exercises. By ALFRED ELWES **1/6**

Spanish-English and English-Spanish Dictionary.
Including a large number of Technical Terms used in Mining, Engineering,
&c., with the proper Accents and the Gender of every Noun. By ALFRED
ELWES **4/0**
※ *Or with the* GRAMMAR, **6 0.**

Portuguese Grammar,

In a Simple and Practical Form. With Exercises. By ALFRED ELWES. **1/6**

Portuguese-English and English-Portuguese Dictionary.

Including a large number of Technical Terms used in Mining, Engineering, &c., with the proper Accents and the Gender of every Noun. By ALFRED ELWES.   Fourth Edition, revised **5/0**

* *Or with the* GRAMMAR, **7/0**.

Animal Physics,

Handbook of. By DIONYSIUS LARDNER, D.C.L. With 520 Illustrations. In One Vol. (732 pages), cloth boards **7/6**

* *Sold also in Two Parts, as follows:—*

ANIMAL PHYSICS. By Dr. LARDNER. Part I., Chapters I.—VII. **4/0**
ANIMAL PHYSICS. By Dr. LARDNER. Part II., Chapters VIII.—XVIII. **3/0**

BRADBURY, AGNEW & CO., LD., PRINTERS, LONDON AND TONBRIDGE.
[58—31.5.]

CROSBY LOCKWOOD & SON'S
Catalogue of
Scientific, Technical and Industrial Books.

MECHANICAL ENGINEERING, ETC.

THE MECHANICAL ENGINEER'S POCKET-BOOK.

Comprising Tables, Formulæ, Rules, and Data: A Handy Book of Reference for Daily Use in Engineering Practice. By D. KINNEAR CLARK, M. Inst. C.E., Fifth Edition, thoroughly Revised and Enlarged. By H. H. P. POWLES, A.M.I C.E., M.I.M.E. Small 8vo, 700 pp., bound in flexible Leather Cover, rounded corners. *[Just Published.* Net **6/0**

SUMMARY OF CONTENTS:—MATHEMATICAL TABLES.—MEASUREMENT OF SURFACES AND SOLIDS.—ENGLISH WEIGHTS AND MEASURES.—FRENCH METRIC WEIGHTS AND MEASURES.—FOREIGN WEIGHTS AND MEASURES.—MONEYS.—SPECIFIC GRAVITY, WEIGHT, AND VOLUME.—MANUFACTURED METALS.—STEEL PIPES.—BOLTS AND NUTS.— SUNDRY ARTICLES IN WROUGHT AND CAST IRON, COPPER, BRASS, LEAD, TIN, ZINC.— STRENGTH OF MATERIALS.—STRENGTH OF TIMBER.—STRENGTH OF CAST IRON.— STRENGTH OF WROUGHT IRON.—STRENGTH OF STEEL.—TENSILE STRENGTH OF COPPER, LEAD, &c.—RESISTANCE OF STONES AND OTHER BUILDING MATERIALS.—RIVETED JOINTS IN BOILER PLATES.—BOILER SHELLS.—WIRE ROPES AND HEMP ROPES.—CHAINS AND CHAIN CABLES.—FRAMING.—HARDNESS OF METALS, ALLOYS, AND STONES.—LABOUR OF ANIMALS.—MECHANICAL PRINCIPLES.—GRAVITY AND FALL OF BODIES.—ACCELERATING AND RETARDING FORCES.—MILL GEARING, SHAFTING, &c.—TRANSMISSION OF MOTIVE POWER.—HEAT.—COMBUSTION: FUELS.—WARMING, VENTILATION, COOKING STOVES.— STEAM.—STEAM ENGINES AND BOILERS.—RAILWAYS.—TRAMWAYS.—STEAM SHIPS.— PUMPING STEAM ENGINES AND PUMPS.—COAL GAS, GAS ENGINES, &c.—AIR IN MOTION. —COMPRESSED AIR.—HOT AIR ENGINES.—WATER POWER.—SPEED OF CUTTING TOOLS. —COLOURS.—ELECTRICAL ENGINEERING.

"Mr. Clark manifests what is an innate perception of what is likely to be useful in a pocket-book, and he is really unrivalled in the art of condensation. It is very difficult to hit upon any mechanical engineering subject concerning which this work supplies no information, and the excellent index at the end adds to its utility. In one word, it is an exceedingly handy and efficient tool, possessed of which the engineer will be saved many a wearisome calculation, or yet more wearisome hunt through various text-books and treatises, and, as such, we can heartily recommend it to our readers."—*The Engineer.*

"It would be found difficult to compress more matter within a similar compass, or produce a book of 700 pages which should be more compact or convenient for pocket reference. . . . Will be appreciated by mechanical engineers of all classes."—*Practical Engineer.*

MR. HUTTON'S PRACTICAL HANDBOOKS.

THE WORKS' MANAGER'S HANDBOOK.

Comprising Modern Rules, Tables, and Data. For Engineers, Millwrights, and Boiler Makers; Tool Makers, Machinists, and Metal Workers; Iron and Brass Founders, &c. By W. S. HUTTON, Civil and Mechanical Engineer, Author of "The Practical Engineer's Handbook." Sixth Edition, carefully Revised, and Enlarged. In One handsome Volume, medium 8vo, strongly bound **15/0**

The Author having compiled Rules and Data for his own use in a great variety of modern engineering work, and having found his notes extremely useful, decided to publish them—revised to date—believing that a practical work, suited to the DAILY REQUIREMENTS OF MODERN ENGINEERS, *would be favourably received.*

"Of this edition we may repeat the appreciative remarks we made upon the first and third. Since the appearance of the latter very considerable modifications have been made, although the total number of pages remains almost the same. It is a very useful collection of rules, tables, and workshop and drawing office data."—*The Engineer,* May 10, 1895.

"The author treats every subject from the point of view of one who has collected workshop notes for application in workshop practice, rather than from the theoretical or literary aspect. The volume contains a great deal of that kind of information which is gained only by practical experience, and is seldom written in books."—*The Engineer,* June 5, 1885.

"The volume is an exceedingly useful one, brimful with engineer's notes, memoranda, and rules, and well worthy of being on every mechanical engineer's bookshelf."—*Mechanical World.*

"The information is precisely that likely to be required in practice. . . . The work forms a desirable addition to the library not only of the works' manager, but of any one connected with general engineering."—*Mining Journal.*

"Brimful of useful information, stated in a concise form, Mr. Hutton's books have met a pressing want among engineers. The book must prove extremely useful to every practical man possessing a copy."—*Practical Engineer.*

THE PRACTICAL ENGINEER'S HANDBOOK.

Comprising a Treatise on Modern Engines and Boilers, Marine, Locomotive, and Stationary. And containing a large collection of Rules and Practical Data relating to Recent Practice in Designing and Constructing all kinds of Engines, Boilers, and other Engineering work. The whole constituting a comprehensive Key to the Board of Trade and other Examinations for Certificates of Competency in Modern Mechanical Engineering. By WALTER S. HUTTON, Civil and Mechanical Engineer, Author of "The Works' Manager's Handbook for Engineers," &c. With upwards of 420 Illustrations. Sixth Edition, Revised and Enlarged. Medium 8vo, nearly 560 pp., strongly bound. **18/0**

This Work is designed as a companion to the Author's ."WORKS' MANAGER'S HANDBOOK." *It possesses many new and original features, and contains, like its predecessor, a quantity of matter not originally intended for publication but collected by the Author for his own use in the construction of a great variety of* MODERN ENGINEERING WORK.

The information is given in a condensed and concise form, and is illustrated by upwards of 420 Engravings; and comprises a quantity of tabulated matter of great value to all engaged in designing, constructing, or estimating for ENGINES, BOILERS, *and* OTHER ENGINEERING WORK.

"We have kept it at hand for several weeks, referring to it as occasion arose, and we have not on a single occasion consulted its pages without finding the information of which we were in quest."—*Athenæum.*

"A thoroughly good practical handbook, which no engineer can go through without learning something that will be of service to him."—*Marine Engineer.*

"An excellent book of reference for engineers, and a valuable text-book for students of engineering."—*Scotsman.*

"This valuable manual embodies the results and experience of the leading authorities on mechanical engineering."—*Building News.*

"The author has collected together a surprising quantity of rules and practical data, and has shown much judgment in the selections he has made. . . . There is no doubt that this book is one of the most useful of its kind published, and will be a very popular compendium."—*Engineer.*

"A mass of information set down in simple language, and in such a form that it can be easily referred to at any time. The matter is uniformly good and well chosen, and is greatly elucidated by the illustrations. The book will find its way on to most engineers' shelves, where it will rank as one of the most useful books of reference."—*Practical Engineer.*

"Full of useful information, and should be found on the office shelf of all practical engineers."—*English Mechanic.*

MR. HUTTON'S PRACTICAL HANDBOOKS—*continued.*

STEAM BOILER CONSTRUCTION.

A Practical Handbook for Engineers, Boiler-Makers, and Steam Users. Containing a large Collection of Rules and Data relating to Recent Practice in the Design, Construction, and Working of all Kinds of Stationary, Loco-motive, and Marine Steam-Boilers. By WALTER S. HUTTON, Civil and Mechanical Engineer, Author of "The Works' Manager's Handbook," "The Practical Engineer's Handbook," &c. With upwards of 500 Illustrations. Fourth Edition. carefully Revised, and Enlarged. Medium 8vo, over 680 pages, cloth, strongly bound. [*Just Published.* **18/0**

☛ THIS WORK *is issued in continuation of the Series of Handbooks written by the Author, viz. :*—"THE WORKS' MANAGER'S HANDBOOK" *and* "THE PRACTICAL ENGINEER'S HANDBOOK," *which are so highly appreciated by engineers for the practical nature of their information ; and is consequently written in the same style as those works.*

The Author believes that the concentration, in a convenient form for easy reference, of such a large amount of thoroughly practical information on Steam-Boilers, will be of considerable service to those for whom it is intended, and he trusts the book may be deemed worthy of as favourable a reception as has been accorded to its predecessors.

" One of the best, if not the best, books on boilers that has ever been published. The infor-mation is of the right kind, in a simple and accessible form. So far as generation is concerned, this is, undoubtedly, the standard book on steam practice."—*Electrical Review.*

" Every detail, both in boiler design and management, is clearly laid before the reader. The volume shows that boiler construction has been reduced to the condition of one of the most exact sciences; and such a book is of the utmost value to the *fin de siècle* Engineer and Works Manager." —*Marine Engineer.*

" There has long been room for a modern handbook on steam boilers; there is not that room now, because Mr. Hutton has filled it. It is a thoroughly practical book for those who are occupied in the construction, design, selection, or use of boilers."—*Engineer.*

" The book is of so important and comprehensive a character that it must find its way into the libraries of every one interested in boiler using or boiler manufacture if they wish to be thoroughly informed. We strongly recommend the book for the intrinsic value of its contents."—*Machinery Market.*

PRACTICAL MECHANICS' WORKSHOP COMPANION.

Comprising a great variety of the most useful Rules and Formulæ in Mechanical Science, with numerous Tables of Practical Data and Calculated Results for Facilitating Mechanical Operations. By WILLIAM TEMPLETON, Author of "The Engineer's Practical Assistant," &c., &c. Eighteenth Edition, Revised, Modernised, and considerably Enlarged by WALTER S. HUTTON, C.E., Author of "The Works' Manager's Handbook," "The Practical Engineer's Hand-book," &c. Fcap. 8vo, nearly 500 pp., with 8 Plates and upwards of 250 Illus-trative Diagrams, strongly bound for workshop or pocket wear and tear . **6/0**

" In its modernised form Hutton's ' Templeton' should have a wide sale, for it contains much valuable information which the mechanic will often find of use, and not a few tables and notes which he might look for in vain in other works. This modernised edition will be appreciated by all who have learned to value the original editions of ' Templeton.'"—*English Mechanic.*

" It has met with great success in the engineering workshop, as we can testify; and there are a great many men who, in a great measure, owe their rise in life to this little book."—*Building News.*

" This familiar text-book—well known to all mechanics and engineers—is of essential service to the every-day requirements of engineers, millwrights, and the various trades connected with engineering and building. The new modernised edition is worth its weight in gold."—*Building News.* (Second Notice.)

" This well-known and largely-used book contains information, brought up to date, of the sort so useful to the foreman and draughtsman. So much fresh information has been introduced as to constitute it practically a new book. It will be largely used in the office and workshop."—*Mechanical World.*

" The publishers wisely entrusted the task of revision of this popular, valuable, and useful book to Mr. Hutton, than whom a more competent man they could not have found."—*Iron.*

ENGINEER'S AND MILLWRIGHT'S ASSISTANT.

A Collection of Useful Tables, Rules, and Data. By WILLIAM TEMPLETON. Eighth Edition, with Additions. 18mo, cloth **2/6**

" Occupies a foremost place among books of this kind. A more suitable present to an apprentice to any of the mechanical trades could not possibly be made."—*Building News.*

" A deservedly popular work. It should be in the ' drawer' of every mechanic."—*English Mechanic.*

THE MECHANICAL ENGINEER'S REFERENCE BOOK.

For Machine and Boiler Construction. In Two Parts. Part I. GENERAL ENGINEERING DATA. Part II. BOILER CONSTRUCTION. With 51 Plates and numerous Illustrations. By NELSON FOLEY, M.I.N.A. Second Edition, Revised throughout and much Enlarged. Folio, half-bound. *Net* **£3 3s.**

PART I.—MEASURES.—CIRCUMFERENCES AND AREAS, &c., SQUARES, CUBES, FOURTH POWERS.—SQUARE AND CUBE ROOTS.—SURFACE OF TUBES.—RECIPROCALS.—LOGARITHMS. — MENSURATION. — SPECIFIC GRAVITIES AND WEIGHTS.—WORK AND POWER. — HEAT. — COMBUSTION. — EXPANSION AND CONTRACTION.—EXPANSION OF GASES.—STEAM.— STATIC FORCES.—GRAVITATION AND ATTRACTION.—MOTION AND COMPUTATION OF RESULTING FORCES.—ACCUMULATED WORK.—CENTRE AND RADIUS OF GYRATION.—MOMENT OF INERTIA.—CENTRE OF OSCILLATION.—ELECTRICITY.—STRENGTH OF MATERIALS.—ELASTICITY.—TRST SHEETS OF METALS.—FRICTION.—TRANSMISSION OF POWER.—FLOW OF LIQUIDS.—FLOW OF GASES.—AIR PUMPS, SURFACE CONDENSERS, &c.—SPEED OF STEAMSHIPS.—PROPELLERS.—CUTTING TOOLS.—FLANGES.—COPPER SHEETS AND TUBES.—SCREWS, NUTS, BOLT HEADS, &c.—VARIOUS RECIPES AND MISCELLANEOUS MATTER.—WITH DIAGRAMS FOR VALVE-GEAR, BELTING AND ROPES, DISCHARGE AND SUCTION PIPES, SCREW PROPELLERS, AND COPPER PIPES.

PART II.—TREATING OF POWER OF BOILERS.—USEFUL RATIOS.—NOTES ON CONSTRUCTION. — CYLINDRICAL BOILER SHELLS. — CIRCULAR FURNACES. — FLAT PLATES.—STAYS. — GIRDERS.—SCREWS. — HYDRAULIC TESTS. — RIVETING. — BOILER SETTING, CHIMNEYS, AND MOUNTINGS.—FUELS, &c.—EXAMPLES OF BOILERS AND SPEEDS OF STEAMSHIPS.—NOMINAL AND NORMAL HORSE POWER.—WITH DIAGRAMS FOR ALL BOILER CALCULATIONS AND DRAWINGS OF MANY VARIETIES OF BOILERS.

" Mr. Foley is well fitted to compile such a work. The diagrams are a great feature of the work. It may be stated that Mr. Foley has produced a volume which will undoubtedly fulfil the desire of the author and become indispensable to all mechanical engineers."—*Marine Engineer.*

" We have carefully examined this work, and pronounce it a most excellent reference book for the use of marine engineers."—*Journal of American Society of Naval Engineers.*

TEXT-BOOK ON THE STEAM ENGINE.

With a Supplement on GAS ENGINES and PART II. on HEAT ENGINES. By T. M. GOODEVE, M.A., Barrister-at-Law, Professor of Mechanics at the Royal College of Science, London ; Author of " The Principles of Mechanics," " The Elements of Mechanism," &c. Fourteenth Edition. Crown 8vo, cloth . **6/0**

" Professor Goodeve has given us a treatise on the steam engine which will bear comparison with anything written by Huxley or Maxwell, and we can award it no higher praise."—*Engineer.*

" Mr. Goodeve's text-book is a work of which every young engineer should possess himself." —*Mining Journal.*

ON GAS ENGINES.

With Appendix describing a Recent Engine with Tube Igniter. By T. M. GOODEVE, M.A. Crown 8vo, cloth **2/6**

" Like all Mr. Goodeve's writings, the present is no exception in point of general excellence. It is a valuable little volume."—*Mechanical World.*

GAS AND OIL ENGINE MANAGEMENT.

A Practical Guide for Users and Attendants, being Notes on Selection, Construction, and Management By M. POWIS BALE, M.I M.E., A.M.I.C.E. Author of " Woodworking Machinery," &c. Crown 8vo, cloth.

[*Just Published. Net* **3/6**

THE GAS-ENGINE HANDBOOK.

A Manual of Useful Information for the Designer and the Engineer. By E. W. ROBERTS, M.E. With Forty Full-page Engravings. Small Fcap. 8vo, leather.

Net **8/6**

A TREATISE ON STEAM BOILERS.

Their Strength, Construction, and Economical Working. By R. WILSON, C.E. Fifth Edition. 12mo, cloth **6/0**

" The best treatise that has ever been published on steam boilers."—*Engineer.*

THE MECHANICAL ENGINEER'S COMPANION

of Areas, Circumferences, Decimal Equivalents, in inches and feet, millimetres, squares, cubes, roots, &c. ; Strength of Bolts, Weight of Iron, &c. ; Weights, Measures, and other Data. Also Practical Rules for Engine Proportions. By R. EDWARDS, M.Inst.C.E. Fcap. 8vo, cloth. **3/6**

" A very useful little volume. It contains many tables, classified data and memoranda generally useful to engineers."—*Engineer.*

" What it professes to be, ' a handy office companion,' giving in a succinct form, a variety of information likely to be required by mechanical engineers in their everyday office work."—*Nature.*

A HANDBOOK ON THE STEAM ENGINE.

With especial Reference to Small and Medium-sized Engines. For the Use of Engine Makers, Mechanical Draughtsmen, Engineering Students, and users of Steam Power. By HERMAN HAEDER, C.E. Translated from the German with additions and alterations, by H. H. P. POWLES, A.M.I.C.E.. M.I.M.E. Third Edition, Revised. With nearly 1,100 Illustrations. Crown 8vo, cloth . . . . *Net* **7/6**

"A perfect encyclopædia of the steam engine and its details, and one which must take a permanent place in English drawing-offices and workshops."—*A Foreman Pattern-maker.*

"This is an excellent book, and should be in the hands of all who are interested in the construction and design of medium-sized stationary engines. . . . A careful study of its contents and the arrangement of the sections leads to the conclusion that there is probably no other book like it in this country. The volume aims at showing the results of practical experience, and it certainly may claim a complete achievement of this idea."—*Nature.*

"There can be no question as to its value. We cordially commend it to all concerned in the design and construction of the steam engine."—*Mechanical World.*

BOILER AND FACTORY CHIMNEYS.

Their Draught-Power and Stability. With a chapter on *Lightning Conductors.* By ROBERT WILSON, A.I.C.E., Author of " A Treatise on Steam Boilers," &c. Crown 8vo, cloth **3/6**

A valuable contribution to the literature of scientific building."—*The Builder.*

BOILER MAKER'S READY RECKONER & ASSISTANT.

With Examples of Practical Geometry and Templating, for the Use of Platers, Smiths, and Riveters. By JOHN COURTNEY, Edited by D. K. CLARK, M.I.C.E. Fourth Edition, 480 pp., with 140 Illustrations. Fcap. 8vo, half-bound **7/0**

" No workman or apprentice should be without this book."—*Iron Trade Circular.*

REFRIGERATION, COLD STORAGE, & ICE-MAKING:

A Practical Treatise on the Art and Science of Refrigeration. By A. J. WALLIS-TAYLER, A.M.Inst.C.E., Author of " Refrigerating and Ice-Making Machinery." 600 pp., with 360 Illustrations. Medium 8vo, cloth. *Net* **15/0**

"The author has to be congratulated on the completion and production of such an important work and it cannot fail to have a large body of readers, for it leaves out nothing that would in any way be of value to those interested in the subject."—*Steamship.*

"No one whose duty it is to handle the mammoth preserving installations of these latter days can afford to be without this valuable book."—*Glasgow Herald.*

THE POCKET BOOK OF REFRIGERATION AND ICE-MAKING.

Edited by A. J. WALLIS-TAYLER, A.M.Inst.C.E. Author of " Refrigerating and Ice-making Machinery," &c. Small Crown 8vo, cloth.
[*Just Published.* Net **3/6**

REFRIGERATING & ICE-MAKING MACHINERY.

A Descriptive Treatise for the Use of Persons Employing Refrigerating and Ice-Making Installations, and others. By A. J. WALLIS-TAYLER, A.-M. Inst. C.E. Third Edition, Enlarged. Crown 8vo, cloth . . **7/6**

" Practical, explicit, and profusely illustrated."—*Glasgow Herald.*

" We recommend the book, which gives the cost of various systems and illustrations showing details of parts of machinery and general arrangements of complete installations."—*Builder.*

" May be recommended as a useful description of the machinery, the processes, and of the facts, figures, and tabulated physics of refrigerating. It is one of the best compilations on the subject."—*Engineer.*

ENGINEERING ESTIMATES, COSTS, AND ACCOUNTS.

A Guide to Commercial Engineering. With numerous examples of Estimates and Costs of Millwright Work, Miscellaneous Productions, Steam Engines and Steam Boilers; and a Section on the Preparation of Costs Accounts. By A GENERAL MANAGER. Second Edition. 8vo, cloth. **12/0**

" This is an excellent and very useful book, covering subject-matter in constant requisition in every factory and workshop. . . . The book is invaluable, not only to the young engineer, but also to the estimate department of every works."—*Builder.*

" We accord the work unqualified praise. The information is given in a plain, straightforward manner, and bears throughout evidence of the intimate practical acquaintance of the author with every phase of commercial engineering."—*Mechanical World.*

HOISTING MACHINERY.

An Elementary Treatise on. Including the Elements of Crane Construction and Descriptions of the Various Types of Cranes in Use. By JOSEPH HORNER, A.M.I.M.E., Author of "Pattern-Making," and other Works. Crown 8vo, with 215 Illustrations, including Folding Plates, cloth.

[Just Published. **Net 7 6**

AËRIAL OR WIRE-ROPE TRAMWAYS.

Their Construction and Management. By A. J. WALLIS-TAYLER, A.M.Inst.C.E. With 81 Illustrations. Crown 8vo, cloth **7/6**

"This is in its way an excellent volume. Without going into the minutiæ of the subject, it yet lays before its readers a very good exposition of the various systems of rope transmission in use, and gives as well not a little valuable information about their working, repair, and management. We can safely recommend it as a useful general treatise on the subject."—*The Engineer.*

MOTOR CARS OR POWER-CARRIAGES FOR COMMON ROADS.

By A. J. WALLIS-TAYLER, A.M. Inst. C.E., Author of "Modern Cycles," &c. 212 pp., with 76 Illustrations. Crown 8vo, cloth **4/6**

"The book is clearly expressed throughout, and is just the sort of work that an engineer thinking of turning his attention to motor-carriage work, would do well to read as a preliminary to starting operations."—*Engineering.*

PLATING AND BOILER MAKING.

A Practical Handbook for Workshop Operations. By JOSEPH G. HORNER, A.M.I.M.E. 380 pp. with 338 Illustrations. Crown 8vo, cloth . . **7/6**

"This work is characterised by that evidence of close acquaintance with workshop methods which will render the book exceedingly acceptable to the practical hand. We have no hesitation in commending the work as a serviceable and practical handbook on a subject which has not hitherto received much attention from those qualified to deal with it in a satisfactory manner."—*Mechanical World.*

PATTERN MAKING.

A Practical Treatise, embracing the Main Types of Engineering Construction, and including Gearing, Engine Work, Sheaves and Pulleys, Pipes and Columns, Screws, Machine Parts, Pumps and Cocks, the Moulding of Patterns in Loam and Greensand, estimating the weight of Castings &c. By JOSEPH G. HORNER, A.M.I.M.E. Third Edition, Enlarged. With 486 Illustrations. Crown 8vo, cloth. *Net* **7/6**

"A well-written technical guide, evidently written by a man who understands and has practised what he has written about. . . . We cordially recommend it to engineering students, young journeymen, and others desirous of being initiated into the mysteries of pattern-making."—*Builder.*
"An excellent *vade mecum* for the apprentice who desires to become master of his trade."—*English Mechanic.*

MECHANICAL ENGINEERING TERMS

(Lockwood's Dictionary of). Embracing those current in the Drawing Office, Pattern Shop, Foundry, Fitting, Turning, Smiths', and Boiler Shops, &c. Comprising upwards of 6,000 Definitions. Edited by J. G. HORNER, A.M.I.M.E. Third Edition, Revised, with Additions. Crown 8vo, cloth . . *Net* **7/6**

"Just the sort of handy dictionary required by the various trades engaged in mechanical engineering. The practical engineering pupil will find the book of great value in his studies, and every foreman engineer and mechanic should have a copy."—*Building News.*

TOOTHED GEARING.

A Practical Handbook for Offices and Workshops. By J. HORNER, A.M.I.M.E. With 184 Illustrations. Crown 8vo, cloth **6/0**

"We give the book our unqualified praise for its thoroughness of treatment, and recommend it to all interested as the most practical book on the subject yet written."—*Mechanical World.*

FIRES, FIRE-ENGINES, AND FIRE BRIGADES.

With a History of Fire-Engines, their Construction, Use, and Management; Foreign Fire Systems; Hints on Fire-Brigades, &c. By C. F. T. YOUNG, C.E. 8vo, cloth . . . . **£1 4s.**

"To such of our readers as are interested in the subject of fires and fire apparatus we can most heartily commend this book."—*Engineering.*

AËRIAL NAVIGATION.

A Practical Handbook on the Construction of Dirigible Balloons, Aërostats, Aëroplanes, and Aëremotors. By FREDERICK WALKER, C.E., Associate Member of the Aëronautic Institute. With 104 Illustrations. Large Crown 8vo, cloth . . . . Net **7/6**

STONE-WORKING MACHINERY.

A Manual dealing with the Rapid and Economical Conversion of Stone. With Hints on the Arrangement and Management of Stone Works. By M. POWIS BALE, M.I.M.E. Second Edition, enlarged. Crown 8vo, cloth . . **9/0**

" The book should be in the hands of every mason or student of stonework."—*Colliery Guardian.*

" A capital handbook for all who manipulate stone for building or ornamental purposes."—*Machinery Market.*

PUMPS AND PUMPING.

A Handbook for Pump Users. Being Notes on Selection, Construction, and Management. By M. POWIS BALE, M.I.M.E. Fourth Edition. Crown 8vo, cloth **3/6**

" The matter is set forth as concisely as possible. . In fact, condensation rather than diffuseness has been the author's aim throughout ; yet he does not seem to have omitted anything likely to be of use."—*Journal of Gas Lighting.*

" Thoroughly practical and clearly written."—*Glasgow Herald.*

MILLING MACHINES AND PROCESSES.

A Practical Treatise on Shaping Metals by Rotary Cutters. Including Information on Making and Grinding the Cutters. By PAUL N. HASLUCK, Author of " Lathe-Work." With upwards of 300 Engravings. Large crown 8vo, cloth **12/6**

" A new departure in engineering literature. . . . We can recommend this work to all interested in milling machines ; it is what it professes to be—a practical treatise."—*Engineer.*

" A capital and reliable book which will no doubt be of considerable service both to those who are already acquainted with the process as well as to those who contemplate its adoption."—*Industries.*

LATHE-WORK.

A Practical Treatise on the Tools, Appliances, and Processes employed in the Art of Turning. By PAUL N. HASLUCK. Eighth Edition. Crown 8vo, cloth **5/0**

" Written by a man who knows not only how work ought to be done, but who also knows how to do it, and how to convey his knowledge to others. To all turners this book would be valuable."—*Engineering.*

" We can safely recommend the work to young engineers. To the amateur it will simply be invaluable. To the student it will convey a great deal of useful information."—*Engineer.*

SCREW-THREADS,

And Methods of Producing Them. With numerous Tables and complete Directions for using Screw-Cutting Lathes. By PAUL N. HASLUCK, Author of " Lathe-Work," &c. Sixth Edition. Waistcoat-pocket size . . **1/6**

" Full of useful information, hints and practical criticism. Taps, dies, and screwing tools generally are illustrated and their action described."—*Mechanical World.*

" It is a complete compendium of all the details of the screw-cutting lathe ; in fact, a *multum in-parvo* on all the subjects it treats upon."—*Carpenter and Builder.*

TABLES AND MEMORANDA FOR ENGINEERS,
MECHANICS, ARCHITECTS, BUILDERS, &c.

Selected and Arranged by FRANCIS SMITH. Seventh Edition, Revised, including ELECTRICAL TABLES, FORMULÆ, and MEMORANDA. Waistcoat-pocket size, limp leather. [*Just Published.* **1/6**

" It would, perhaps, be as difficult to make a small pocket-book selection of notes and formulæ to suit ALL engineers as it would be to make a universal medicine ; but Mr. Smith's waistcoat-pocket collection may be looked upon as a successful attempt."—*Engineer.*

" The best example we have ever seen of 270 pages of useful matter packed into the dimensions of a card-case."—*Building News.* " A veritable pocket treasury of knowledge."—*Iron.*

POCKET GLOSSARY OF TECHNICAL TERMS.

English-French, French-English ; with Tables suitable for the Architectural, Engineering, Manufacturing, and Nautical Professions. By JOHN JAMES FLETCHER. Third Edition, 200 pp. Waistcoat-pocket size, limp leather **1/6**

" It is a very great advantage for readers and correspondents in France and England to have so large a number of the words relating to engineering and manufacturers collected in a lilliputian volume. The little book will be useful both to students and travellers."—*Architect.*

" The glossary of terms is very complete, and many of the Tables are new and well arranged We cordially commend the book."—*Mechanical World.*

THE ENGINEER'S YEAR BOOK FOR 1904.

Comprising Formulæ, Rules, Tables, Data and Memoranda in Civil, Mechanical, Electrical, Marine and Mine Engineering. By H. R. KEMPE, A.M. Inst. C.E., M. I. E. E., Principal Technical Officer, Engineer-in-Chief's Office, General Post Office, London, Author of "A Handbook of Electrical Testing," "The Electrical Engineer's Pocket-Book," &c. With 1,000 Illustrations, specially Engraved for the work. Crown 8vo, 900 pp., leather. [*Just Published.* **8/0**

"Kempe's Year Book really requires no commendation. Its sphere of usefulness is widely known, and it is used by engineers the world over."—*The Engineer.*
"The volume is distinctly in advance of most similar publications in this country."—*Engineering.*
"This valuable and well-designed book of reference meets the demands of all descriptions of engineers."—*Saturday Review.*
"Teems with up-to-date information in every branch of engineering and construction."—*Building News.*
"The needs of the engineering profession could hardly be supplied in a more admirable, complete and convenient form. To say that it more than sustains all comparisons is praise of the highest sort, and that may justly be said of it."—*Mining Journal.*
"There is certainly room for the newcomer, which supplies explanations and directions, as well as formulæ and tables. It deserves to become one of the most successful of the technical annuals."—*Architect.*
"Brings together with great skill all the technical information which an engineer has to use day by day. It is in every way admirably equipped, and is sure to prove successful."—*Scotsman.*
"The up-to-dateness of Mr. Kempe's compilation is a quality that will not be lost on the busy people for whom the work is intended."—*Glasgow Herald.*

THE PORTABLE ENGINE.

A Practical Manual on its Construction and Management. For the use of Owners and Users of Steam Engines generally. By WILLIAM DYSON WANSBROUGH. Crown 8vo, cloth **3/6**

"This is a work of value to those who use steam machinery. . . . Should be read by every one who has a steam engine, on a farm or elsewhere."—*Mark Lane Express.*

IRON AND STEEL.

A Work for the Forge, Foundry, Factory, and Office. Containing ready, useful, and trustworthy Information for Ironmasters and their Stock-takers; Managers of Bar, Rail, Plate, and Sheet Rolling Mills; Iron and Metal Founders; Iron Ship and Bridge Builders; Mechanical, Mining, and Consulting Engineers; Architects, Contractors, Builders, &c. By CHARLES HOARE, Author of "The Slide Rule," &c. Ninth Edition. 32mo, leather   .   **6/0**

CONDENSED MECHANICS.

A Selection of Formulæ, Rules, Tables, and Data or the Use of Engineering Students, &c. By W. G. C. HUGHES, A.M.I.C.E. Crown 8vo, cloth . **2/6**

"The book is well fitted for those who are preparing for examination and wish to refresh their knowledge by going through their formulæ again.'—*Marine Engineer.*

THE SAFE USE OF STEAM.

Containing Rules for Unprofessional Steam Users. By an ENGINEER. Seventh Edition. Sewed   .   .   .   .   **6p.**

"If steam-users would but learn this little book by heart, boiler explosions would become sensations by their rarity."—*English Mechanic.*

THE CARE AND MANAGEMENT OF STATIONARY ENGINES.

A Practical Handbook for Men-in-charge. By C. HURST. Crown 8vo. *Net* **1/0**

THE LOCOMOTIVE ENGINE.

The Autobiography of an Old Locomotive Engine. By ROBERT WEATHERBURN, M.I.M.E. With Illustrations and Portraits of GEORGE and ROBERT STEPHENSON. Crown 8vo, cloth.   *Net* **2/6**

THE LOCOMOTIVE ENGINE AND ITS DEVELOPMENT.

A Popular Treatise on the Gradual Improvements made in Railway Engines between 1803 and 1903. By CLEMENT E. STRETTON, C.E. Sixth Edition, Revised and Enlarged. Crown 8vo, cloth.   [*Just Published. Net* **4/6**

"Students of railway history and all who are interested in the evolution of the modern locomotive will find much to attract and entertain in this volume."—*The Times.*

MODERN MACHINE SHOP TOOLS,
THEIR CONSTRUCTION, OPERATION, AND MANIPULATION.

Including both Hand and Machine Tools. An entirely New and Fully Illustrated Work, treating this Subject in a Concise and Comprehensive Manner. A Book of Practical Instruction in all Classes of Machine Shop Practice. Including Chapters on Filing, Fitting, and Scraping Surfaces ; on Drills, Reamers, Taps, and Dies ; the Lathe and its Tools ; Planers, Shapers, and their Tools ; Milling Machines and Cutters ; Gear Cutters and Gear Cutting ; Drilling Machines and Drill Work ; Grinding Machines and their Work ; Hardening and Tempering, Gearing, Belting, and Transmission Machinery; Useful Data and Tables. By WILLIAM H. VAN DERVOORT, M.E. Illustrated by 673 Engravings of Latest Tools and Methods, all of which are fully described. Medium 8vo, cloth. *[Just Published. Net* **21/0**

LOCOMOTIVE ENGINE DRIVING.

A Practical Manual for Engineers in Charge of Locomotive Engines. By MICHAEL REYNOLDS, formerly Locomotive Inspector, L. B. & S. C. R. Eleventh Edition. Including a KEY TO THE LOCOMOTIVE ENGINE. Crown 8vo, cloth **4/6**

" Mr. Reynolds has supplied a want, and has supplied it well. We can confidently recommend the book not only to the practical driver, but to everyone who takes an interest in the performance of locomotive engines."—*The Engineer.*

"Mr. Reynolds has opened a new chapter in the literature of the day. This admirable practical treatise, of the practical utility of which we have to speak in terms of warm commendation." —*Athenæum.*

THE MODEL LOCOMOTIVE ENGINEER,

Fireman, and Engine-Boy. Comprising a Historical Notice of the Pioneer Locomotive Engines and their Inventors. By MICHAEL REYNOLDS. Second Edition, with Revised Appendix. Crown 8vo, cloth. **4/6**

" We should be glad to see this book in the possession of everyone in the kingdom who has ever laid, or is to lay, hands on a locomotive engine."—*Iron.*

CONTINUOUS RAILWAY BRAKES.

A Practical Treatise on the several Systems in Use in the United Kingdom : their Construction and Performance. By MICHAEL REYNOLDS. 8vo, cloth **9/0**

" A popular explanation of the different brakes. It will be of great assistance in forming public opinion, and will be studied with benefit by those who take an interest in the brake."—*English Mechanic.*

STATIONARY ENGINE DRIVING.

A Practical Manual for Engineers in Charge of Stationary Engines. By MICHAEL REYNOLDS. Sixth Edition. With Plates and Woodcuts. Crown 8vo, cloth **4/6**

" The author's advice on the various points treated is clear and practical."—*Engineering.*
" Our author leaves no stone unturned. He is determined that his readers shall not only know something about the stationary engine, but all about it."—*Engineer.*

ENGINE-DRIVING LIFE.

Stirring Adventures and Incidents in the Lives of Locomotive Engine-Drivers. By MICHAEL REYNOLDS. Third Edition. Crown 8vo, cloth . **1/6**

" From first to last perfectly fascinating. Wilkie Collins's most thrilling conceptions are thrown into the shade by true incidents, endless in their variety, related in every page."—*North British Mail.*

THE ENGINEMAN'S POCKET COMPANION,

And Practical Educator for Enginemen, Boiler Attendants, and Mechanics. By MICHAEL REYNOLDS. With 45 Illustrations and numerous Diagrams Fourth Edition, Revised. Royal 18mo, strongly bound for pocket wear. **3/6**

" A most meritorious work, giving in a succinct and practical form all the information an engine-minder desirous of mastering the scientific principles of his daily calling would require."— *The Miller.*

CIVIL ENGINEERING, SURVEYING, ETC.

PIONEER IRRIGATION.

A Manual of Information for Farmers in the Colonies. By E. O. MAWSON, Executive Engineer, Public Works Department, Bombay. With Numerous Plates and Diagrams. Medium 8vo, cloth.

[Just ready. Price about **7/6** *net.*

SUMMARY OF CONTENTS:—VALUE OF IRRIGATION, AND SOURCES OF WATER SUPPLY.—DAMS AND WEIRS.—CANALS.—UNDERGROUND WATER.—METHODS OF IRRIGATION.—SEWAGE IRRIGATION.—IMPERIAL AUTOMATIC SLUICE GATES.—THE CULTIVATION OF IRRIGATED CROPS, VEGETABLES, AND FRUIT TREES.—LIGHT RAILWAYS FOR HEAVY TRAFFIC.—USEFUL MEMORANDA, TABLES, AND DATA.

TUNNELLING.

A Practical Treatise. By CHARLES PRELINI, C.E. With additions by CHARLES S. HILL, C.E. With 150 Diagrams and Illustrations. Royal 8vo, cloth . . . . *Net* **16 0**

PRACTICAL TUNNELLING.

Explaining in detail Setting-out the Works, Shaft-sinking, and Heading-driving, Ranging the Lines and Levelling underground, Sub-Excavating, Timbering and the Construction of the Brickwork of Tunnels. By F. W. SIMMS, M. Inst. C.E. Fourth Edition, Revised and Further Extended, including the most recent (1895) Examples of Sub-aqueous and other Tunnels, by D. KINNEAR CLARK, M. Inst. C.E. With 34 Folding Plates. Imperial 8vo, cloth **£2 2s.**

" The present (1896) edition has been brought right up to date, and is thus rendered a work to which civil engineers generally should have ready access, and to which engineers who have construction work can hardly afford to be without, but which to the younger members of the profession is invaluable, as from its pages they can learn the state to which the science of tunnelling has attained."—*Railway News.*

THE WATER SUPPLY OF TOWNS AND THE CONSTRUCTION OF WATER-WORKS.

A Practical Treatise for the Use of Engineers and Students of Engineering. By W. K. BURTON, A.M. Inst. C.E., Consulting Engineer to the Tokyo Water-works. Second Edition, Revised and Extended. With numerous Plates and Illustrations. Super-royal 8vo, buckram. *[Just Published.* **25/0**

I. INTRODUCTORY. — II. DIFFERENT QUALITIES OF WATER. — III. QUANTITY OF WATER TO BE PROVIDED.—IV. ON ASCERTAINING WHETHER A PROPOSED SOURCE OF SUPPLY IS SUFFICIENT.—V. ON ESTIMATING THE STORAGE CAPACITY REQUIRED TO BE PROVIDED.—VI. CLASSIFICATION OF WATER-WORKS.—VII. IMPOUNDING RESERVOIRS.—VIII. EARTHWORK DAMS.—IX. MASONRY DAMS.—X. THE PURIFICATION OF WATER.—XI. SETTLING RESERVOIRS.—XII. SAND FILTRATION.—XIII. PURIFICATION OF WATER BY ACTION OF IRON, SOFTENING OF WATER BY ACTION OF LIME, NATURAL FILTRATION.—XIV. SERVICE OR CLEAN WATER RESERVOIRS—WATER TOWERS—STAND PIPES.—XV. THE CONNECTION OF SETTLING RESERVOIRS, FILTER BEDS AND SERVICE RESERVOIRS.—XVI. PUMPING MACHINERY.—XVII. FLOW OF WATER IN CONDUITS—PIPES AND OPEN CHANNELS.—XVIII. DISTRIBUTION SYSTEMS.—XIX. SPECIAL PROVISIONS FOR THE EXTINCTION OF FIRE.—XX. PIPES FOR WATER-WORKS.—XXI. PREVENTION OF WASTE OF WATER.—XXII. VARIOUS APPLIANCES USED IN CONNECTION WITH WATER-WORKS.

APPENDIX I. By PROF. JOHN MILNE, F.R.S.—CONSIDERATIONS CONCERNING THE PROBABLE EFFECTS OF EARTHQUAKES ON WATER-WORKS, AND THE SPECIAL PRECAUTIONS TO BE TAKEN IN EARTHQUAKE COUNTRIES.

APPENDIX II. By JOHN DE RIJKE, C.E.—ON SAND DUNES AND DUNE SAND AS A SOURCE OF WATER SUPPLY.

" The chapter upon filtration of water is very complete, and the details of construction well illustrated. . . . The work should be specially valuable to civil engineers engaged in work in Japan, but the interest is by no means confined to that locality."—*Engineer.*

" We congratulate the author upon the practical commonsense shown in the preparation of this work. . . . The plates and diagrams have evidently been prepared with great care, and cannot fail to be of great assistance to the student."—*Builder.*

RURAL WATER SUPPLY.

A Practical Handbook on the Supply of Water and Construction of Water-works for small Country Districts. By ALLAN GREENWELL, A.M.I.C.E., and W. T. CURRY, A.M.I.C.E., F.G.S. With Illustrations. Second Edition, Revised. Crown 8vo, cloth **5/0**

" We conscientiously recommend it as a very useful book for those concerned in obtaining water for small districts, giving a great deal of practical information in a smal lcompass."—*Builder.*

" The volume contains valuable information upon all matters connected with water supply. . . . It is full of details on points which are continually before water-works engineers."—*Nature.*

THE WATER SUPPLY OF CITIES AND TOWNS.

By WILLIAM HUMBER, A. M. Inst. C.E., and M. Inst. M.E., Author of "Cast and Wrought Iron Bridge Construction," &c., &c. Illustrated with 50 Double Plates, 1 Single Plate, Coloured Frontispiece, and upwards of 250 Woodcuts, and containing 400 pp. of Text. Imp. 4to, elegantly and substantially half-bound in morocco . . . . *Net* **£6 6s.**

LIST OF CONTENTS. I. HISTORICAL SKETCH OF SOME OF THE MEANS THAT HAVE BEEN ADOPTED FOR THE SUPPLY OF WATER TO CITIES AND TOWNS.—II. WATER AND THE FOREIGN MATTER USUALLY ASSOCIATED WITH IT.—III. RAINFALL AND EVAPORATION.—IV. SPRINGS AND THE WATER-BEARING FORMATIONS OF VARIOUS DISTRICTS.—V. MEASUREMENT AND ESTIMATION OF THE FLOW OF WATER.—VI. ON THE SELECTION OF THE SOURCE OF SUPPLY.—VII. WELLS.—VIII. RESERVOIRS.—IX. THE PURIFICATION OF WATER.—X. PUMPS.—XI. PUMPING MACHINERY.—XII. CONDUITS.—XIII. DISTRIBUTION OF WATER.—XIV. METERS, SERVICE PIPES, AND HOUSE FITTINGS.—XV. THE LAW AND ECONOMY OF WATER-WORKS.—XVI. CONSTANT AND INTERMITTENT SUPPLY.—XVII. DESCRIPTION OF PLATES.—APPENDICES, GIVING TABLES OF RATES OF SUPPLY, VELOCITIES, &c., &c., TOGETHER WITH SPECIFICATIONS OF SEVERAL WORKS ILLUSTRATED, AMONG WHICH WILL BE FOUND: ABERDEEN, BIDEFORD, CANTERBURY, DUNDEE, HALIFAX, LAMBETH, ROTHERHAM, DUBLIN, AND OTHERS.

"The most systematic and valuable work upon water supply hitherto produced in English, or in any other language. Mr. Humber's work is characterised almost throughout by an exhaustiveness much more distinctive of French and German than of English technical treatises."—*Engineer.*

THE PROGRESS OF ENGINEERING (1863-6).

By WM. HUMBER, A.M.Inst.C.E. Complete in Four Vols. Containing 148 Double Plates, with Portraits and Copious Descriptive Letterpress. Impl. 4to, half-morocco. Price, complete, **£12 12s.** ; or each Volume sold separately at **£3 3s.** per Volume. *Descriptive List of Contents on application.*

HYDRAULIC POWER ENGINEERING.

A Practical Manual on the Concentration and Transmission of Power by Hydraulic Machinery. By G. CROYDON MARKS, A.M. Inst. C.E. With nearly 200 Illustrations. 8vo, cloth. *Net* **9/0**

SUMMARY OF CONTENTS. PRINCIPLES OF HYDRAULICS.—THE FLOW OF WATER.—HYDRAULIC PRESSURES, MATERIAL.—TEST LOAD PACKINGS FOR SLIDING SURFACES.—PIPE JOINTS.—CONTROLLING VALVES.—PLATFORM LIFTS.—WORKSHOP AND FOUNDRY CRANES.—WAREHOUSE AND DOCK CRANES.—HYDRAULIC ACCUMULATORS.—PRESSES FOR BALING AND OTHER PURPOSES.—SHEET METAL WORKING AND FORGING MACHINERY.—HYDRAULIC RIVETTERS.—HAND, POWER, AND STEAM PUMPS.—TURBINES.—IMPULSE TURBINES.—REACTION TURBINES.—DESIGN OF TURBINES IN DETAIL.—WATER WHEELS.—HYDRAULIC ENGINES.—RECENT ACHIEVEMENTS.—PRESSURE OF WATER.—ACTION OF PUMPS, &c.

"We have nothing but praise for this thoroughly valuable work. The author has succeeded in rendering his subject interesting as well as instructive."—*Practical Engineer.*
"Can be unhesitatingly recommended as a useful and up-to-date manual on hydraulic transmission and utilisation of power."—*Mechanical World*

HYDRAULIC TABLES, CO-EFFICIENTS, & FORMULÆ.

For Finding the Discharge of Water from Orifices, Notches, Weirs, Pipes, and Rivers. With New Formulæ, Tables, and General Information on Rain-fall, Catchment-Basins, Drainage, Sewerage, Water Supply for Towns and Mill Power. By JOHN NEVILLE, C.E., M.R.I.A. Third Edition, revised, with additions. Numerous Illustrations. Crown 8vo, cloth . . . **14/0**

"It is, of all English books on the subject, the one nearest to completeness."—*Architect.*

HYDRAULIC MANUAL.

Consisting of Working Tables and Explanatory Text. Intended as a Guide in Hydraulic Calculations and Field Operations. By LOWIS D'A. JACKSON, Author of "Aid to Survey Practice," "Modern Metrology," &c. Fourth Edition, Enlarged. Large crown 8vo, cloth **16/0**

"The author has constructed a manual which may be accepted as a trustworthy guide to this branch of the engineer's profession."—*Engineering.*

WATER ENGINEERING.

A Practical Treatise on the Measurement, Storage, Conveyance, and Utilisation of Water for the Supply of Towns, for Mill Power, and for other Purposes. By CHARLES SLAGG, A.M.Inst.C.E. Second Edition. Crown 8vo, cloth . **7/6**

"As a small practical treatise on the water supply of towns, and on some applications of water-power, the work is in many respects excellent."—*Engineering.*

THE RECLAMATION OF LAND FROM TIDAL WATERS.

A Handbook for Engineers, Landed Proprietors, and others interested in Works of Reclamation. By ALEX. BEAZELEY, M.Inst. C.E. 8vo, cloth.

Net **10/6**

" The book shows in a concise way what has to be done in reclaiming land from the sea, and the best way of doing it. The work contains a great deal of practical and useful information which cannot fail to be of service to engineers entrusted with the enclosure of salt marshes, and to land-owners intending to reclaim land from the sea."—*The Engineer.*

" The author has carried out his task efficiently and well, and his book contains a large amount of information of great service to engineers and others interested in works of reclamation." —*Nature.*

MASONRY DAMS FROM INCEPTION TO COMPLETION.

Including numerous Formulæ, Forms of Specification and Tender, Pocket Diagram of Forces, &c. For the use of Civil and Mining Engineers. By C. F. COURTNEY, M. Inst. C.E. 8vo, cloth **9/0**

" The volume contains a good deal of valuable data. Many useful suggestions will be found in the remarks on site and position, location of dam, foundations and construction.'—*Building News.*

RIVER BARS.

The Causes of their Formation, and their Treatment by " Induced Tidal Scour "; with a Description of the Successful Reduction by this Method of the Bar at Dublin. By I. J. MANN, Assist. Eng. to the Dublin Port and Docks Board. Royal 8vo, cloth **7/6**

" We recommend all interested in harbour works—and, indeed, those concerned in the improvements of rivers generally—to read Mr. Mann's interesting work."—*Engineer.*

TRAMWAYS: THEIR CONSTRUCTION AND WORKING.

Embracing a Comprehensive History of the System; with an exhaustive Analysis of the Various Modes of Traction, including Horse Power, Steam, Cable Traction, Electric Traction, &c.; a Description of the Varieties of Rolling Stock; and ample Details of Cost and Working Expenses. New Edition, Thoroughly Revised, and Including the Progress recently made in Tramway Construction, &c., &c. By D. KINNEAR CLARK, M. Inst. C.E. With 400 Illustrations. 8vo, 780 pp., buckram.   **28/0**

" The new volume is one which will rank, among tramway engineers and those interested in tramway working, with the Author's world-famed book on railway machinery."—*The Engineer.*

SURVEYING AS PRACTISED BY CIVIL ENGINEERS AND SURVEYORS.

Including the Setting-out of Works for Construction and Surveys Abroad, with many Examples taken from Actual Practice. A Handbook for use in the Field and the Office, intended also as a Text-book for Students. By JOHN WHITE-LAW, Jun., A.M. Inst. C.E., Author of " Points and Crossings." With about 260 Illustrations. Demy 8vo, cloth   .   .   .   .   *Net* **10/6**

" This work is written with admirable lucidity, and will certainly be found of distinct value both to students and to those engaged in actual practice."—*The Builder.*

PRACTICAL SURVEYING.

A Text-Book for Students preparing for Examinations or for Survey-work in the Colonies. By GEORGE W. USILL, A.M.I.C.E. With 4 Lithographic Plates and upwards of 330 Illustrations. Seventh Edition. Including Tables of Natural Sines, Tangents, Secants, &c. Crown 8vo, **7/6** cloth; or, on THIN PAPER, leather, gilt edges, rounded corners, for pocket use .   .   **12/6**

" The best forms of instruments are described as to their construction, uses and modes of employment, and there are innumerable hints on work and equipment such as the author, in his experience as surveyor, draughtsman and teacher, has found necessary, and which the student in his inexperience will find most serviceable."—*Engineer.*

" The first book which should be put in the hands of a pupil of Civil Engineering."— *Architect.*

AID TO SURVEY PRACTICE.

For Reference in Surveying, Levelling, and Setting-out; and in Route Sur-veys of Travellers by Land and Sea. With Tables, Illustrations, and Records. By LOWIS D'A. JACKSON, A.M.I.C.E. Second Edition, Enlarged. 8vo, cloth **12/6**

" Mr. Jackson has produced a valuable *vade-mecum* for the surveyor. We can recommend this book as containing an admirable supplement to the teaching of the accomplished surveyor."— *Athenæum.*

" The author brings to his work a fortunate union of theory and practical experience which, aided by a clear and lucid style of writing, renders the book a very useful one."—*Builder.*

SURVEYING WITH THE TACHEOMETER.

A practical Manual for the use of Civil and Military Engineers and Surveyors. Including two series of Tables specially computed for the Reduction of Readings in Sexagesimal and in Centesimal Degrees. By NEIL KENNEDY, M. Inst. C. E. With Diagrams and Plates. Demy 8vo, cloth. *Net* **10/6**

" The work is very clearly written, and should remove all difficulties in the way of any surveyor desirous of making use of this useful and rapid instrument."—*Nature.*

ENGINEER'S & MINING SURVEYOR'S FIELD BOOK.

Consisting of a Series of Tables, with Rules, Explanations of Systems, and use of Theodolite for Traverse Surveying and plotting the work with minute accuracy by means of Straight Edge and Set Square only ; Levelling with the Theodolite, Setting-out Curves with and without the Theodolite, Earthwork Tables, &c. By W. DAVIS HASKOLL, C.E. With numerous Woodcuts. Fourth Edition, Enlarged. Crown 8vo, cloth **12/0**

" The book is very handy ; the separate tables of sines and tangents to every minute will make it useful for many other purposes, the genuine traverse tables existing all the same."—*Athenæum.*

LAND AND MARINE SURVEYING.

In Reference to the Preparation of Plans for Roads and Railways ; Canals, Rivers, Towns' Water Supplies ; Docks and Harbours. With Description and Use of Surveying Instruments. By W. DAVIS HASKOLL, C.E. Second Edition, Revised, with Additions. Large crown 8vo, cloth . . . **9/0**

" This book must prove of great value to the student. We have no hesitation in recommending it, feeling assured that it will more than repay a careful study."—*Mechanical World.*
" A most useful book for the student. We can strongly recommend it as a carefully-written and valuable text-book. It enjoys a well-deserved repute among surveyors."—*Builder.*

PRINCIPLES AND PRACTICE OF LEVELLING.

Showing its Application to Purposes of Railway and Civil Engineering in the Construction of Roads ; with Mr. TELFORD'S Rules for the same. By FREDERICK W. SIMMS, M. Inst. C.E. Eighth Edition, with LAW'S Practical Examples for Setting-out Railway Curves, and TRAUTWINE'S Field Practice of Laying-out Circular Curves. With 7 Plates and numerous Woodcuts. 8vo **8/6**

" The text-book on levelling in most of our engineering schools and colleges."—*Engineer.*
" The publishers have rendered a substantial service to the profession, especially to the younger members, by bringing out the present edition of Mr. Simms's useful work."—*Engineering.*

AN OUTLINE OF THE METHOD OF CONDUCTING A TRIGONOMETRICAL SURVEY.

For the Formation of Geographical and Topographical Maps and Plans, Military Reconnaissance, LEVELLING, &c., with Useful Problems, Formulæ, and Tables. By Lieut.-General FROME, R.E. Fourth Edition, Revised and partly Re-written by Major-General Sir CHARLES WARREN, G.C.M.G., R.E. With 19 Plates and 115 Woodcuts, royal 8vo, cloth **16/0**

" No words of praise from us can strengthen the position so well and so steadily maintained by this work. Sir Charles Warren has revised the entire work, and made such additions as were necessary to bring every portion of the contents up to the present date."—*Broad Arrow.*

TABLES OF TANGENTIAL ANGLES AND MULTIPLES FOR SETTING-OUT CURVES.

From 5 to 200 Radius. By A. BEAZELEY, M. Inst. C.E. 6th Edition, Revised. With an Appendix on the use of the Tables for Measuring up Curves. Printed on 50 Cards, and sold in a cloth box, waistcoat-pocket size. **3/6**

" Each table is printed on a small card, which, placed on the theodolite, leaves the hands free to manipulate the instrument—no small advantage as regards the rapidity of work."—*Engineer.*
" Very handy : a man may know that all his day's work must fall on two of these cards, which he puts into his own card-case, and leaves the rest behind."—*Athenæum.*

HANDY GENERAL EARTH-WORK TABLES.

Giving the Contents in Cubic Yards of Centre and Slopes of Cuttings and Embankments from 3 inches to 80 feet in Depth or Height, for use with either 66 feet Chain or 100 feet Chain. By J. H. WATSON BUCK, M. Inst. C.E. On a Sheet mounted in cloth case **3/6**

EARTHWORK TABLES.

Showing the Contents in Cubic Yards of Embankments, Cuttings, &c., of Heights or Depths up to an average of 80 feet. By JOSEPH BROADBENT, C.E., and FRANCIS CAMPIN, C.E.   Crown 8vo, cloth **5/0**

" The way in which accuracy is attained, by a simple division of each cross section into three elements, two in which are constant and one variable, is ingenious."—*Athenæum.*

A MANUAL ON EARTHWORK.

By ALEX. J. GRAHAM, C.E.   With numerous Diagrams.   Second Edition. 18mo, cloth **2/6**

THE CONSTRUCTION OF LARGE TUNNEL SHAFTS.

A Practical and Theoretical Essay.   By J. H. WATSON BUCK, M. Inst. C.E., Resident Engineer, L. and N. W. R.   With Folding Plates, 8vo, cloth **12/0**

" Many of the methods given are of extreme practical value to the mason, and the observations on the form of arch, the rules for ordering the stone, and the construction of the templates, will be found of considerable use.   We commend the book to the engineering profession."— *Building News.*

" Will be regarded by civil engineers as of the utmost value, and calculated to save much time and obviate many mistakes."—*Colliery Guardian.*

CAST & WROUGHT IRON BRIDGE CONSTRUCTION

(A Complete and Practical Treatise on), including Iron Foundations. In Three Parts.—Theoretical, Practical, and Descriptive.   By WILLIAM HUMBER, A. M. Inst. C.E., and M. Inst. M.E.   Third Edition, revised and much improved, with 115 Double Plates (20 of which now first appear in this edition), and numerous Additions to the Text.   In 2 vols., imp. 4to, half-bound in morocco   .   .   .   .   **£6 16s. 6D.**

" A very valuable contribution to the standard literature of civil engineering.   In addition to elevations, plans, and sections, large scale details are given, which very much enhance the instructive worth of those illustrations."—*Civil Engineer and Architect's Journal.*

" Mr. Humber's stately volumes, lately issued—in which the most important bridges erected during the last five years, under the direction of the late Mr. Brunel, Sir W. Cubitt, Mr. Hawkshaw, Mr. Page, Mr. Fowler, Mr. Hemans, and others among our most eminent engineers, are drawn and specified in great detail."—*Engineer.*

ESSAY ON OBLIQUE BRIDGES

(Practical and Theoretical).   With 13 large Plates.   By the late GEORGE WATSON BUCK, M.I.C.E.   Fourth Edition, revised by his Son, J. H. WATSON BUCK, M.I.C.E. ; and with the addition of Description to Diagrams for Facilitating the Construction of Oblique Bridges, by W. H. BARLOW, M.I.C.E. Royal 8vo, cloth **12/0**

" The standard text-book for all engineers regarding skew arches is Mr. Buck's treatise, and it would be impossible to consult a better."—*Engineer.*

" Mr. Buck's treatise is recognised as a standard text-book, and his treatment has divested the subject of many of the intricacies supposed to belong to it.   As a guide to the engineer and architect, on a confessedly difficult subject, Mr. Buck's work is unsurpassed."—*Building News.*

THE CONSTRUCTION OF OBLIQUE ARCHES

(A Practical Treatise on).   By JOHN HART.   Third Edition, with Plates. Imperial 8vo, cloth **8/0**

GRAPHIC AND ANALYTIC STATICS.

In their Practical Application to the Treatment of Stresses in Roofs Solid Girders, Lattice, Bowstring, and Suspension Bridges, Braced Iron Arches and Piers, and other Frameworks.   By R. HUDSON GRAHAM, C.E.   Containing Diagrams and Plates to Scale.   With numerous Examples, many taken from existing Structures.   Specially arranged for Class-work in Colleges and Universities.   Second Edition, Revised and Enlarged.   8vo, cloth   .   **16/0**

" Mr. Graham's book will find a place wherever graphic and analytic statics are used or studied."—*Engineer.*

" The work is excellent from a practical point of view, and has evidently been prepared with much care.   The directions for working are ample, and are illustrated by an abundance of well-selected examples.   It is an excellent text-book for the practical draughtsman."—*Athenæum.*

WEIGHTS OF WROUGHT IRON & STEEL GIRDERS.

A Graphic Table for Facilitating the Computation of the Weights of Wrought Iron and Steel Girders, &c., for Parliamentary and other Estimates.   By J. H. WATSON BUCK, M. Inst. C.E.   On a Sheet   .   .   .   **2/6**

GEOMETRY FOR TECHNICAL STUDENTS.

An Introduction to Pure and Applied Geometry and the Mensuration of Surfaces and Solids, including Problems in Plane Geometry useful in Drawing. By E. H. Sprague, A.M.I.C.E. Crown 8vo, cloth. [*Just Published. Net* **1/0**

PRACTICAL GEOMETRY.

For the Architect, Engineer, and Mechanic. Giving Rules for the Delineation and Application of various Geometrical Lines, Figures, and Curves. By E. W. Tarn, M.A., Architect. 8vo, cloth **9/0**

"No book with the same objects in view has ever been published in which the clearness of the rules laid down and the illustrative diagrams have been so satisfactory."—*Scotsman.*

THE GEOMETRY OF COMPASSES.

Or, Problems Resolved by the mere Description of Circles and the Use of Coloured Diagrams and Symbols. By Oliver Byrne. Coloured Plates. Crown 8vo, cloth **3/6**

EXPERIMENTS ON THE FLEXURE OF BEAMS

Resulting in the Discovery of New Laws of Failure by Buckling. By Albert E. Guy. Medium 8vo, cloth. [*Just Published. Net* **9/0**

HANDY BOOK FOR THE CALCULATION OF STRAINS

In Girders and Similar Structures and their Strength. Consisting of Formulæ and Corresponding Diagrams, with numerous details for Practical Application, &c. By William Humber, A. M. Inst. C.E., &c. Fifth Edition. Crown 8vo, with nearly 100 Woodcuts and 3 Plates, cloth . . **7/6**

"The formulæ are neatly-expressed, and the diagrams good."—*Athenæum.*
"We heartily commend this really *handy* book to our engineer and architect readers."—*English Mechanic.*

TRUSSES OF WOOD AND IRON.

Practical Applications of Science in Determining the Stresses, Breaking Weights, Safe Loads, Scantlings, and Details of Construction. With Complete Working Drawings. By William Griffiths, Surveyor. 8vo, cloth **4/6**

"This handy little book enters so minutely into every detail connected with the construction of roof trusses that no student need be ignorant of these matters."—*Practical Engineer.*

THE STRAINS ON STRUCTURES OF IRONWORK.

With Practical Remarks on Iron Construction. By F. W. Sheilds, M.I.C.E. 8vo, cloth **5/0**

A TREATISE ON THE STRENGTH OF MATERIALS.

With Rules for Application in Architecture, the Construction of Suspension Bridges, Railways, &c. By Peter Barlow, F.R.S. A new Edition, revised by his Sons, P. W. Barlow, F.R.S., and W. H. Barlow, F.R.S. ; to which are added, Experiments by Hodgkinson, Fairbairn, and Kirkaldv; and Formulæ for calculating Girders, &c. Edited by Wm. Humber, A.M.I.C.E. 8vo, 400 pp., with 19 Plates and numerous Woodcuts, cloth. . . **18/0**

"Valuable alike to the student, tyro, and the experienced practitioner, it will always rank in future as it has hitherto done, as the standard treatise on that particular subject."—*Engineer.*

SAFE RAILWAY WORKING.

A Treatise on Railway Accidents, their Cause and Prevention; with a Description of Modern Appliances and Systems. By Clement E. Stretton, C.E. With Illustrations and Coloured Plates. Third Edition, Enlarged. Crown 8vo, cloth **3/6**

"A book for the engineer, the directors, the managers; and, in short, all who wish for information on railway matters will find a perfect encyclopædia in 'Safe Railway Working.'"—*Railway Review.*

EXPANSION OF STRUCTURES BY HEAT.

By John Keily, C.E., late of the Indian Public Works Department. Crown 8vo, cloth **3/6**

"The aim the author has set before him, viz., to show the effects of heat upon metallic and other structures, is a laudable one, for this is a branch of physics upon which the engineer or architect can find but little reliable and comprehensive data in books."—*Builder.*

PUBLICATIONS OF THE ENGINEERING
STANDARDS COMMITTEE.

THE ENGINEERING STANDARDS COMMITTEE is the outcome of a Committee appointed by the Institution of Civil Engineers at the instance of Sir John Wolfe Barry, K.C.B., to inquire into the advisability of Standardising Rolled Iron and Steel Sections.

The Committee is supported by the Institution of Civil Engineers, the Institution of Mechanical Engineers, the Institution of Naval Architects, the Iron and Steel Institute, and the Institution of Electrical Engineers; and the value and importance of its labours has been emphatically recognised by his Majesty's Government, who have made a liberal grant from the Public Funds by way of contribution to the financial resources of the Committee.

The subjects already dealt with, or under consideration by the Committee, include not only Rolled Iron and Steel Sections, but Tests for Iron and Steel Material used in the Construction of Ships and their Machinery, Bridges and General Building Construction, Railway Rolling Stock Underframes, Component Parts of Locomotives, Railway and Tramway Rails, Electrical Plant, Insulating Materials, Screw Threads and Limit Gauges, Pipe Flanges, Cement, &c.

Reports already Published :—

1. **BRITISH STANDARD SECTIONS.**
   List 1. EQUAL ANGLES —List 2. UNEQUAL ANGLES.—List 3. BULB ANGLES. List 4. BULB TEES.—List 5. BULB PLATES.—List 7. CHANNELS.—List 8. BEAMS. F'cap. folio, sewed.   [*Just Published. Net* 1/0

2. **BRITISH STANDARD TRAMWAY RAILS AND FISH**
   PLATES : STANDARD SECTIONS AND SPECIFICATION.
   F'cap. folio, sewed.   [*Just Published. Net* 21/0

3. **REPORT ON THE INFLUENCE OF GAUGE LENGTH**
   AND SECTION OF TEST BAR ON THE PERCENTAGE OF
   ELONGATION.
   By Professor W. C. UNWIN, F.R.S.   F'cap. folio, sewed.
                                         [*Just Published. Net* 2/6

4. **PROPERTIES OF STANDARD BEAMS.**
   Demy 8vo, sewed.   [*Just Published. Net* 1/0

MARINE ENGINEERING, SHIPBUILDING, NAVIGATION, ETC.

THE NAVAL ARCHITECT'S AND SHIPBUILDER'S

POCKET-BOOK of Formulæ, Rules, and Tables, and Marine Engineer's and Surveyor's Handy Book of Reference. By CLEMENT MACKROW, M.I.N.A. Eighth Edition, Carefully Revised and Enlarged. Fcap., leather. *Net* **12/6**

SUMMARY OF CONTENTS:—SIGNS AND SYMBOLS, DECIMAL FRACTIONS.—TRIGO-NOMETRY.—PRACTICAL GEOMETRY.—MENSURATION.—CENTRES AND MOMENTS OF FIGURES.—MOMENTS OF INERTIA AND RADII OF GYRATION.—ALGEBRAICAL EXPRESSIONS FOR SIMPSON'S RULES.—MECHANICAL PRINCIPLES.—CENTRE OF GRAVITY.—LAWS OF MOTION.—DISPLACEMENT, CENTRE OF BUOYANCY.—CENTRE OF GRAVITY OF SHIP'S HULL—STABILITY CURVES AND METACENTRES.—SEA AND SHALLOW-WATER WAVES.—ROLLING OF SHIPS.—PROPULSION AND RESISTANCE OF VESSELS.—SPEED TRIALS.—SAILING CENTRE OF EFFORT.—DISTANCES DOWN RIVERS, COAST LINES.—STEERING AND RUDDERS OF VESSELS.—LAUNCHING CALCULATIONS AND VELOCITIES.—WEIGHT OF MATERIAL AND GEAR.—GUN PARTICULARS AND WEIGHT.—STANDARD GAUGES.—RIVETED JOINTS AND RIVETING.—STRENGTH AND TESTS OF MATERIALS.—BINDING AND SHEARING STRESSES, &c.—STRENGTH OF SHAFTING, PILLARS, WHEELS, &c.—HYDRAULIC DATA, &c.—CONIC SECTIONS, CATENARIAN CURVES.—MECHANICAL POWERS, WORK.—BOARD OF TRADE REGULATIONS FOR BOILERS AND ENGINES.—BOARD OF TRADE REGULATIONS FOR SHIPS.—LLOYD'S RULES FOR BOILERS.—LLOYD'S WEIGHT OF CHAINS.—LLOYD'S SCANTLINGS FOR SHIPS.—DATA OF ENGINES AND VESSELS.—SHIPS' FITTINGS AND TESTS.—SEASONING PRESERVING TIMBER.—MEASUREMENT OF TIMBER.—ALLOYS, PAINTS, VARNISHES.—DATA FOR STOWAGE.—ADMIRALTY TRANS-PORT REGULATIONS.—RULES FOR HORSE-POWER, SCREW PROPELLERS, &c.—PER-CENTAGES FOR BUTT STRAPS, &c.—PARTICULARS OF YACHTS.—MASTING AND RIGGING VESSELS.—DISTANCES OF FOREIGN PORTS.—TONNAGE TABLES.—VOCABULARY OF FRENCH AND ENGLISH TERMS.—ENGLISH WEIGHTS AND MEASURES.—FOREIGN WEIGHTS AND MEASURES.—DECIMAL EQUIVALENTS.—FOREIGN MONEY.—DISCOUNT AND WAGES TABLES.—USEFUL NUMBERS AND READY RECKONERS.—TABLES OF CIRCULAR MEASURES.—TABLES OF AREAS OF AND CIRCUMFERENCES OF CIRCLES.—TABLES OF AREAS OF SEGMENTS OF CIRCLES.—TABLES OF SQUARES AND CUBES AND ROOTS OF NUMBERS.—TABLES OF LOGARITHMS OF NUMBERS.—TABLES OF HYPER-BOLIC LOGARITHMS.—TABLES OF NATURAL SINES, TANGENTS, &c.—TABLES OF LOGARITHMIC SINES, TANGENTS, &c.

" In these days of advanced knowledge a work like this is of the greatest value. It contains a vast amount of information. We unhesitatingly say that it is the most valuable compilation for its specific purpose that has ever been printed. No naval architect, engineer, surveyor, seaman, wood or iron shipbuilder, can afford to be without this work."—*Nautical Magazine.*

" Should be used by all who are engaged in the construction or design of vessels. . . . Will be found to contain the most useful tables and formulæ required by shipbuilders, carefully collected from the best authorities, and put together in a popular and simple form. The book is one of exceptional merit."—*Engineer.*

" The professional shipbuilder has now, in a convenient and accessible form, reliable data for solving many of the numerous problems that present themselves in the course of his work."—*Iron.*

" There is no doubt that a pocket-book of this description must be a necessity in the ship-building trade. . . The volume contains a mass of useful information clearly expressed and presented in a handy form."—*Marine Engineer.*

WANNAN'S MARINE ENGINEER'S GUIDE

To Board of Trade Examinations for Certificates of Competency. Containing all Latest Questions to Date, with Simple, Clear, and Correct Solutions; 302 Elementary Questions with Illustrated Answers, and Verbal Questions and Answers; complete Set of Drawings with Statements completed. By A. C. WANNAN, C.E., Consulting Engineer, and E. W. I. WANNAN, M.I.M.E., Certificated First Class Marine Engineer. With numerous Engravings. Third Edition, Enlarged. 500 pages. Large crown 8vo, cloth . . *Net* **10/6**

" The book is clearly and plainly written and avoids unnecessary explanations and formulas, and we consider it a valuable book for students of marine engineering."—*Nautical Magazine.*

WANNAN'S MARINE ENGINEER'S POCKET-BOOK.

Containing Latest Board of Trade Rules and Data for Marine Engineers. By A. C. WANNAN. Third Edition, Revised, Enlarged, and Brought up to Date. Square 18mo, with thumb Index, leather. [*Just Published.* **5/0**

" There is a great deal of useful information in this little pocket-book. It is of the rule-of thumb order, and is, on that account, well adapted to the uses of the sea-going engineer."—*Engineer.*

THE SHIPBUILDING INDUSTRY OF GERMANY.

Compiled and Edited by G. LEHMANN-FELSKOWSKI. With Coloured Prints, Art Supplements, and numerous Illustrations throughout the text. Super-royal 4to, cloth. [*Just Published.* *Net* **10/6**

SEA TERMS, PHRASES, AND WORDS

(Technical Dictionary of) used in the English and French Languages (English-French, French-English). For the Use of Seamen, Engineers, Pilots, Shipbuilders, Shipowners, and Ship-brokers. Compiled by W. PIRRIE, late of the African Steamship Company. Fcap. 8vo, cloth limp   .   .   **5/0**

"This volume will be highly appreciated by seamen, engineers, pilots, shipbuilders and shipowners. It will be found wonderfully accurate and complete."—*Scotsman.*
"A very useful dictionary, which has long been wanted by French and English engineers, masters, officers and others."—*Shipping World.*

ELECTRIC SHIP-LIGHTING.

A Handbook on the Practical Fitting and Running of Ships' Electrical Plant, for the Use of Shipowners and Builders, Marine Electricians and Sea-going Engineers in Charge. By J. W. URQUHART, Author of "Electric Light," "Dynamo Construction," &c.   Second Edition, Revised and Extended. With numerous Illustrations. Crown 8vo, cloth **7/6**

MARINE ENGINEER'S POCKET-BOOK.

Consisting of useful Tables and Formulæ. By FRANK PROCTOR, A.I.N.A. Third Edition. Royal 32mo, leather. **4/0**

"We recommend it to our readers as going far to supply a long-felt want."—*Naval Science.*
"A most useful companion to all marine engineers."—*United Service Gazette.*

ELEMENTARY MARINE ENGINEERING.

A Manual for Young Marine Engineers and Apprentices. In the Form of Questions and Answers on Metals, Alloys, Strength of Materials, Construction and Management of Marine Engines and Boilers, Geometry, &c. With an Appendix of Useful Tables. By J. S. BREWER. Crown 8vo, cloth   **1/6**

"Contains much valuable information for the class for whom it is i tended, especially in the chapters on the management of boilers and engines."—*Nautical Magazine.*

MARINE ENGINES AND STEAM VESSELS.

A Treatise on. By ROBERT MURRAY, C.E. Eighth Edition, thoroughly Revised, with considerable Additions by the Author and by GEORGE CARLISLE, C.E., Senior Surveyor to the Board of Trade at Liverpool. Crown 8vo, cloth **4/6**

PRACTICAL NAVIGATION.

Consisting of THE SAILOR'S SEA-BOOK, by JAMES GREENWOOD and W. H. ROSSER; together with the requisite Mathematical and Nautical Tables for the Working of the Problems, by HENRY LAW, C.E., and Professor J. R. YOUNG. Illustrated. 12mo, strongly half-bound   .   .   .   **7/0**

THE ART AND SCIENCE OF SAILMAKING.

By SAMUEL B. SADLER, Practical Sailmaker, late in the employment of Messrs. Ratsey and Lapthorne, of Cowes and Gosport. With Plates and other Illustrations. Small 4to, cloth **12/6**

"This extremely practical work gives a complete education in all the branches of the manufacture, cutting out, roping, seaming, and goring. It is copiously illustrated, and will form a first-rate text-book and guide."—*Portsmouth Times.*

CHAIN CABLES AND CHAINS.

Comprising Sizes and Curves of Links, Studs, &c., Iron for Cables and Chains, Chain Cable and Chain Making, Forming and Welding Links, Strength of Cables and Chains, Certificates for Cables, Marking Cables, Prices of Chain Cables and Chains, Historical Notes, Acts of Parliament, Statutory Tests, Charges for Testing, List of Manufacturers of Cables, &c., &c. By THOMAS W. TRAILL, F.E.R.N., M.Inst.C.E., Engineer-Surveyor-in-Chief, Board of Trade, Inspector of Chain Cable and Anchor Proving Establishments, and General Superintendent Lloyd's Committee on Proving Establishments. With numerous Tables, Illustrations, and Lithographic Drawings. Folio, cloth, bevelled boards .   .   .   .   **£2 2s.**

"It contains a vast amount of valuable information. Nothing seems to be wanting to make it a complete and standard work of reference on the subject."—*Nautical Magazine.*

MINING, METALLURGY, AND COLLIERY WORKING.

THE OIL FIELDS OF RUSSIA AND THE RUSSIAN PETROLEUM INDUSTRY.

A Practical Handbook on the Exploration, Exploitation, and Management of Russian Oil Properties, including Notes on the Origin of Petroleum in Russia, a Description of the Theory and Practice of Liquid Fuel, and a Translation of the Rules and Regulations concerning Russian Oil Properties. By A. BEEBY THOMPSON, A.M.I.M.E., late Chief Engineer and Manager of the European Petroleum Company's Russian Oil Properties. About 500 pp. With numerous Illustrations and Photographic Plates, and a Map of the Balakhany-Saboontchy-Romany Oil Field. Super-royal 8vo, cloth.
[*Just Published.* Net **£3 3s.**

MACHINERY FOR METALLIFEROUS MINES.

A Practical Treatise for Mining Engineers, Metallurgists, and Managers of Mines. By E. HENRY DAVIES, M.E., F.G.S. 600 pp. With Folding Plates and other Illustrations. Medium 8vo, cloth . . . . *Net* **25/0**
" Deals exhaustively with the many and complex details which go to make up the sum total of machinery and other requirements for the successful working of metalliferous mines, and as a book of ready reference is of the highest value to mine managers and directors."—*Mining Journal.*

THE DEEP LEVEL MINES OF THE RAND,

And their Future Development, considered from the Commercial Point of View. By G. A. DENNY (of Johannesburg), M.N.E.I.M.E., Consulting Engineer to the General Mining and Finance Corporation, Ltd., of London, Berlin, Paris, and Johannesburg. Fully Illustrated with Diagrams and Folding Plates. Royal 8vo, buckram . . . . *Net* **25/0**
"Mr. Denny by confining himself to the consideration of the future of the deep-level mines of the Rand breaks new ground, and by dealing with the subject rather from a commercial stand-point than from a scientific one, appeals to a wide circle of readers. The book cannot fail to prove of very great value to investors in South African mines."—*Mining Journal.*

PROSPECTING FOR GOLD.

A Handbook of Information and Hints for Prospectors based on Personal Experience. By DANIEL J. RANKIN, F.R.S.G.S., M.R.A.S., formerly Manager of the Central African Company, and Leader of African Gold Pros-pecting Expeditions. With Illustrations specially Drawn and Engraved for the Work F'cap. 8vo, leather . . . . *Net* **7/6**
"This well-compiled book contains a collection of the richest gems of useful knowledge for the prospector's benefit. A special table is given to accelerate the spotting at a glance of minerals associated with gold."—*Mining Journal.*

THE METALLURGY OF GOLD.

A Practical Treatise on the Metallurgical Treatment of Gold-bearing Ores. Including the Assaying, Melting, and Refining of Gold. By M. EISSLER, M. Inst. M.M. Fifth Edition, Enlarged. With over 300 Illustrations and numerous Folding Plates. Medium 8vo, cloth . . . . *Net* **21/0**
" This book thoroughly deserves its title of a 'Practical Treatise.' The whole process of gold mining, from the breaking of the quartz to the assay of the bullion, is described in clear and orderly narrative and with much, but not too much, fulness of detail."—*Saturday Review.*

THE CYANIDE PROCESS OF GOLD EXTRACTION.

And its Practical Application on the Witwatersrand Gold Fields and elsewhere. By M. EISSLER, M. Inst. M.M. With Diagrams and Working Drawings. Third Edition, Revised and Enlarged. 8vo, cloth . . . . *Net* **7/6**
"This book is just what was needed to acquaint mining men with the actual working of a process which is not only the most popular, but is, as a general rule, the most successful for the extraction of gold from tailings."—*Mining Journal.*

DIAMOND DRILLING FOR GOLD & OTHER MINERALS.

A Practical Handbook on the Use of Modern Diamond Core Drills in Pro-specting and Exploiting Mineral-Bearing Properties, including Particulars of the Costs of Apparatus and Working. By G. A. DENNY, M.N.E. Inst. M.E., M. Inst. M.M. Medium 8vo, 168 pp., with Illustrative Diagrams . **12/6**
" There is certainly scope for a work on diamond drilling, and Mr. Denny deserves grateful recognition for supplying a decided want."—*Mining Journal.*

GOLD ASSAYING.

A Practical Handbook for Assayers, Bankers, Chemists, Bullion Smelters, Goldsmiths, Mining and Metallurgical Engineers, Prospectors, Students, and others. By H. JOSHUA PHILLIPS, F.I.C., F.C.S., A.I.C.E., Author of "Engineering Chemistry," etc. Large Crown 8vo, cloth.

[Just ready, price about **7/6** *net.*

FIELD TESTING FOR GOLD AND SILVER.

A Practical Manual for Prospectors and Miners. By W. H. MERRITT, M.N.E. Inst. M.E., A.R.S.M., &c. With Photographic Plates and other Illustrations. Fcap. 8vo, leather . . . . . *Net* **5/0**

"As an instructor of prospectors' classes Mr. Merritt has the advantage of knowing exactly the information likely to be most valuable to the miner in the field. The contents cover all the details of sampling and testing gold and silver ores. A useful addition to a prospector's kit."—*Mining Journal.*

THE PROSPECTOR'S HANDBOOK.

A Guide for the Prospector and Traveller in search of Metal-Bearing or other Valuable Minerals. By J. W. ANDERSON, M.A. (Camb.), F.R.G.S. Ninth Edition. Small crown 8vo, **3/6** cloth ; or, leather **4/6**

"Will supply a much-felt want, especially among Colonists, in whose way are so often thrown many mineralogical specimens the value of which it is difficult to determine."—*Engineer.*

"How to find commercial minerals, and how to identify them when they are found, are the leading points to which attention is directed. The author has managed to pack as much practical detail into his pages as would supply material for a book three times its size."—*Mining Journal.*

THE METALLURGY OF SILVER.

A Practical Treatise on the Amalgamation, Roasting, and Lixiviation of Silver Ores. Including the Assaying, Melting, and Refining of Silver Bullion. By M. EISSLER, M. Inst. M.M. Third Edition. Crown 8vo, cloth . **10/6**

"A practical treatise, and a technical work which we are convinced will supply a long-felt want amongst practical men, and at the same time be of value to students and others indirectly connected with the industries."—*Mining Journal.*

THE HYDRO-METALLURGY OF COPPER.

Being an Account of Processes Adopted in the Hydro-Metallurgical Treatment of Cupriferous Ores, Including the Manufacture of Copper Vitriol, with Chapters on the Sources of Supply of Copper and the Roasting of Copper Ores. By M. EISSLER, M. Inst. M.M. 8vo, cloth . . . . *Net* **12/9**

"In this volume the various processes for the extraction of copper by wet methods are fully detailed. Costs are given when available, and a great deal of useful information about the copper industry of the world is presented in an interesting and attractive manner."—*Mining Journal.*

THE METALLURGY OF ARGENTIFEROUS LEAD.

A Practical Treatise on the Smelting of Silver-Lead Ores and the Refining of Lead Bullion. Including Reports on various Smelting Establishments and Descriptions of Modern Smelting Furnaces and Plants in Europe and America. By M. EISSLER, M. Inst. M.M., Author of "The Metallurgy of Gold," &c. Crown 8vo, 400 pp., with 183 Illustrations, cloth **12/6**

"The numerous metallurgical processes, which are fully and extensively treated of, embrace all the stages experienced in the passage of the lead from the various natural states to its issue from the refinery as an article of commerce."—*Practical Engineer.*

METALLIFEROUS MINERALS AND MINING.

By D. C. DAVIES, F.G.S. Sixth Edition, thoroughly Revised and much Enlarged by his Son, E. HENRY DAVIES, M.E., F.G.S. 600 pp., with 173 Illustrations. Large crown 8vo, cloth . . . . . *Net* **12/6**

"Neither the practical miner nor the general reader, interested in mines, can have a better book for his companion and his guide."—*Mining Journal.*

EARTHY AND OTHER MINERALS AND MINING.

By D. C. DAVIES, F.G.S., Author of "Metalliferous Minerals," &c. Third Edition, Revised and Enlarged by his Son, E. HENRY DAVIES, M.E., F.G.S. With about 100 Illustrations. Crown 8vo, cloth **12/6**

"We do not remember to have met with any English work on mining matters that contains the same amount of information packed in equally convenient form."—*Academy.*

BRITISH MINING.

A Treatise on the History, Discovery, Practical Development, and Future Prospects of Metalliferous Mines in the United Kingdom. By ROBERT HUNT, F.R.S., late Keeper of Mining Records. Upwards of 950 pp., with 230 Illustrations. Second Edition, Revised. Super-royal 8vo, cloth **£2 2s.**

POCKET-BOOK FOR MINERS AND METALLURGISTS.

Comprising Rules, Formulæ, Tables, and Notes for Use in Field and Office Work. By F. DANVERS POWER, F.G.S., M.E. Second Edition, Corrected. Fcap. 8vo, leather **9/0**

"This excellent book is an admirable example of its kind, and ought to find a large sale amongst English-speaking prospectors and mining engineers."—*Engineering.*

THE MINER'S HANDBOOK.

A Handy Book of Reference on the subjects of Mineral Deposits, Mining Operations, Ore Dressing, &c. For the Use of Students and others interested in Mining Matters. Compiled by JOHN MILNE, F.R.S., Professor of Mining in the Imperial University of Japan. Third Edition. Fcap. 8vo, leather **7/6**

"Professor Milne's handbook is sure to be received with favour by all connected with mining, and will be extremely popular among students."—*Athenæum.*

IRON ORES of GREAT BRITAIN and IRELAND.

Their Mode of Occurrence, Age and Origin, and the Methods of Searching for and Working Them. With a Notice of some of the Iron Ores of Spain. By J. D. KENDALL, F.G.S., Mining Engineer. Crown 8vo, cloth . . **16/0**

MINE DRAINAGE.

A Complete Practical Treatise on Direct-Acting Underground Steam Pumping Machinery. By STEPHEN MICHELL. Second Edition, Re-written and Enlarged. With 250 Illustrations. Royal 8vo, cloth . *Net* **25/0**

HORIZONTAL PUMPING ENGINES.—ROTARY AND NON-ROTARY HORIZONTAL ENGINES.—SIMPLE AND COMPOUND STEAM PUMPS.—VERTICAL PUMPING ENGINES.—ROTARY AND NON-ROTARY VERTICAL ENGINES.—SIMPLE AND COMPOUND STEAM PUMPS. — TRIPLE-EXPANSION STEAM PUMPS. — PULSATING STEAM PUMPS. — PUMP VALVES.—SINKING PUMPS, &c., &c.

"This volume contains an immense amount of important and interesting new matter. The book should undoubtedly prove of great use to all who wish for information on the subject."—*The Engineer.*

ELECTRICITY AS APPLIED TO MINING.

By ARNOLD LUPTON, M.Inst.C.E., M.I.M.E., M.I.E.E., late Professor of Coal Mining at the Yorkshire College, Victoria University, Mining Engineer and Colliery Manager; G. D. ASPINALL PARR, M.I.E.E., A.M.I.M.E., Associate of the Central Technical College, City and Guilds of London, Head of the Electrical Engineering Department, Yorkshire College, Victoria University; and HERBERT PERKIN, M.I.M.E., Certified Colliery Manager, Assistant Lecturer in the Mining Department of the Yorkshire College, Victoria University. With about 170 Illustrations. Medium 8vo, cloth.

Net **9/0**

(For SUMMARY OF CONTENTS, see page 23.)

THE COLLIERY MANAGER'S HANDBOOK.

A Comprehensive Treatise on the Laying-out and Working of Collieries, Designed as a Book of Reference for Colliery Managers, and for the Use of Coal Mining Students preparing for First-class Certificates. By CALEB PAMELY, Mining Engineer and Surveyor; Member of the North of England Institute of Mining and Mechanical Engineers; and Member of the South Wales Institute of Mining Engineers. With 700 Plans, Diagrams, and other Illustrations. Fourth Edition, Revised and Enlarged. 964 pp. Medium 8vo, cloth **£1 5s.**

GEOLOGY.—SEARCH FOR COAL.—MINERAL LEASES AND OTHER HOLDINGS.—SHAFT SINKING.—FITTING UP THE SHAFT AND SURFACE ARRANGEMENTS.—STEAM BOILERS AND THEIR FITTINGS.—TIMBERING AND WALLING.—NARROW WORK AND METHODS OF WORKING. — UNDERGROUND CONVEYANCE. — DRAINAGE.—THE GASES MET WITH IN MINES; VENTILATION. — ON THE FRICTION OF AIR IN MINES. — THE PRIESTMAN OIL ENGINE; PETROLEUM AND NATURAL GAS. — SURVEYING AND PLANNING.—SAFETY LAMPS AND FIREDAMP DETECTORS.—SUNDRY AND INCIDENTAL OPERATIONS AND APPLIANCES.—COLLIERY EXPLOSIONS.—MISCELLANEOUS QUESTIONS AND ANSWERS.—*Appendix:* SUMMARY OF REPORT OF H.M. COMMISSIONERS ON ACCIDENTS IN MINES.

"Mr. Pamely's work is eminently suited to the purpose or which it is intended, being clear, interesting, exhaustive, rich in detail, and up to date, giving descriptions of the latest machines in every department. A mining engineer could scarcely go wrong who followed this work."—*Colliery Guardian.*

"Mr. Pamely has not only given us a comprehensive reference book of a very high order suitable to the requirements of mining engineers and colliery managers, but has also provided mining students with a class-book that is as interesting as it is instructive."—*Colliery Manager.*

"This is the most complete 'all-round work on coal-mining published in the English language. . . . No library of coal-mining books is complete without it."—*Colliery Engineer* (Scranton, Pa., U.S.A.).

COLLIERY WORKING AND MANAGEMENT.

Comprising the Duties of a Colliery Manager, the Oversight and Arrangement of Labour and Wages, and the different Systems of Working Coal
Seams. By H. F. BULMAN and R. A. S. REDMAYNE. 350 pp., with
28 Plates and other Illustrations, including Underground Photographs.
Medium 8vo, cloth.   **15/0**

" This is, indeed, an admirable Handbook for Colliery Managers, in fact it is an indispensable
adjunct to a Colliery Manager's education, as well as being a most useful and interesting work
on the subject for all who in any way have to do with coal mining. The underground photographs
are an attractive feature of the work, being very lifelike and necessarily true representations of the
scenes they depict."—*Colliery Guardian.*

" Mr. Bulman and Mr. Redmayne, who are both experienced Colliery Managers of great
literary ability, are to be congratulated on having supplied an authoritative work dealing with a side
of the subject of coal mining which has hitherto received but scant treatment. The authors
elucidate their text by 119 woodcuts and 28 plates, most of the latter being admirable reproductions
of photographs taken underground with the aid of the magnesium flash-light. These illustrations
are excellent."—*Nature.*

COAL AND COAL MINING.

By the late Sir WARINGTON W. SMYTH, M.A., F.R.S., Chief Inspector of the
Mines of the Crown and of the Duchy of Cornwall. Eighth Edition, Revised
and Extended by T. FORSTER BROWN, Mining and Civil Engineer, Chief
Inspector of the Mines of the Crown and of the Duchy of Cornwall. Crown
8vo, cloth.   **3/6**

" As an outline is given of every known coal-field in this and other countries, as well as of the
principal methods of working, the book will doubtless interest a very large number of readers."—
Mining Journal.

NOTES AND FORMULÆ FOR MINING STUDENTS.

By JOHN HERMAN MERIVALE, M.A., Late Professor of Mining in the Durham
College of Science, Newcastle-upon-Tyne. Fourth Edition, Revised and
Enlarged. By H. F. BULMAN, A.M.Inst.C.E. Small crown 8vo, cloth. **2/6**

" The author has done his work in a creditable manner, and has produced a book that will
be of service to students and those who are practically engaged in mining operations."—*Engineer.*

INFLAMMABLE GAS AND VAPOUR IN THE AIR

(The Detection and Measurement of). By FRANK CLOWES, D.Sc., Lond.,
F.I.C. With a Chapter on THE DETECTION AND MEASUREMENT OF PETRO-
LEUM VAPOUR by BOVERTON REDWOOD, F.R.S.E., Consulting Adviser to the
Corporation of London under the Petroleum Acts. Crown 8vo, cloth. *Net* **5/0**

" Professor Clowes has given us a volume on a subject of much industrial importance.
Those interested in these matters may be recommended to study this book, which is easy of compre-
hension and contains many good things."—*The Engineer.*

COAL & IRON INDUSTRIES of the UNITED KINGDOM.

Comprising a Description of the Coal Fields, and of the Principal Seams of
Coal, with Returns of their Produce and its Distribution, and Analyses of
Special Varieties. Also, an Account of the Occurrence of Iron Ores in Veins or
Seams ; Analyses of each Variety ; and a History of the Rise and Progress of
Pig Iron Manufacture. By RICHARD MEADE. 8vo, cloth . . **£1 8s.**

" Of this book we may unreservedly say that it is the best of its class which we have ever
met. . . . A book of reference which no one engaged in the iron or coal trades should omit from
his library."—*Iron and Coal Trades Review.*

ASBESTOS AND ASBESTIC.

Their Properties, Occurrence, and Use. By ROBERT H. JONES, F.S.A.,
Mineralogist, Hon. Mem. Asbestos Club, Black Lake, Canada. With
Ten Collotype Plates and other Illustrations. Demy 8vo, cloth.   . **16/0**

" An interesting and invaluable work."—*Colliery Guardian.*

GRANITES AND OUR GRANITE INDUSTRIES.

By GEORGE F. HARRIS, F.G.S. With Illustrations. Crown 8vo, cloth **2/6**

TRAVERSE TABLES.

For use in Mine Surveying.   By WILLIAM LINTERN, C.E. With two plates.
Small crown 8vo, cloth   .   .   .   . *Net* **3/0**

ELECTRICITY, ELECTRICAL ENGINEERING, ETC.

THE ELEMENTS OF ELECTRICAL ENGINEERING.

A First Year's Course for Students. By TYSON SEWELL, A.I.E.E., Assistant Lecturer and Demonstrator in Electrical Engineering at the Polytechnic, Regent Street, London. Second Edition, Revised, with Additional Chapters on Alternating Current Working, and Appendix of Questions and Answers. 450 pages, with 274 Illustrations. Demy 8vo, cloth. [*Just Published.* Net **7/6**

OHM'S LAW.—UNITS EMPLOYED IN ELECTRICAL ENGINEERING.—SERIES AND PARALLEL CIRCUITS; CURRENT DENSITY AND POTENTIAL DROP IN THE CIRCUIT.—THE HEATING EFFECT OF THE ELECTRIC CURRENT.—THE MAGNETIC EFFECT OF AN ELECTRIC CURRENT.—THE MAGNETISATION OF IRON.—ELECTRO-CHEMISTRY; PRIMARY BATTERIES.—ACCUMULATORS.—INDICATING INSTRUMENTS; AMMETERS, VOLTMETERS, OHMMETERS.—ELECTRICITY SUPPLY METERS.—MEASURING INSTRUMENTS, AND THE MEASUREMENT OF ELECTRICAL RESISTANCE. — MEASUREMENT OF POTENTIAL DIFFERENCE, CAPACITY, CURRENT STRENGTH, AND PERMEABILITY.—ARC LAMPS.—INCANDESCENT LAMPS; MANUFACTURE AND INSTALLATION; PHOTOMETRY. — THE CONTINUOUS CURRENT DYNAMO.—DIRECT CURRENT MOTORS.—ALTERNATING CURRENTS. —TRANSFORMERS, ALTERNATORS, SYNCHRONOUS MOTORS.—POLYPHASE WORKING.—APPENDIX OF QUESTIONS AND ANSWERS.

"An excellent treatise for students of the elementary facts connected with electrical engineering."—*The Electrician.*

"One of the best books for those commencing the study of electrical engineering. Everything is explained in simple language which even a beginner cannot fail to understand."—*Engineer.*

"One welcomes this book, which is sound in its treatment, and admirably calculated to give students the knowledge and information they most require."—*Nature.*

CONDUCTORS FOR ELECTRICAL DISTRIBUTION.

Their Materials and Manufacture, The Calculation of Circuits, Pole-Line Construction, Underground Working, and other Uses. By F. A. C. PERRINE, A.M., D.Sc.; formerly Professor of Electrical Engineering, Leland Stanford, Jr., University; M.Amer.I.E.E. 8vo, cloth. [*Just Published.* Net **20/-**

CONDUCTOR MATERIALS—ALLOYED CONDUCTORS—MANUFACTURE OF WIRE—WIRE-FINISHING—WIRE INSULATION—CABLES—CALCULATION OF CIRCUITS—KELVIN'S LAW OF ECONOMY IN CONDUCTORS—MULTIPLE ARC DISTRIBUTION—ALTERNATING CURRENT CALCULATION—OVERHEAD LINES—POLE LINE—LINE INSULATORS—UNDERGROUND CONDUCTORS.

WIRELESS TELEGRAPHY;

Its Origins, Development, Inventions, and Apparatus. By CHARLES HENRY SEWALL. With 85 Diagrams and Illustrations. Demy 8vo, cloth.
[*Just Published.* Net **10/6**

ARMATURE WINDINGS OF DIRECT CURRENT DYNAMOS.

Extension and Application of a General Winding Rule. By E. ARNOLD, Engineer. Assistant Professor in Electrotechnics and Machine Design at the Riga Polytechnic School Translated from the Original German by FRANCIS B. DE GRESS, M.E, Chief of Testing Department, Crocker-Wheeler Company. With 146 Illustrations. Medium 8vo, cloth . . . . *Net* **12/-**

ELECTRICITY AS APPLIED TO MINING.

By ARNOLD LUPTON, M.Inst C.E., M.I M.E., M.I.E E., late Professor o Coal Mining at the Yorkshire College, Victoria University, Mining Engineer and Colliery Manager; G. D. ASPINALL PARR, M.I.E.E., A M.I.M.E., Associate of the Central Technical College, City and Guilds of London, Head of the Electrical Engineering Department, Yorkshire College, Victoria University; and HERBERT PERKIN, M I.M E. Certificated Colliery Manager, Assistant Lecturer in the Mining Department of the Yorkshire College, Victoria University. With about 170 Illustrations. Medium 8vo, cloth *Net* **9/-**

INTRODUCTORY. — DYNAMIC ELECTRICITY. — DRIVING OF THE DYNAMO. — THE STEAM TURBINE.—DISTRIBUTION OF ELECTRICAL ENERGY.—STARTING AND STOPPING ELECTRICAL GENERATORS AND MOTORS.—ELECTRIC CABLES.—CENTRAL ELECTRICAL PLANTS.—ELECTRICITY APPLIED TO PUMPING AND HAULING.—ELECTRICITY APPLIED TO COAL-CUTTING.—TYPICAL ELECTRIC PLANTS RECENTLY ERECTED. — ELECTRIC LIGHTING BY ARC AND GLOW LAMPS—MISCELLANEOUS APPLICATIONS OF ELECTRICITY —ELECTRICITY AS COMPARED WITH OTHER MODES OF TRANSMITTING POWER.—DANGERS OF ELECTRICITY.

DYNAMO, MOTOR AND SWITCHBOARD CIRCUITS
FOR ELECTRICAL ENGINEERS.

A Practical Handbook dealing with Direct, Alternating and Polyphase Currents. By WILLIAM R. BOWKER, C.E., M.E., E.E., Lecturer on Physics and Electrical Engineering at the Municipal Technical School, Bury. 8vo, cloth. *[Just ready, price about* **6 0** *net.*

DYNAMO ELECTRIC MACHINERY: its CONSTRUC-
TION, DESIGN, and OPERATION.

By SAMUEL SHELDON, A.M., Ph.D, Professor of Physics and Electrical Engineering at the Polytechnic Institute of Brooklyn, assisted by HOBART MASON, B.S.

In two volumes, sold separately, as follows:—

Vol. I.—DIRECT CURRENT MACHINES. Third Edition, Revised. Large crown 8vo. 280 pages, with 200 Illustrations . . *Net* **12/0**

Vol. II.—ALTERNATING CURRENT MACHINES. Large crown 8vo. 260 pages, with 184 Illustrations . . . . . . *Net* **12/0**

Designed as Text-books for use in Technical Educational Institutions, and by Engineers whose work includes the handling of Direct and Alternating Current Machines respectively, and for Students proficient in mathematics.

ELECTRICAL AND MAGNETIC CALCULATIONS.

For the Use of Electrical Engineers and Artisans, Teachers, Students, and all others interested in the Theory and Application of Electricity and Magnetism. By A. A. ATKINSON, Professor of Electricity in Ohio University. Crown 8vo, cloth . . . . *Net* **9/0**
"To teachers and those who already possess a fair knowledge of their subject we can recommend this book as being useful to consult when requiring data or formulæ which it is neither convenient nor necessary to retain by memory."—*The Electrician.*

SUBMARINE TELEGRAPHS.

Their History, Construction, and Working. Founded in part on WÜNSCHEN-DORFF's " Traité de Télégraphie Sous-Marine," and Compiled from Authoritative and Exclusive Sources. By CHARLES BRIGHT, F.R.S.E., A.M.Inst.C.E., M.I.E.E. 780 pp., fully Illustrated, including Maps and Folding Plates. Royal 8vo, cloth . . . . *Net* **£3 3s.**
"There are few, if any, persons more fitted to write a treatise on submarine telegraphy than Mr. Charles Bright. He has done his work admirably, and has written in a way which will appeal as much to the layman as to the engineer. This admirable volume must, for many years to come, hold the position of the English classic on submarine telegraphy."—*Engineer.*
"This book is full of information. It makes a book of reference which should be in every engineer's library."—*Nature.*
"Mr. Bright's interestingly written and admirably illustrated book will meet with a welcome reception from cable men."—*Electrician.*
"The author deals with his subject from all points of view—political and strategical as well as scientific. The work will be of interest, not only to men of science, but to the general public. We can strongly recommend it."—*Athenæum.*

THE ELECTRICAL ENGINEER'S POCKET-BOOK.

Consisting of Modern Rules, Formulæ, Tables, and Data. By H. R. KEMPE, M.I.E.E., A.M.Inst.C.E., Technical Officer Postal Telegraphs, Author of "A Handbook of Electrical Testing," &c. Second Edition, thoroughly Revised, with Additions. With numerous Illustrations. Royal 32mo, oblong, leather **5/0**
"It is the best book of its kind."—*Electrical Engineer.*
"The Electrical Engineer's Pocket-Book is a good one."—*Electrician.*
"Strongly recommended to those engaged in the electrical industries."—*Electrical Review.*

POWER TRANSMITTED BY ELECTRICITY.

And applied by the Electric Motor, including Electric Railway Construction. By P. ATKINSON, A.M., Ph.D. Third Edition, Fully Revised, and New Matter added. With 94 Illustrations. Crown 8vo, cloth . . *Net* **9/0**

DYNAMIC ELECTRICITY AND MAGNETISM.

By PHILIP ATKINSON, A.M., Ph.D., Author of "Elements of Static Electricity," &c. Crown 8vo, 417 pp., with 120 Illustrations, cloth . **10/6**

THE MANAGEMENT OF DYNAMOS.

A Handybook of Theory and Practice for the Use of Mechanics, Engineers, Students, and others in Charge of Dynamos. By G. W. LUMMIS-PATERSON. Third Edition, Revised. Crown 8vo, cloth **4/6**
" An example which deserves to be taken as a model by other authors. The subject is treated in a manner which any intelligent man who is fit to be entrusted with charge of an engine should be able to understand. It is a useful book to all who make, tend, or employ electric machinery."
—*Architect.*

THE STANDARD ELECTRICAL DICTIONARY.

A Popular Encyclopædia of Words and Terms Used in the Practice of Electrical Engineering. Containing upwards of 3,000 definitions, By T. O'CONOR SLOANE, A.M., Ph.D. Third Edition, with Appendix. Crown 8vo, 690 pp., 390 Illustrations, cloth . . . . *Net* **7/6**
" The work has many attractive features in it, and is, beyond doubt, a well put together and useful publication. The amount of ground covered may be gathered from the fact that in the index about 5,000 references will be found."—*Electrical Review.*

ELECTRIC LIGHT FITTING.

A Handbook for Working Electrical Engineers, embodying Practical Notes on Installation Management. By J. W. URQUHART, Electrician, Author of " Electric Light," &c. With numerous Illustrations. Third Edition, Revised, with Additions. Crown 8vo, cloth **5/0**
" This volume deals with the mechanics of electric lighting, and is addressed to men who are already engaged in the work, or are training for it. The work traverses a great deal of ground, and may be read as a sequel to the author's useful work on ' Electric Light.' "—*Electrician.*
" The book is well worth the perusal of the workman, for whom it is written."—*Electrical Review.*

ELECTRIC LIGHT.

Its Production and Use, Embodying Plain Directions for the Treatment of Dynamo-Electric Machines, Batteries, Accumulators, and Electric Lamps. By J. W. URQUHART, C.E. Sixth Edition, Enlarged. Crown 8vo, cloth.
7/6
" The whole ground of electric lighting is more or less covered and explained in a very clear and concise manner."—*Electrical Review.*
" A *vade-mecum* of the salient facts connected with the science of electric lighting."—*Electrician.*

DYNAMO CONSTRUCTION.

A Practical Handbook for the Use of Engineer-Constructors and Electricians-in-Charge. Embracing Framework Building, Field Magnet and Armature Winding and Grouping, Compounding, &c. By J. W. URQUHART. Second Edition, Enlarged, with 114 Illustrations. Crown 8vo, cloth . . **7/6**
" Mr. Urquhart's book is the first one which deals with these matters in such a way that the engineering student can understand them. The book is very readable, and the author leads his readers up to difficult subjects by reasonably simple tests."—*Engineering Review.*

ELECTRIC SHIP-LIGHTING.

A Handbook on the Practical Fitting and Running of Ships' Electrical Plant. For the Use of Shipowners and Builders, Marine Electricians, and Seagoing Engineers-in-Charge. By J. W. URQUHART, C.E. Second Edition, Revised and Extended. With 88 Illustrations, Crown 8vo, cloth . . . **7/6**
" The subject of ship electric lighting is one of vast importance and Mr. Urquhart is to be highly complimented for placing such a valuable work at the service of marine electricians."—*The Steamship.*

ELECTRIC LIGHTING (ELEMENTARY PRINCIPLES OF).

By ALAN A. CAMPBELL SWINTON, M.Inst.C.E., M.I.E.E. Fifth Edition. With 16 Illustrations. Crown 8vo, cloth **1/6**

ELECTRIC LIGHT FOR COUNTRY HOUSES.

A Practical Handbook on the Erection and Running of Small Installations, with Particulars of the Cost of Plant and Working. By J. H. KNIGHT. Third Edition, Revised. Crown 8vo, wrapper **1/0**

HOW TO MAKE A DYNAMO.

A Practical Treatise for Amateurs. Containing Illustrations and Detailed Instructions for Constructing a Small Dynamo to Produce the Electric Light. By ALFRED CROFTS. Sixth Edition, Revised. Crown 8vo, cloth . **2/0**

THE STUDENT'S TEXT-BOOK OF ELECTRICITY.

By H. M. NOAD, F.R.S. 650 pp., with 470 Illustrations. Crown 8vo, cloth.
9/0

ARCHITECTURE, BUILDING, ETC.

PRACTICAL BUILDING CONSTRUCTION.

A Handbook for Students Preparing for Examinations, and a Book of Reference for Persons Engaged in Building. By JOHN PARNELL ALLEN, Surveyor, Lecturer on Building Construction at the Durham College of Science, Newcastle-on-Tyne. Fourth Edition, Revised and Enlarged. Medium 8vo, 570 pp., with 1,000 Illustrations, cloth.

[*Just Published. Net* **7/6**

"The most complete exposition of building construction we have seen. It contains all that is necessary to prepare students for the various examinations in building construction."—*Building News.*

"The author depends nearly as much on his diagrams as on his type. The pages suggest the hand of a man of experience in building operations—and the volume must be a blessing to many teachers as well as to students."—*The Architect.*

"The work is sure to prove a formidable rival to great and small competitors alike, and bids fair to take a permanent place as a favourite student's text-book. The large number of illustrations deserve particular mention for the great merit they possess for purposes of reference in exactly corresponding to convenient scales."—*Journal of the Royal Institute of British Architects.*

PRACTICAL MASONRY.

A Guide to the Art of Stone Cutting. Comprising the Construction, Setting Out, and Working of Stairs, Circular Work, Arches, Niches, Domes, Pendentives, Vaults, Tracery Windows, &c.; to which are added Supplements relating to Masonry Estimating and Quantity Surveying, and to Building Stones, and a Glossary of Terms. For the Use of Students, Masons, and other Workmen. By WILLIAM R. PURCHASE, Building Inspector to the Borough of Hove. Fourth Edition, Enlarged. Royal 8vo, 210 pp., with 52 Lithographic Plates, comprising over 400 Diagrams, cloth.

[*Just Published. Net* **7/6**

"Mr. Purchase's 'Practical Masonry' will undoubtedly be found useful to all interested in this important subject, whether theoretically or practically. Most of the examples given are from actual work carried out, the diagrams being carefully drawn. The book is a practical treatise on the subject, the author himself having commenced as an operative mason, and afterwards acted as foreman mason on many large and important buildings prior to the attainment of his present position. It should be found of general utility to architectural students and others, as well as to those to whom it is specially addressed."—*Journal of the Royal Institute of British Architects.*

MODERN PLUMBING, STEAM AND HOT WATER HEATING.

A New Practical Work for the Plumber, the Heating Engineer, the Architect, and the Builder. By J. J. LAWLER, Author of "American Sanitary Plumbing," &c. With 284 Illustrations and Folding Plates. 4to, cloth   .   *Net* **21/-**

HEATING BY HOT WATER,

VENTILATION AND HOT WATER SUPPLY.

By WALTER JONES, M.I.M.E. 340 pages, with 140 Illustrations. Royal 8vo, cloth.   [*Just Published. Net* **6/0**

CONCRETE : ITS NATURE AND USES.

A Book for Architects, Builders, Contractors, and Clerks of Works. By GEORGE L. SUTCLIFFE, A.R.I.B.A. 350 pp., with Illustrations. Crown 8vo, cloth **7/6**

"The author treats a difficult subject in a lucid manner. The manual fills a long-felt gap. It is careful and exhaustive; equally useful as a student's guide and an architect's book of reference."—*Journal of the Royal Institute of British Architects.*

LOCKWOOD'S BUILDER'S PRICE BOOK for 1904.

A Comprehensive Handbook of the Latest Prices and Data for Builders, Architects, Engineers, and Contractors. Re-constructed, Re-written, and Greatly Enlarged. By FRANCIS T. W. MILLER. 800 closely-printed pages, crown 8vo, cloth **4/0**

"This book is a very useful one, and should find a place in every English office connected with the building and engineering professions."—*Industries.*

"An excellent book of reference."—*Architect.*

"In its new and revised form this Price Book is what a work of this kind should be—comprehensive, reliable, well arranged, legible, and well bound."—*British Architect.*

DECORATIVE PART OF CIVIL ARCHITECTURE.

By Sir WILLIAM CHAMBERS, F.R.S. With Portrait, Illustrations, Notes, and an EXAMINATION OF GRECIAN ARCHITECTURE, by JOSEPH GWILT, F.S.A. Revised and Edited by W. H. LEEDS. 66 Plates, 4to, cloth   .   .   **21/0**

THE MECHANICS OF ARCHITECTURE.

A Treatise on Applied Mechanics, especially Adapted to the Use of Architects. By E. W. TARN, M.A., Author of "The Science of Building," &c. Second Edition, Enlarged. Illustrated with 125 Diagrams. Crown 8vo, cloth **7/6**

"The book is a very useful and helpful manual of architectural mechanics."—*Builder.*

A HANDY BOOK OF VILLA ARCHITECTURE.

Being a Series of Designs for Villa Residences in various Styles. With Outline Specifications and Estimates. By C. WICKES, Architect, Author of "The Spires and Towers of England," &c. 61 Plates, 4to, half-morocco, gilt edges . . . . **£1 11s. 6d.**

"The whole of the designs bear evidence of their being the work of an artistic architect, and they will prove very valuable and suggestive."—*Building News.*

THE ARCHITECT'S GUIDE.

Being a Text-book of Useful Information for Architects, Engineers, Surveyors, Contractors, Clerks of Works, &c., &c. By F. ROGERS. Crown 8vo, cloth. **3/6**

ARCHITECTURAL PERSPECTIVE.

The whole Course and Operations of the Draughtsman in Drawing a Large House in Linear Perspective. Illustrated by 43 Folding Plates. By F. O. FERGUSON. Third Edition. 8vo, boards **3/6**

"It is the most intelligible of the treatises on this ill-treated subject that I have met with."— E. INGRESS BELL, ESQ., in the *R.I.B.A. Journal.*

PRACTICAL RULES ON DRAWING.

For the Operative Builder and Young Student in Architecture. By GEORGE PYNE. 14 Plates, 4to, boards **7/6**

MEASURING AND VALUING ARTIFICERS' WORK

(The Student's Guide to the Practice of). Containing Directions for taking Dimensions, Abstracting the same, and bringing the Quantities into Bill, with Tables of Constants for Valuation of Labour, and for the Calculation of Areas and Solidities. Originally edited by E. DOBSON, Architect. With Additions by E. W. TARN, M.A. Seventh Edition, Revised. With 8 Plates and 63 Woodcuts. Crown 8vo, cloth. **7/6**

"This edition will be found the most complete treatise on the principles of measuring and valuing artificers' work that has yet been published."—*Building News.*

TECHNICAL GUIDE, MEASURER, AND ESTIMATOR.

For Builders and Surveyors. Containing Technical Directions for Measuring Work in all the Building Trades, Complete Specifications for Houses, Roads, and Drains, and an Easy Method of Estimating the parts of a Building collectively. By A. C. BEATON. Ninth Edition. Waistcoat-pocket size, gilt edges **1/6**

"No builder, architect, surveyor, or valuer should be without his ' Beaton.'"—*Building News.*

SPECIFICATIONS FOR PRACTICAL ARCHITECTURE.

A Guide to the Architect, Engineer, Surveyor, and Builder. With an Essay on the Structure and Science of Modern Buildings. Upon the Basis of the Work by ALFRED BARTHOLOMEW, thoroughly Revised, Corrected, and greatly added to by FREDERICK ROGERS, Architect. Third Edition, Revised. 8vo, cloth **15/0**

"The work is too well known to need any recommendation from us. It is one of the books with which every young architect must be equipped."—*Architect.*

THE HOUSE-OWNER'S ESTIMATOR.

Or, What will it Cost to Build, Alter, or Repair? A Price Book for Unprofessional People as well as the Architectural Surveyor and Builder. By J. D. SIMON. Edited by F. T. W. MILLER, A.R.I.B.A. Fifth Edition. Carefully Revised. Crown 8vo, cloth. . . . . . . *Net* **3/6**

"In two years it will repay its cost a hundred times over."—*Field.*

SANITATION AND WATER SUPPLY.

THE HEALTH OFFICER'S POCKET-BOOK.

A Guide to Sanitary Practice and Law. For Medical Officers of Health, Sanitary Inspectors, Members of Sanitary Authorities, &c. By EDWARD F. WILLOUGHBY, M.D. (Lond.), &c. Second Edition, Revised and Enlarged. Fcap. 8vo, leather . . . . *Net* **10/6**

"It is a mine of condensed information of a pertinent and useful kind on the various subjects of which it treats. The different subjects are succinctly but fully and scientifically dealt with."—*The Lancet.*

"We recommend all those engaged in practical sanitary work to furnish themselves with a copy for reference."—*Sanitary Journal.*

THE BACTERIAL PURIFICATION OF SEWAGE:

Being a Practical Account of the Various Modern Biological Methods of Purifying Sewage. By SIDNEY BARWISE, M.D. (Lond.), D.P.H. (Camb.), etc. With 10 Page Plates and 2 Folding Diagrams. Royal 8vo, cloth.

Net **6/0**

THE PURIFICATION OF SEWAGE.

Being a Brief Account of the Scientific Principles of Sewage Purification, and their Practical Application. By SIDNEY BARWISE, M.D. (Lond.), M.R.C.S., D.P.H. (Camb.), Fellow of the Sanitary Institute, Medical Officer of Health to the Derbyshire County Council. Crown 8vo, cloth **5/0**

WATER AND ITS PURIFICATION.

A Handbook for the Use of Local Authorities, Sanitary Officers, and others interested in Water Supply. By S. RIDEAL, D.Sc. Lond., F.I.C. Second Edition, Revised, with Additions, including numerous Illustrations and Tables. Large Crown 8vo, cloth . . . . *Net* **9/0**

RURAL WATER SUPPLY.

A Practical Handbook on the Supply of Water and Construction of Water-works for Small Country Districts. By ALLAN GREENWELL, A.M.I.C.E., and W. T. CURRY, A.M.I.C.E. Revised Edition. Crown 8vo, cloth **5/0**

THE WATER SUPPLY OF CITIES AND TOWNS.

By WILLIAM HUMBER, A.M. Inst. C.E., and M.Inst. M.E. Imp. 4to, half-bound morocco. (See page 11.) . . . . *Net* **£6 6s.**

THE WATER SUPPLY OF TOWNS AND THE CON-STRUCTION OF WATER-WORKS.

By PROFESSOR W. K. BURTON, A.M. Inst. C.E. Second Edition, Revised and Extended. Royal 8vo, cloth. (See page 10.) . . . . **£1 5s.**

WATER ENGINEERING.

A Practical Treatise on the Measurement, Storage, Conveyance, and Utilisation of Water for the Supply of Towns. By C. SLAGG, A.M. Inst. C.E. **7/6**

SANITARY WORK IN SMALL TOWNS AND VILLAGES.

By CHARLES SLAGG, A. M. Inst. C.E. Crown 8vo, cloth . . . **3/0**

PLUMBING.

A Text-book to the Practice of the Art or Craft of the Plumber. By W. P. BUCHAN. Ninth Edition, Enlarged, with 500 Illustrations. Crown 8vo, **3/6**

VENTILATION.

A Text-book to the Practice of the Art of Ventilating Buildings. By W. P. BUCHAN, R.P. Crown 8vo, cloth **3/6**

CARPENTRY, TIMBER, ETC.

THE ELEMENTARY PRINCIPLES OF CARPENTRY.

A Treatise on the Pressure and Equilibrium of Timber Framing, the Resistance of Timber, and the Construction of Floors, Arches, Bridges, Roofs, Uniting Iron and Stone with Timber, &c. To which is added an Essay on the Nature and Properties of Timber, &c., with Descriptions of the kinds of Wood used in Building; also numerous Tables of the Scantlings of Timber for different purposes, the Specific Gravities of Materials, &c. By THOMAS TREDGOLD, C.E. With an Appendix of Specimens of Various Roofs of Iron and Stone, Illustrated. Seventh Edition, thoroughly Revised and considerably Enlarged by E. WYNDHAM TARN, M.A., Author of "The Science of Building," &c. With 61 Plates, Portrait of the Author, and several Woodcuts. In One large Vol., 4to, cloth . . . . **£1 5s.**

"Ought to be in every architect's and every builder's library."—*Builder.*

"A work whose monumental excellence must commend it wherever skilful carpentry is concerned. The author's principles are rather confirmed than impaired by time. The additional plates are of great intrinsic value."—*Building News.*

WOODWORKING MACHINERY.

Its Rise, Progress, and Construction. With Hints on the Management of Saw Mills and the Economical Conversion of Timber. Illustrated with Examples of Recent Designs by leading English, French, and American Engineers. By M. POWIS BALE, A.M.Inst.C.E., M.I.M.E. Second Edition, Revised, with large Additions, large crown 8vo, 440 pp., cloth **9/0**

"Mr. Bale is evidently an expert on the subject, and he has collected so much information that his book is all-sufficient for builders and others engaged in the conversion of timber."—*Architect.*

"The most comprehensive compendium of wood-working machinery we have seen. The author is a thorough master of his subject."—*Building News.*

SAW MILLS.

Their Arrangement and Management, and the Economical Conversion of Timber. By M. POWIS BALE, A.M.Inst.C.E. Second Edition, Revised. Crown 8vo, cloth. **10/6**

"The *administration* of a large sawing establishment is discussed, and the subject examined from a financial standpoint. Hence the size, shape, order, and disposition of saw mills and the like are gone into in detail, and the course of the timber is traced from its reception to its delivery in its converted state. We could not desire a more complete or practical treatise."—*Builder.*

THE CARPENTER'S GUIDE.

Or, Book of Lines for Carpenters; comprising all the Elementary Principles essential for acquiring a knowledge of Carpentry. Founded on the late PETER NICHOLSON's standard work. A New Edition, Revised by ARTHUR ASHPITEL, F.S.A. Together with Practical Rules on Drawing, by GEORGE PYNE. With 74 Plates, 4to, cloth . . . . **£1 1s.**

A PRACTICAL TREATISE ON HANDRAILING.

Showing New and Simple Methods for Finding the Pitch of the Plank, Drawing the Moulds, Bevelling, Jointing-up, and Squaring the Wreath. By GEORGE COLLINGS. Revised and Enlarged, to which is added A TREATISE ON STAIR-BUILDING. Third Edition. With Plates and Diagrams. 12mo, cloth. **2/6**

"Will be found of practical utility in the execution of this difficult branch of joinery."—*Builder.*

"Almost every difficult phase of this somewhat intricate branch of joinery is elucidated by the aid of plates and explanatory letterpress."—*Furniture Gazette.*

CIRCULAR WORK IN CARPENTRY AND JOINERY.

A Practical Treatise on Circular Work of Single and Double Curvature. By GEORGE COLLINGS. With Diagrams. Fourth Edition, 12mo, cloth . **2/6**

"An excellent example of what a book of this kind should be. Cheap in price, clear in definition, and practical in the examples selected."—*Builder.*

THE CABINET-MAKER'S GUIDE TO THE ENTIRE CONSTRUCTION OF CABINET WORK.

By RICHARD BITMEAD. Illustrated with Plans, Sections and Working Drawings. Crown 8vo, cloth **2/6**

HANDRAILING COMPLETE IN EIGHT LESSONS.

On the Square-Cut System. By J. S. GOLDTHORP, Teacher of Geometry and Building Construction at the Halifax Mechanics' Institute. With Eight Plates and over 150 Practical Exercises. 4to, cloth   .   .   .   **3/6**

"Likely to be of considerable value to joiners and others who take a pride in good work. The arrangement of the book is excellent. We heartily commend it to teachers and students."— *Timber Trades Journal.*

TIMBER MERCHANT'S and BUILDER'S COMPANION.

Containing New and Copious Tables of the Reduced Weight and Measurement of Deals and Battens, of all sizes, and other Useful Tables for the use of Timber Merchants and Builders. By WILLIAM DOWSING. Fourth Edition, Revised and Corrected. Crown 8vo, cloth **3/0**

"We are glad to see a fourth edition of these admirable tables, which for correctness and simplicity of arrangement leave nothing to be desired."— *Timber Trades Journal.*

THE PRACTICAL TIMBER MERCHANT.

Being a Guide for the Use of Building Contractors, Surveyors, Builders, &c., comprising useful Tables for all purposes connected with the Timber Trade, Marks of Wood, Essay on the Strength of Timber, Remarks on the Growth of Timber, &c. By W. RICHARDSON. Second Edition. Fcap. 8vo, cloth   .   **3/6**

"This handy manual contains much valuable information for the use of timber merchants, builders, foresters, and all others connected with the growth, sale, and manufacture of timber."— *Journal of Forestry.*

PACKING-CASE TABLES.

Showing the number of Superficial Feet in Boxes or Packing-Cases, from six inches square and upwards. By W. RICHARDSON, Timber Broker. Third Edition. Oblong 4to, cloth **3/6**

"Invaluable labour-saving tables."— *Ironmonger.*
"Will save much labour and calculation."— *Grocer.*

GUIDE TO SUPERFICIAL MEASUREMENT.

Tables calculated from 1 to 200 inches in length by 1 to 108 inches in breadth. For the use of Architects, Surveyors, Engineers, Timber Merchants, Builders, &c. By JAMES HAWKINGS. Fifth Edition. Fcap., cloth . **3/6**

"These tables will be found of great assistance to all who require to make calculations of superficial measurement."— *English Mechanic.*

PRACTICAL FORESTRY.

And its Bearing on the Improvement of Estates. By CHARLES E. CURTIS, F.S.I., Professor of Forestry, Field Engineering, and General Estate Management, at the College of Agriculture, Downton. Second Edition, Revised. Crown 8vo, cloth **3/6**

PREFATORY REMARKS. — OBJECTS OF PLANTING. — CHOICE OF A FORESTER. — CHOICE OF SOIL AND SITE.—LAYING OUT OF LAND FOR PLANTATIONS.—PREPARATION OF THE GROUND FOR PLANTING.—DRAINAGE.—PLANTING.—DISTANCES AND DISTRIBUTION OF TREES IN PLANTATIONS.—TREES AND GROUND GAME.—ATTENTION AFTER PLANTING.—THINNING OF PLANTATIONS. — PRUNING OF FOREST TREES.—REALIZATION. —METHODS OF SALE.—MEASUREMENT OF TIMBER.—MEASUREMENT AND VALUATION OF LARCH PLANTATION.—FIRE LINES.—COST OF PLANTING.

"Mr. Curtis has in the course of a series of short pithy chapters afforded much information of a useful and practical character on the planting and subsequent treatment of trees."— *Illustrated Carpenter and Builder.*

THE ELEMENTS OF FORESTRY.

Designed to afford Information concerning the Planting and Care of Forest Trees for Ornament or Profit, with suggestions upon the Creation and Care of Woodlands. By F. B. HOUGH. Large crown 8vo, cloth   .   .   .   **10/0**

TIMBER IMPORTER'S, TIMBER MERCHANT'S, AND BUILDER'S STANDARD GUIDE.

By RICHARD E. GRANDY. Comprising:—An Analysis of Deal Standards, Home and Foreign, with Comparative Values and Tabular Arrangements for fixing Net Landed Cost on Baltic and North American Deals, including all intermediate Expenses, Freight, Insurance, &c.; together with copious Information for the Retailer and Builder. Third Edition, Revised. 12mo, cloth **2/0**

"Everything it pretends to be: built up gradually, it leads one from a forest to a treenail, and throws in, as a makeweight, a host of material concerning bricks, columns, cisterns, &c."— *English Mechanic.*

DECORATIVE ARTS, ETC.

SCHOOL OF PAINTING FOR THE IMITATION OF WOODS AND MARBLES.

As Taught and Practised by A. R. VAN DER BURG and P. VAN DER BURG, Directors of the Rotterdam Painting Institution. Royal folio, 18½ by 12½ in., Illustrated with 24 full-size Coloured Plates ; also 12 plain Plates, comprising 154 Figures. Fourth Edition cloth .   [*Just Published. Net* **£1 5s.**

LIST OF PLATES.

1. VARIOUS TOOLS REQUIRED FOR WOOD PAINTING.—2, 3. WALNUT; PRELIMINARY STAGES OF GRAINING AND FINISHED SPECIMEN. — 4. TOOLS USED FOR MARBLE PAINTING AND METHOD OF MANIPULATION.—5, 6. ST. REMI MARBLE; EARLIER OPERATIONS AND FINISHED SPECIMEN. — 7. METHODS OF SKETCHING DIFFERENT GRAINS, KNOTS, &c.—8, 9. ASH: PRELIMINARY STAGES AND FINISHED SPECI-MEN. — 10. METHODS OF SKETCHING MARBLE GRAINS. — 11, 12. BRECHE MARBLE; PRELIMINARY STAGES OF WORKING AND FINISHED SPECIMEN.—13. MAPLE ; METHODS OF PRODUCING THE DIFFERENT GRAINS.—14, 15. BIRD'S-EYE MAPLE; PRELIMINARY STAGES AND FINISHED SPECIMEN.—16. METHODS OF SKETCHING THE DIFFERENT SPECIES OF WHITE MARBLE.—17, 18. WHITE MARBLE ; PRELIMINARY STAGES OF PROCESS AND FINISHED SPECIMEN.—19. MAHOGANY; SPECIMENS OF VARIOUS GRAINS AND METHODS OF MANIPULATION.—20, 21. MAHOGANY ; EARLIER STAGES AND FINISHED SPECIMEN.—22, 23, 24. SIENNA MARBLE; VARIETIES OF GRAIN, PRELIMINARY STAGES AND FINISHED SPECIMEN.—25, 26, 27. JUNIPER WOOD; METHODS OF PRO-DUCING GRAIN, &c.; PRELIMINARY STAGES AND FINISHED SPECIMEN.—28, 29, 30. VERT DE MER MARBLE; VARIETIES OF GRAIN AND METHODS OF WORKING, UNFINISHED AND FINISHED SPECIMENS.—31, 32, 33. OAK ; VARIETIES OF GRAIN, TOOLS EMPLOYED AND METHODS OF MANIPULATION, PRELIMINARY STAGES AND FINISHED SPECIMEN.—34, 35, 36. WAULSORT MARBLE; VARIETIES OF GRAIN, UNFINISHED AND FINISHED SPECIMENS.

" Those who desire to attain skill in the art of painting woods and marbles will find advantage in consulting this book.   .   .   .   Some of the Working Men's Clubs should give their young men the opportunity to study it."—*Builder.*

" A comprehensive guide to the art. The explanations of the processes, the manipulation and management of the colours, and the beautifully executed plates will not be the least valuable to the student who aims at making his work a faithful transcript of nature."—*Building News.*

" Students and novices are fortunate who are able to become the possessors of so noble a work."—*The Architect.*

ELEMENTARY DECORATION.

A Guide to the Simpler Forms of Everyday Art. Together with PRACTICAL HOUSE DECORATION. By JAMES W. FACEY. With numerous Illus-trations. In One Vol., strongly half-bound **5/0**

HOUSE PAINTING, GRAINING, MARBLING, AND SIGN WRITING.

A Practical Manual of. By ELLIS A. DAVIDSON. Eighth Edition. With Coloured Plates and Wood Engravings. Crown 8vo, cloth .   .   . **6/0**

" A mass of information of use to the amateur and of value to the practical man."—*English Mechanic.*

THE DECORATOR'S ASSISTANT.

A Modern Guide for Decorative Artists and Amateurs, Painters, Writers, Gilders, &c. Containing upwards of 600 Receipts, Rules, and Instructions ; with a variety of Information for General Work connected with every Class of Interior and Exterior Decorations, &c. Eighth Edition. Cr. 8vo   . **1/0**

" Full of receipts of value to decorators, painters, gilders, &c. The book contains the gist of larger treatises on colour and technical processes. It would be difficult to meet with a work so full of varied information on the painter's art."—*Building News.*

MARBLE DECORATION

And the Terminology of British and Foreign Marbles. A Handbook for Students. By GEORGE H. BLAGROVE, Author of " Shoring and its Applica-tion," &c. With 28 Illustrations. Crown 8vo, cloth **3/6**

" This most useful and much wanted handbook should be in the hands of every architect and builder."—*Building World.*

" A carefully and usefully written treatise ; the work is essentially practical."—*Scotsman.*

DELAMOTTE'S WORKS ON ILLUMINATION AND ALPHABETS.

ORNAMENTAL ALPHABETS, ANCIENT & MEDIÆVAL.

From the Eighth Century, with Numerals; including Gothic, Church-Text, large and small, German, Italian, Arabesque, Initials for Illumination, Monograms, Crosses, &c., &c., for the use of Architectural and Engineering Draughtsmen, Missal Painters, Masons, Decorative Painters, Lithographers, Engravers, Carvers, &c. Collected and Engraved by F. DELAMOTTE, and printed in Colours. New and Cheaper Edition. Royal 8vo, oblong, ornamental boards **2/6**

" For those who insert enamelled sentences round gilded chalices, who blazon shop legends over shop-doors, who letter church walls with pithy sentences from the Decalogue, this book will be useful."—*Athenæum.*

MODERN ALPHABETS, PLAIN AND ORNAMENTAL.

Including German, Old English, Saxon, Italic, Perspective, Greek, Hebrew, Court Hand, Engrossing, Tuscan, Riband, Gothic, Rustic, and Arabesque; with several Original Designs, and an Analysis of the Roman and Old English Alphabets, large and small, and Numerals, for the use of Draughtsmen, Surveyors, Masons, Decorative Painters, Lithographers, Engravers, Carvers, &c. Collected and Engraved by F. DELAMOTTE, and printed in Colours. New and Cheaper Edition. Royal 8vo, ohlong, ornamental boards . **2/6**

" There is comprised in it every possible shape into which the letters of the alphabet and numerals can be formed, and the talent which has been expended in the conception of the various plain and ornamental letters is wonderful."—*Standard.*

MEDIÆVAL ALPHABETS AND INITIALS.

By F. G. DELAMOTTE. Containing 21 Plates and Illuminated Title, printed in Gold and Colours. With an Introduction by J. WILLIS BROOKS. Fifth Edition. Small 4to, ornamental boards . . . . . *Net* **5/0**

"A volume in which the letters of the alphabet come forth glorified in gilding and all the colours of the prism interwoven and intertwined and intermingled."—*Sun.*

A PRIMER OF THE ART OF ILLUMINATION.

For the Use of Beginners; with a Rudimentary Treatise on the Art, Practical Directions for its Exercise, and Examples taken from Illuminated MSS., printed in Gold and Colours. By F. DELAMOTTE. New and Cheaper Edition. Small 4to, ornamental boards **6/0**

" The examples of ancient MSS. recommended to the student, which, with much good sense, the author chooses from collections accessible to all, are selected with judgment and knowledge as well as taste."—*Athenæum.*

THE EMBROIDERER'S BOOK OF DESIGN.

Containing Initials, Emblems, Cyphers, Monograms, Ornamental Borders, Ecclesiastical Devices, Mediæval and Modern Alphabets, and National Emblems. Collected hy F. DELAMOTTE, and printed in Colours. Oblong royal 8vo, ornamental wrapper . . . . *Net* **2/0**

" The book will be of great assistance to ladies and young children who are endowed with the art of plying the needle in this most ornamental and useful pretty work."—*East Anglian Times.*

WOOD-CARVING FOR AMATEURS.

With Hints on Design. By A LADY. With 10 Plates. New and Cheaper Edition. Crown 8vo, in emhlematic wrapper **2/0**

" The handicraft of the wood-carver, so well as a book can impart it, may be learnt from ' A Lady's ' publication."—*Athenæum.*

PAINTING POPULARLY EXPLAINED.

By THOMAS JOHN GULLICK, Painter, and JOHN TIMBS, F.S.A. Including Fresco, Oil, Mosaic, Water-Colour, Water-Glass, Tempera, Encaustic, Miniature, Painting on Ivory, Vellum, Pottery, Enamel, Glass, &c. Fifth Edition. Crown 8vo, cloth **5/0**

*** *Adopted as a Prize Book at South Kensington.*

" Much may be learned, even by those who fancy they do not require to be taught, from the careful perusal of this unpretending but comprehensive treatise."—*Art Journal.*

NATURAL SCIENCE, ETC.

THE VISIBLE UNIVERSE.

Chapters on the Origin and Construction of the Heavens. By J. E. Gore, F.R.A.S., Author of "Star Groups," &c. Illustrated by 6 Stellar Photographs and 12 Plates. Demy 8vo, cloth **16/0**

STAR GROUPS.

A Student's Guide to the Constellations. By J. Ellard Gore, F.R.A.S., M.R.I.A., &c., Author of "The Visible Universe," "The Scenery of the Heavens," &c. With 30 Maps. Small 4to, cloth **5/0**

AN ASTRONOMICAL GLOSSARY.

Or, Dictionary of Terms used in Astronomy. With Tables of Data and Lists of Remarkable and Interesting Celestial Objects. By J. Ellard Gore, F.R.A.S., Author of " The Visible Universe," &c. Small crown 8vo, cloth.

2/6

THE MICROSCOPE.

Its Construction and Management. Including Technique, Photo-micrography, and the Past and Future of the Microscope. By Dr. Henri van Heurck. Re-Edited and Augmented from the Fourth French Edition, and Translated by Wynne E. Baxter, F.G.S. Imp. 8vo, cloth **18/0**

A MANUAL OF THE MOLLUSCA.

A Treatise on Recent and Fossil Shells. By S. P. Woodward, A.L.S., F.G.S. With an Appendix on Recent and Fossil Conchological Discoveries, by Ralph Tate, A.L.S., F.G.S. With 23 Plates and upwards of 300 Woodcuts. Reprint of Fourth Edition (1880). Crown 8vo, cloth **7/6**

THE TWIN RECORDS OF CREATION.

Or, Geology and Genesis, their Perfect Harmony and Wonderful Concord. By G. W. V. le Vaux. 8vo, cloth **5/0**

LARDNER'S HANDBOOKS OF SCIENCE.

HANDBOOK OF MECHANICS.

Enlarged and re-written by B. Loewy, F.R.A.S. Post 8vo, cloth . **6/0**

HANDBOOK OF HYDROSTATICS AND PNEUMATICS.

Revised and Enlarged by B. Loewy, F.R.A.S. Post 8vo, cloth . **5/0**

HANDBOOK OF HEAT.

Edited and re-written by B. Loewy, F.R.A.S. Post 8vo, cloth . **6/0**

HANDBOOK OF OPTICS.

New Edition. Edited by T. Olver Harding, B.A. Small 8vo, cloth **5/0**

ELECTRICITY, MAGNETISM, AND ACOUSTICS.

Edited by Geo. C. Foster, B.A. Small 8vo, cloth **5/0**

HANDBOOK OF ASTRONOMY.

Revised and Edited by Edwin Dunkin, F.R.A.S. 8vo, cloth . . **9/6**

MUSEUM OF SCIENCE AND ART.

With upwards of 1,200 Engravings. In Six Double Volumes, **£1 1s.** Cloth, or half-morocco . . . . **£1 11s. 6d.**

NATURAL PHILOSOPHY FOR SCHOOLS . . **3/6**

ANIMAL PHYSIOLOGY FOR SCHOOLS . . **3/6**

THE ELECTRIC TELEGRAPH.

Revised by E. B. Bright, F.R.A.S. Fcap. 8vo, cloth . . . **2/6**

CHEMICAL MANUFACTURES, CHEMISTRY, ETC.

THE OIL FIELDS OF RUSSIA AND THE RUSSIAN PETROLEUM INDUSTRY.

A Practical Handbook on the Exploration, Exploitation, and Management of Russian Oil Properties, including Notes on the Origin of Petroleum in Russia, a Description of the Theory and Practice of Liquid Fuel, and a Translation of the Rules and Regulations concerning Russian Oil Properties. By A. BEEBY THOMPSON, A.M.I.M.E., late Chief Engineer and Manager of the European Petroleum Company's Russian Oil Properties. About 500 pp., with numerous Illustrations and Photographic Plates, and a Map of the Balakhany-Saboontchy-Romany Oil Field. Super-royal 8vo, cloth.

[Just Published. Net **£3 3s.**

THE ANALYSIS OF OILS AND ALLIED SUBSTANCES.

By A. C. WRIGHT, M.A.Oxon., B.Sc.Lond., formerly Assistant Lecturer in Chemistry at the Yorkshire College, Leeds, and Lecturer in Chemistry at the Hull Technical School. Demy 8vo, cloth.   .   .   .   .   .   *Net* **9/0**

THE GAS ENGINEER'S POCKET-BOOK.

Comprising Tables, Notes and Memoranda relating to the Manufacture, Distribution and Use of Coal Gas and the Construction of Gas Works. By H. O'CONNOR, A.M.Inst.C.E. Second Edition, Revised. 470 pp., crown 8vo, fully Illustrated, leather **10/6**

"The book contains a vast amount of information. The author goes consecutively through the engineering details and practical methods involved in each of the different processes or parts of a gas-works. He has certainly succeeded in making a compilation of hard matters of fact absolutely interesting to read."—*Gas World.*

"The volume contains a great quantity of specialised information, compiled, we believe, from trustworthy sources, which should make it of considerable value to those for whom it is specifically produced."—*Engineer.*

LIGHTING BY ACETYLENE

Generators, Burners, and Electric Furnaces. By WILLIAM E. GIBBS, M.E. With 66 Illustrations. Crown 8vo, cloth. **7/6**

ENGINEERING CHEMISTRY.

A Practical Treatise for the Use of Analytical Chemists, Engineers, Iron Masters, Iron Founders, Students and others. Comprising Methods of Analysis and Valuation of the Principal Materials used in Engineering Work, with numerous Analyses, Examples and Suggestions. By H. JOSHUA PHILLIPS F.I.C., F.C.S. Third Edition, Revised and Enlarged. Crown 8vo, 420 pp., with Plates and other Illustrations, cloth.   .   .   .   *Net* **10/6**

"In this work the author has rendered no small service to a numerous body of practical men. . . . The analytical methods may be pronounced most satisfactory, being as accurate as the despatch required of engineering chemists permits."—*Chemical News.*

"The analytical methods given are, as a whole, such as are likely to give rapid and trust-worthy results in experienced hands. . . . There is much excellent descriptive matter in the work, the chapter on 'Oils and Lubrication' being specially noticeable in this respect."—*Engineer.*

NITRO-EXPLOSIVES.

A Practical Treatise concerning the Properties, Manufacture, and Analysis of Nitrated Substances, including the Fulminates, Smokeless Powders, and Celluloid. By P. GERALD SANFORD, F.I.C., Consulting Chemist to the Cotton Powder Company, Limited, &c. With Illustrations. Crown 8vo, cloth. **9/0**

"One of the very few text-books in which can be found just what is wanted. Mr. Sanford goes steadily through the whole list of explosives commonly used, he names any given explosive, and tells us of what it is composed and how it is manufactured. The book is excellent."—*Engineer.*

A HANDBOOK ON MODERN EXPLOSIVES.

A Practical Treatise on the Manufacture and Use of Dynamite, Gun-Cotton, Nitro-Glycerine and other Explosive Compounds, including Collodion-Cotton. With Chapters on Explosives in Practical Application. By M. EISSLER, M.E. Second Edition, Enlarged. Crown 8vo, cloth   .   .   **12/6**

" A veritable mine of information on the subject of explosives employed for military, mining and blasting purposes."—*Army and Navy Gazette.*

A MANUAL OF THE ALKALI TRADE.

Including the Manufacture of Sulphuric Acid, Sulphate of Soda, and Bleaching Powder. By JOHN LOMAS, Alkali Manufacturer. With 232 Illustrations and Working Drawings, Second Edition, with Additions. Super-royal 8vo, cloth . . . . : . . . **£1 10s.**

" We find not merely a sound and luminous explanation of the chemical principles of the trade, but a notice of numerous matters which have a most important bearing on the successful conduct of alkali works, but which are generally overlooked by even experienced technological authors."—*Chemical Review.*

DANGEROUS GOODS.

Their Sources and Properties, Modes of Storage and Transport. With Notes and Comments on Accidents arising therefrom. A Guide for the Use of Government and Railway Officials, Steamship Owners, &c. By H. JOSHUA PHILLIPS, F.I.C., F.C.S. Crown 8vo, 374 pp., cloth . . . **9/0**

" Merits a wide circulation, and an intelligent, appreciative study."—*Chemical News.*

THE BLOWPIPE IN CHEMISTRY, MINERALOGY, Etc.

Containing all known Methods of Anhydrous Analysis, many Working Examples, and Instructions for Making Apparatus. By Lieut.-Colonel W. A. Ross, R.A., F.G.S. Second Edition, Enlarged. Crown 8vo, cloth . **5/0**

" The student who goes conscientiously through the course of experimentation here laid down will gain a better insight into inorganic chemistry and mineralogy than if he had ' got up ' any of the best text-books of the day, and passed any number of examinations in their contents "—*Chemical News.*

THE MANUAL OF COLOURS AND DYE-WARES.

Their Properties, Applications, Valuations, Impurities and Sophistications. For the Use of Dyers, Printers, Drysalters, Brokers, &c. By J. W. SLATER. Second Edition, Revised and greatly Enlarged. Crown 8vo, cloth . **7/6**

" There is no other work which covers precisely the same ground. To students preparing for examinations in dyeing and printing it will prove exceedingly useful."—*Chemical News.*

A HANDY BOOK FOR BREWERS.

Being a Practical Guide to the Art of Brewing and Malting. Embracing the Conclusions of Modern Research which bear upon the Practice of Brewing. By HERBERT EDWARDS WRIGHT, M.A. Second Edition, Enlarged. Crown 8vo, 530 pp., cloth **12/6**

" May be consulted with advantage by the student who is preparing himself for examinational tests, while the scientific brewer will find in it a *résumé* of all the most important discoveries of modern times. The work is written throughout in a clear and concise manner, and the author takes great care to discriminate between vague theories and well-established facts "—*Brewers' Journal.*

" We have great pleasure in recommending this handy book, and have no hesitation in saying that it is one of the best—if not the best—which has yet been written on the subject of beer-brewing in this country; it should have a place on the shelves of every brewer's library."—*Brewers' Guardian.*

FUELS: SOLID, LIQUID, AND GASEOUS.

Their Analysis and Valuation. For the Use of Chemists and Engineers. By H. J. PHILLIPS, F.C.S., formerly Analytical and Consulting Chemist to the G.E. Rlwy. Fourth Edition. Crown 8vo, cloth **2/0**

" Ought to have its place in the laboratory of every metallurgical establishment and wherever fuel is used on a large scale."—*Chemical News.*

THE ARTISTS' MANUAL OF PIGMENTS.

Showing their Composition, Conditions of Permanency, Non-Permanency, and Adulterations, &c., with Tests of Purity. By H. C. STANDAGE. Third Edition. Crown 8vo, cloth **2/6**

" This work is indeed *multum-in-parvo*, and we can, with good conscience, recommend it to all who come in contact with pigments, whether as makers, dealers, or users."—*Chemical Review.*

A POCKET-BOOK OF MENSURATION AND GAUGING.

Containing Tables, Rules, and Memoranda for Revenue Officers, Brewers, Spirit Merchants, &c. By J. B. MANT, Inland Revenue. Second Edition, Revised. 18mo, leather **4/0**

" Should be in the hands of every practical brewer."—*Brewers' Journal.*

INDUSTRIAL ARTS, TRADES, AND MANUFACTURES.

TEA MACHINERY AND TEA FACTORIES.

A Descriptive Treatise on the Mechanical Appliances required in the Cultivation of the Tea Plant and the Preparation of Tea for the Market. By A. J. WALLIS-TAYLER, A. M. Inst. C.E. Medium 8vo, 468 pp. With 218 Illustrations . . . . *Net* **25/0**

SUMMARY OF CONTENTS.

MECHANICAL CULTIVATION OR TILLAGE OF THE SOIL.—PLUCKING OR GATHERING THE LEAF.—TEA FACTORIES.—THE DRESSING, MANUFACTURE, OR PREPARATION OF TEA BY MECHANICAL MEANS. — ARTIFICIAL WITHERING OF THE LEAF.— MACHINES FOR ROLLING OR CURLING THE LEAF.—FERMENTING PROCESS.— MACHINES FOR THE AUTOMATIC DRYING OR FIRING OF THE LEAF.—MACHINES FOR NON-AUTOMATIC DRYING OR FIRING OF THE LEAF.—DRYING OR FIRING MACHINES. —BREAKING OR CUTTING, AND SORTING MACHINES.—PACKING THE TEA.—MEANS OF TRANSPORT ON TEA PLANTATIONS.—MISCELLANEOUS MACHINERY AND APPARATUS. —FINAL TREATMENT OF THE TEA.—TABLES AND MEMORANDA.

"The subject of tea machinery is now one of the first interest to a large class of people, to whom we strongly commend the volume."—*Chamber of Commerce Journal.*
"When tea planting was first introduced into the British possessions little, if any, machinery was employed, but now its use is almost universal. This volume contains a very full account of the machinery necessary for the proper outfit of a factory, and also a description of the processes best carried out by this machinery."—*Journal Society of Arts.*

FLOUR MANUFACTURE.

A Treatise on Milling Science and Practice. By FRIEDRICH KICK, Imperial Regierungsrath, Professor of Mechanical Technology in the Imperial German Polytechnic Institute, Prague. Translated from the Second Enlarged and Revised Edition with Supplement. By H. H. P. POWLES, Assoc. Memb. Institution of Civil Engineers. Nearly 400 pp. Illustrated with 28 Folding Plates, and 167 Woodcuts. Royal 8vo, cloth . . . . . **£1 5s.**

"This invaluable work is, and will remain, the standard authority on the science of milling. . . . The miller who has read and digested this work will have laid the foundation, so to speak, of a successful career; he will have acquired a number of general principles which he can proceed to apply. In this handsome volume we at last have the accepted text-book of modern milling in good, sound English, which has little, if any, trace of the German idiom."—*The Miller.*
"The appearance of this celebrated work in English is very opportune, and British millers will, we are sure, not be slow in availing themselves of its pages."—*Millers' Gazette.*

COTTON MANUFACTURE.

A Manual of Practical Instruction of the Processes of Opening, Carding, Combing, Drawing, Doubling and Spinning of Cotton, the Methods of Dyeing, &c. For the Use of Operatives, Overlookers, and Manufacturers. By JOHN LISTER, Technical Instructor, Pendleton. 8vo, cloth . . **7/6**

"This invaluable volume is a distinct advance in the literature of cotton manufacture."— *Machinery.*
"It is thoroughly reliable, fulfilling nearly all the requirements desired."—*Glasgow Herald.*

MODERN CYCLES.

A Practical Handbook on their Construction and Repair. By A. J. WALLIS-TAYLER, A. M. Inst. C. E., Author of "Refrigerating Machinery," &c. With upwards of 300 Illustrations. Crown 8vo, cloth **10/6**

"The large trade that is done in the component parts of bicycles has placed in the way of men mechanically inclined extraordinary facilities for building bicycles for their own use. . . . The book will prove a valuable guide for all those who aspire to the manufacture or repair of their own machines."—*The Field.*
"A most comprehensive and up-to-date treatise."—*The Cycle.*
"A very useful book, which is quite entitled to rank as a standard work for students of cycle construction."—*Wheeling.*

MOTOR CARS OR POWER CARRIAGES FOR COMMON ROADS.

By A. J. WALLIS-TAYLER, Assoc. Memb. Inst. C.E., Author of "Modern Cycles," &c. 212 pp., with 76 Illustrations. Crown 8vo, cloth . . **4/6**

"The book is clearly expressed throughout, and is just the sort of work that an engineer, thinking of turning his attention to motor-carriage work, would do well to read as a preliminary to starting operations."—*Engineering.*

PRACTICAL TANNING.

A Handbook of Modern Processes, Receipts, and Suggestions for the Treatment of Hides, Skins, and Pelts of every Description. By L. A. FLEMMING, American Tanner. 472 pages. 8vo, cloth. [*Just Published. Net* **25/0**

THE ART OF LEATHER MANUFACTURE.

Being a Practical Handbook, in which the Operations of Tanning, Currying, and Leather Dressing are fully Described, and the Principles of Tanning Explained, and many Recent Processes Introduced ; as also Methods for the Estimation of Tannin, and a Description of the Arts of Glue Boiling, Gut Dressing, &c. By ALEXANDER WATT. Fourth Edition. Crown 8vo cloth.

9/0

"A sound, comprehensive treatise on tanning and its accessories. The book is an eminently valuable production, which redounds to the credit of both author and publishers."—*Chemical Review.*

THE ART OF SOAP-MAKING.

A Practical Handbook of the Manufacture of Hard and Soft Soaps, Toilet Soaps, &c. Including many New Processes, and a Chapter on the Recovery of Glycerine from Waste Leys. By ALEXANDER WATT. Sixth Edition, including an Appendix on Modern Candlemaking. Crown 8vo, cloth . **7/6**

"The work will prove very useful, not merely to the technological student, but to the practical soap boiler who wishes to understand the theory of his art."—*Chemical News.*
"A thoroughly practical treatise. We congratulate the author on the success of his endeavour to fill a void in English technical literature."—*Nature.*

PRACTICAL PAPER-MAKING.

A Manual for Paper-Makers and Owners and Managers of Paper-Mills. With Tables, Calculations, &c. By G. CLAPPERTON, Paper-Maker. With Illustrations of Fibres from Micro-Photographs. Crown 8vo, cloth . **5/0**

" The author caters for the requirements of responsible mill hands, apprentices, &c., whilst his manual will be found of great service to students of technology, as well as to veteran paper-makers and mill owners. The illustrations form an excellent feature."—*The World's Paper Trade Review.*

THE ART OF PAPER-MAKING.

A Practical Handbook of the Manufacture of Paper from Rags, Esparto, Straw, and other Fibrous Materials. Including the Manufacture of Pulp from Wood Fibre, with a Description of the Machinery and Appliances used. To which are added Details of Processes for Recovering Soda from Waste Liquors. By ALEXANDER WATT. With Illustrations. Crown 8vo, cloth . **7/6**

"It may be regarded as the standard work on the subject. The book is full of valuable Information. The 'Art of Paper-Making' is in every respect a model of a text-book, either for a technical class, or for the private student."—*Paper and Printing Trades Journal.*

A TREATISE ON PAPER.

For Printers and Stationers. With an Outline of Paper Manufacture ; Complete Tables of Sizes, and Specimens of Different Kinds of Paper. By RICHARD PARKINSON, late of the Manchester Technical School. Demy 8vo, cloth **3/6**

CEMENTS, PASTES, GLUES, AND GUMS.

A Practical Guide to the Manufacture and Application of the various Agglutinants required in the Building, Metal-Working, Wood-Working, and Leather-Working Trades, and for Workshop and Office Use. With upwards of 900 Recipes. By H. C. STANDAGE. Third Edition. Crown 8vo, cloth . **2/0**

"We have pleasure in speaking favourably of this volume. So far as we have had experience, which is not inconsiderable, this manual is trustworthy."—*Athenæum.*

THE CABINET-MAKER'S GUIDE
TO THE ENTIRE CONSTRUCTION OF CABINET WORK.

Including Veneering, Marquetrie, Buhlwork, Mosaic, Inlaying, &c. By RICHARD BITMEAD. Illustrated with Plans, Sections, and Working Drawings. Small crown 8vo, cloth **2/6**

FRENCH POLISHING AND ENAMELLING.

A Practical Work of Instruction. Including Numerous Recipes for making Polishes, Varnishes, Glaze-Lacquers, Revivers, &c. By RICHARD BITMEAD, Author of "The Cabinet-Maker's Guide." Small crown 8vo, cloth . **1/6**

WATCH REPAIRING, CLEANING, AND ADJUSTING.

A Practical Handbook dealing with the Materials and Tools Used, and the Methods of Repairing, Cleaning, Altering, and Adjusting all kinds of English and Foreign Watches, Repeaters, Chronographs and Marine Chronometers. By F. J. GARRARD, Springer and Adjuster of Marine Chronometers and Deck Watches for the Admiralty. With over 200 Illustrations. Crown 8vo, cloth.
[Just Published. Net **4/6**

MODERN HOROLOGY, IN THEORY AND PRACTICE.

Translated from the French of CLAUDIUS SAUNIER, ex-Director of the School of Horology at Macon, by JULIEN TRIPPLIN, F.R.A.S., Besançon Watch Manufacturer, and EDWARD RIGG, M.A., Assayer in the Royal Mint. With Seventy-eight Woodcuts and Twenty-two Coloured Copper Plates. Second Edition. Super-royal 8vo, **£2 2s.** cloth; half-calf   .   .   . **£2 10s.**

" There is no horological work in the English language at all to be compared to this production of M. Saunier's for clearness and completeness. It is alike good as a guide for the student and as a reference for the experienced horologist and skilled workman."—*Horological Journal.*

" The latest, the most complete, and the most reliable of those literary productions to which continental watchmakers are indebted for the mechanical superiority over their English brethren —in fact, the Book of Books is M. Saunier's ' Treatise.'"—*Watchmaker, Jeweller, and Silversmith.*

THE WATCH ADJUSTER'S MANUAL.

A Practical Guide for the Watch and Chronometer Adjuster in Making, Springing, Timing and Adjusting for Isochronism, Positions and Temperatures. By C. E. FRITTS. 370 pp., with Illustrations, 8vo, cloth   .   .   . **16/0**

THE WATCHMAKER'S HANDBOOK.

Intended as a Workshop Companion for those engaged in Watchmaking and the Allied Mechanical Arts. Translated from the French of CLAUDIUS SAUNIER, and enlarged by JULIEN TRIPPLIN, F.R.A.S., and EDWARD RIGG, M.A., Assayer in the Royal Mint. Third Edition. Cr. 8vo, cloth.   . **9/0**

" Each part is truly a treatise in itself. The arrangement is good and the language is clear and concise. It is an admirable guide for the young watchmaker."—*Engineering.*

HISTORY OF WATCHES & OTHER TIMEKEEPERS.

By JAMES F. KENDAL, M.B.H. Inst. **1/6** boards ; or cloth, gilt   . **2/6**

" The best which has yet appeared on this subject in the English language."—*Industries.*

" Open the book where you may, there is interesting matter in it concerning the ingenious devices of the ancient or modern horologer."—*Saturday Review.*

ELECTRO-PLATING & ELECTRO-REFINING OF METALS.

Being a new edition of ALEXANDER WATT's "ELECTRO-DEPOSITION." Revised and Largely Rewritten by ARNOLD PHILIP, B.Sc., A.I.E.E., Principal Assistant to the Admiralty Chemist. Large Crown 8vo, cloth.   . *Net* **12/6**

" Altogether the work can be highly recommended to every electro-plater, and is of undoubted interest to every electro-metallurgist."—*Electrical Review.*

" Eminently a book for the practical worker in electro-deposition. It contains practical descriptions of methods, processes and materials, as actually pursued and used in the workshop."—*Engineer.*

ELECTRO-METALLURGY.

Practically Treated. By ALEXANDER WATT. Tenth Edition, including the most recent Processes. 12mo, cloth **3/6**

" From this book both amateur and artisan may learn everything necessary for the successful prosecution of electroplating."—*Iron.*

JEWELLER'S ASSISTANT IN WORKING IN GOLD.

A Practical Treatise for Masters and Workmen, Compiled from the Experience of Thirty Years' Workshop Practice. By GEORGE E. GEE. Crown 8vo. **7/6**

" This manual of technical education is apparently destined to be a valuable auxiliary to a handicraft which is certainly capable of great improvement."—*The Times.*

ELECTROPLATING.

A Practical Handbook on the Deposition of Copper, Silver, Nickel, Gold, Aluminium, Brass, Platinum, &c., &c. By J. W. URQUHART, C.E. Fourth Edition, Revised. Crown 8vo, cloth **5/0**

" An excellent practical manual."—*Engineering.*

" An excellent work, giving the newest information."—*Horological Journal.*

ELECTROTYPING.

The Reproduction and Multiplication of Printing Surfaces and Works of Art by the Electro-Deposition of Metals. By J. W. URQUHART, C.E. Crown 8vo, cloth **5/0**

"The book is thoroughly practical; the reader is, therefore, conducted through the leading laws of electricity, then through the metals used by electrotypers, the apparatus, and the depositing processes, up to the final preparation of the work."—*Art Journal.*

GOLDSMITH'S HANDBOOK.

By GEORGE E. GEE, Jeweller, &c. Fifth Edition. 12mo, cloth . . **3/0**

"A good, sound educator."—*Horological Journal.*

SILVERSMITH'S HANDBOOK.

By GEORGE E. GEE, Jeweller, &c. Third Edition, with numerous Illustrations. 12mo, cloth **3/0**

"The chief merit of the work is its practical character. . . . The workers in the trade will speedily discover its merits when they sit down to study it."—*English Mechanic.*

⁎⁎⁎ The above two works together, strongly half-bound, price 7s.

SHEET METAL WORKER'S INSTRUCTOR.

Comprising a Selection of Geometrical Problems and Practical Rules for Describing the Various Patterns Required by Zinc, Sheet-Iron, Copper, and Tin-Plate Workers. By REUBEN HENRY WARN, Practical Tin-Plate Worker. New Edition, Revised and greatly Enlarged by JOSEPH G. HORNER, A.M.I.M.E. Crown 8vo, 254 pp., with 430 Illustrations, cloth . . **7/6**

SAVOURIES AND SWEETS

Suitable for Luncheons and Dinners. By Miss M. L. ALLEN (Mrs. A. MACAIRE), Author of "Breakfast Dishes," &c. Twenty-ninth Edition. F'cap 8vo, sewed **1/0**

BREAKFAST DISHES

For Every Morning of Three Months. By Miss ALLEN (Mrs A. MACAIRE), Author of "Savouries and Sweets," &c. Twenty-second Edition. F'cap 8vo, sewed **1/0**

BREAD & BISCUIT BAKER'S & SUGAR-BOILER'S ASSISTANT.

Including a large variety of Modern Recipes. With Remarks on the Art of Bread-making. By ROBERT WELLS. Third Edition. Crown 8vo . . **1/0**

"A large number of wrinkles for the ordinary cook, as well as the baker."—*Saturday Review.*

PASTRYCOOK & CONFECTIONER'S GUIDE.

For Hotels, Restaurants, and the Trade in general, adapted also for Family Use. By R. WELLS, Author of "The Bread and Biscuit Baker" . . **1/0**

"We cannot speak too highly of this really excellent work. In these days of keen competition our readers cannot do better than purchase this book."—*Bakers' Times.*

ORNAMENTAL CONFECTIONERY.

A Guide for Bakers, Confectioners and Pastrycooks; including a variety of Modern Recipes, and Remarks on Decorative and Coloured Work. With 129 Original Designs. By ROBERT WELLS. Crown 8vo, cloth . . . **5/0**

"A valuable work, practical, and should be in the hands of every baker and confectioner. The illustrative designs are worth treble the amount charged for the work."—*Bakers' Times.*

MODERN FLOUR CONFECTIONER.

Containing a large Collection of Recipes for Cheap Cakes, Biscuits, &c. With remarks on the Ingredients Used in their Manufacture. By R. WELLS. **1/0**

"The work is of a decidedly practical character, and in every recipe regard is had to economical working."—*North British Daily Mail.*

RUBBER HAND STAMPS

And the Manipulation of Rubber. A Practical Treatise on the Manufacture of Indiarubber Hand Stamps, Small Articles of Indiarubber, The Hektograph, Special Inks, Cements, and Allied Subjects. By T. O'CONOR SLOANE, A.M., Ph.D. With numerous Illustrations. Square 8vo, cloth. . . . **5/0**

HANDYBOOKS FOR HANDICRAFTS.

BY PAUL N. HASLUCK.

Editor of "Work" (New Series), Author of "Lathe Work," "Milling Machines," &c.
Crown 8vo, 144 pp., price 1s. each.

☞ *These* HANDYBOOKS *have been written to supply information for* WORKMEN, STUDENTS, *and* AMATEURS *in the several Handicrafts, on the actual* PRACTICE *of the* WORKSHOP, *and are intended to convey in plain language* TECHNICAL KNOWLEDGE *of the several* CRAFTS. *In describing the processes employed, and the manipulation of material, workshop terms are used ; workshop practice is fully explained ; and the text is freely illustrated with drawings of modern tools, appliances, and processes.*

METAL TURNER'S HANDYBOOK.

A Practical Manual for Workers at the Foot-Lathe. With 100 Illustrations.
1/O

"The book will be of service alike to the amateur and the artisan turner. It displays thorough knowledge of the subject."—*Scotsman.*

WOOD TURNER'S HANDYBOOK.

A Practical Manual for Workers at the Lathe. With over 100 Illustrations.
1/O

"We recommend the book to young turners and amateurs. A multitude of workmen have hitherto sought in vain for a manual of this special industry."—*Mechanical World.*

WATCH JOBBER'S HANDYBOOK.

A Practical Manual on Cleaning, Repairing, and Adjusting. With upwards of 100 Illustrations **1/O**
"We strongly advise all young persons connected with the watch trade to acquire and study this inexpensive work."—*Clerkenwell Chronicle.*

PATTERN MAKER'S HANDYBOOK.

A Practical Manual on the Construction of Patterns for Founders. With upwards of 100 Illustrations **1/O**
"A most valuable, if not indispensable manual for the pattern maker."—*Knowledge.*

MECHANIC'S WORKSHOP HANDYBOOK.

A Practical Manual on Mechanical Manipulation, embracing Information on various Handicraft Processes. With Useful Notes and Miscellaneous Memoranda. Comprising about 200 Subjects **1/O**
"A very clever and useful book, which should be found in every workshop; and it should certainly find a place in all technical schools."—*Saturday Review.*

MODEL ENGINEER'S HANDYBOOK.

A Practical Manual on the Construction of Model Steam Engines. With upwards of 100 Illustrations. **1/O**
"Mr. Hasluck has produced a very good little book."—*Builder.*

CLOCK JOBBER'S HANDYBOOK.

A Practical Manual on Cleaning, Repairing, and Adjusting. With upwards of 100 Illustrations **1/O**
"It is of inestimable service to those commencing the trade."—*Coventry Standard.*

CABINET WORKER'S HANDYBOOK.

A Practical Manual on the Tools, Materials, Appliances, and Processes employed in Cabinet Work. With upwards of 100 Illustrations . . **1/O**
"Mr. Hasluck's thorough-going little Handybook is amongst the most practical guides we have seen for beginners in cabinet-work."—*Saturday Review.*

WOODWORKER'S HANDYBOOK.

Embracing Information on the Tools, Materials, Appliances and Processes Employed in Woodworking. With 104 Illustrations. **1/O**
"Written by a man who knows, not only how work ought to be done, but how to do it, and how to convey his knowledge to others."—*Engineering.*
"Mr. Hasluck writes admirably, and gives complete instructions."—*Engineer.*
"Mr. Hasluck combines the experience of a practical teacher with the manipulative skill and scientific knowledge of processes of the trained mechanician, and the manuals are marvels of what can be produced at a popular price."—*Schoolmaster.*
"Helpful to workmen of all ages and degrees of experience."—*Daily Chronicle.*
"Concise, clear, and practical."—*Saturday Review.*

COMMERCE, COUNTING-HOUSE WORK, TABLES, ETC.

LESSONS IN COMMERCE.

By Professor R. GAMBARO, of the Royal High Commercial School at Genoa. Edited and Revised by JAMES GAULT, Professor of Commerce and Commercial Law in King's College, London. Fourth Edition. Crown 8vo, cloth . **3/6**

"The publishers of this work have rendered considerable service to the cause of commercial education by the opportune production of this volume. . . . The work is peculiarly acceptable to English readers and an admirable addition to existing class books. In a phrase, we think the work attains its object in furnishing a brief account of those laws and customs of British trade with which the commercial man interested therein should be familiar."—*Chamber of Commerce Journal.*

"An invaluable guide in the hands of those who are preparing for a commercial career, and, in fact, the information it contains on matters of business should be impressed on every one."—*Counting House.*

THE FOREIGN COMMERCIAL CORRESPONDENT.

Being Aids to Commercial Correspondence in Five Languages—English, French, German, Italian, and Spanish. By CONRAD E. BAKER. Third Edition, Carefully Revised Throughout. Crown 8vo, cloth . . . **4/6**

"Whoever wishes to correspond in all the languages mentioned by Mr. Baker cannot do better than study this work, the materials of which are excellent and conveniently arranged. They consist not of entire specimen letters, but—what are far more useful—short passages, sentences, or phrases expressing the same general idea in various forms."—*Athenæum.*

"A careful examination has convinced us that it is unusually complete, well arranged and reliable. The book is a thoroughly good one."—*Schoolmaster.*

FACTORY ACCOUNTS: their PRINCIPLES & PRACTICE.

A Handbook for Accountants and Manufacturers, with Appendices on the Nomenclature of Machine Details; the Income Tax Acts; the Rating of Factories; Fire and Boiler Insurance; the Factory and Workshop Acts, &c., including also a Glossary of Terms and a large number of Specimen Rulings. By EMILE GARCKE and J. M. FELLS. Fifth Edition, Revised and Enlarged. Demy 8vo, cloth **7/6**

"A very interesting description of the requirements of Factory Accounts. . . . The principle of assimilating the Factory Accounts to the general commercial books is one which we thoroughly agree with."—*Accountants' Journal.*

"Characterised by extreme thoroughness. There are few owners of factories who would not derive great benefit from the perusal of this most admirable work."—*Local Government Chronicle.*

MODERN METROLOGY.

A Manual of the Metrical Units and Systems of the present Century. With an Appendix containing a proposed English System. By LOWIS D. A. JACKSON, A. M. Inst. C. E., Author of "Aid to Survey Practice," &c. Large crown 8vo, cloth **12/6**

"We recommend the work to all interested in the practical reform of our weights and measures."—*Nature.*

A SERIES OF METRIC TABLES.

In which the British Standard Measures and Weights are compared with those of the Metric System at present in Use on the Continent. By C. H. DOWLING, C.E. 8vo, cloth **10/6**

"Mr. Dowling's Tables are well put together as a ready reckoner for the conversion of one system into the other."—*Athenæum.*

IRON AND METAL TRADES' COMPANION.

For Expeditiously Ascertaining the Value of any Goods bought or sold by Weight, from 1s. per cwt. to 112s. per cwt., and from one farthing per pound to one shilling per pound. By THOMAS DOWNIE. Strongly bound in leather, 396 pp. **9/0**

"A most useful set of tables, nothing like them before existed."—*Building News.*

"Although specially adapted to the iron and metal trades, the tables will be found useful in every other business in which merchandise is bought and sold by weight."—*Railway News.*

NUMBER, WEIGHT, AND FRACTIONAL CALCULATOR.

Containing upwards of 250,000 Separate Calculations, showing at a Glance the Value at 422 Different Rates, ranging from ₁/₁₆th of a Penny to 20s. each, or per cwt., and £20 per ton, of any number of articles consecutively, from 1 to 470. Any number of cwts., qrs., and lbs., from 1 cwt. to 470 cwts. Any number of tons, cwts., qrs., and lbs., from 1 to 1,000 tons. By WILLIAM CHADWICK, Public Accountant. Third Edition, Revised and Improved. 8vo, strongly bound **18/0**

"It is as easy of reference for any answer or any number of answers as a dictionary. For making up accounts or estimates the book must prove invaluable to all who have any considerable quantity of calculations involving price and measure in any combination to do."—*Engineer*.
"The most perfect work of the kind yet prepared."—*Glasgow Herald*.

THE WEIGHT CALCULATOR.

Being a Series of Tables upon a New and Comprehensive Plan, exhibiting at one Reference the exact Value of any Weight from 1 lb. to 15 tons, at 300 Progressive Rates, from 1d. to 168s. per cwt., and containing 186,000 Direct Answers, which, with their Combinations, consisting of a single addition (mostly to be performed at sight), will afford an aggregate of 10,266,000 Answers; the whole being calculated and designed to ensure correctness and promote despatch. By HENRY HARBEN, Accountant. Sixth Edition, carefully Corrected. Royal 8vo, strongly half-bound.   [*Just Published*. **£1 5s.**

"A practical and useful work of reference for men of business generally."—*Ironmonger*.
"Of priceless value to business men. It is a necessary book in all mercantile offices."—*Sheffield Independent*.

THE DISCOUNT GUIDE.

Comprising several Series of Tables for the Use of Merchants, Manufacturers, Ironmongers, and Others, by which may be ascertained the Exact Profit arising from any mode of using Discounts, either in the Purchase or Sale of Goods, and the method of either Altering a Rate of Discount, or Advancing a Price, so as to produce, by one operation, a sum that will realise any required Profit after allowing one or more Discounts: to which are added Tables of Profit or Advance from 1¼ to 90 per cent., Tables of Discount from 1¼ to 98¾ per cent., and Tables of Commission, &c., from ⅛ to 10 per cent. By HENRY HARBEN, Accountant. New Edition, Corrected. Demy 8vo, half-bound   .   **£1 5s.**

"A book such as this can only be appreciated by business men, to whom the saving of time means saving of money. The work must prove of great value to merchants, manufacturers, and general traders."—*British Trade Journal*.

TABLES OF WAGES.

At 54, 52, 50 and 48 Hours per Week. Showing the Amounts of Wages from One quarter of an hour to Sixty-four hours, in each case at Rates of Wages advancing by One Shilling from 4s. to 55s. per week. By THOS. GARBUTT, Accountant. Square crown 8vo, half-bound **6/0**

IRON-PLATE WEIGHT TABLES.

For Iron Shipbuilders, Engineers, and Iron Merchants. Containing the Calculated Weights of upwards of 150,000 different sizes of Iron Plates from 1 foot by 6 in. by ¼ in. to 10 feet by 5 feet by 1 in. Worked out on the Basis of 40 lbs. to the square foot of Iron of 1 inch in thickness. By H. BURLINSON and W. H. SIMPSON. 4to, half-bound   .   .   .   .   .   **£1 5s.**

AGRICULTURE, FARMING, GARDENING, ETC.

THE COMPLETE GRAZIER AND FARMER'S AND CATTLE BREEDER'S ASSISTANT.

A Compendium of Husbandry. Originally Written by WILLIAM YOUATT. Fourteenth Edition, entirely Re-written, considerably Enlarged, and brought up to Present Requirements, by WILLIAM FREAM, LL.D., Assistant Commissioner, Royal Commission on Agriculture, Author of "The Elements of Agriculture," &c. Royal 8vo, 1,100 pp., 450 Illustrations, handsomely bound.
£1 11s. 6D.

BOOK I. ON THE VARIETIES, BREEDING, REARING, FATTENING AND MANAGEMENT OF CATTLE.
BOOK II. ON THE ECONOMY AND MANAGEMENT OF THE DAIRY.
BOOK III. ON THE BREEDING, REARING, AND MANAGEMENT OF HORSES.
BOOK IV. ON THE BREEDING, REARING, AND FATTENING OF SHEEP.
BOOK V. ON THE BREEDING, REARING, AND FATTENING OF SWINE.
BOOK VI. ON THE DISEASES OF LIVE STOCK.

BOOK VII. ON THE BREEDING, REARING, AND MANAGEMENT OF POULTRY.
BOOK VIII. ON FARM OFFICES AND IMPLEMENTS OF HUSBANDRY.
BOOK IX. ON THE CULTURE AND MANAGEMENT OF GRASS LANDS.
BOOK X. ON THE CULTIVATION AND APPLICATION OF GRASSES, PULSE AND ROOTS.
BOOK XI. ON MANURES AND THEIR APPLICATION TO GRASS LAND AND CROPS.
BOOK XII. MONTHLY CALENDARS OF FARMWORK.

" Dr. Fream is to be congratulated on the successful attempt he has made to give us a work which will at once become the standard classic of the farm practice of the country. We believe that it will be found that it has no compeer among the many works at present in existence. . . . The illustrations are admirable, while the frontispiece, which represents the well-known bull, New Year's Gift, bred by the Queen, is a work of art."—*The Times.*

"The book must be recognised as occupying the proud position of the most exhaustive work of reference in the English language on the subject with which it deals."—*Athenæum.*

"The most comprehensive guide to modern farm practice that exists in the English language to-day. . . . The book is one that ought to be on every farm and in the library of every land owner."—*Mark Lane Express.*

"In point of exhaustiveness and accuracy the work will certainly hold a pre-eminent and unique position among books dealing with scientific agricultural practice. It is, in fact, an agricultural library of itself."—*North British Agriculturist.*

FARM LIVE STOCK OF GREAT BRITAIN.

By ROBERT WALLACE, F.L.S., F.R.S.E., &c., Professor of Agriculture and Rural Economy in the University of Edinburgh. Third Edition, thoroughly Revised and considerably Enlarged. With over 120 Phototypes of Prize Stock. Demy 8vo, 384 pp., with 79 Plates and Maps, cloth. **12/6**

"A really complete work on the history, breeds, and management of the farm stock of Great Britain, and one which is likely to find its way to the shelves of every country gentleman's library."—*The Times.*

"The 'Farm Live Stock of Great Britain' is a production to be proud of, and its issue not the least of the services which its author has rendered to agricultural science."—*Scottish Farmer.*

NOTE-BOOK OF AGRICULTURAL FACTS & FIGURES FOR FARMERS AND FARM STUDENTS.

By PRIMROSE McCONNELL, B.Sc., Fellow of the Highland and Agricultural Society, Author of "Elements of Farming." Sixth Edition, Re-written, Revised, and greatly Enlarged. Fcap. 8vo, 480 pp., leather, gilt edges **6/O**

CONTENTS.—SURVEYING AND LEVELLING.—WEIGHTS AND MEASURES.—MACHINERY AND BUILDINGS. — LABOUR. — OPERATIONS. — DRAINING. — EMBANKING. — GEOLOGICAL MEMORANDA. — SOILS. — MANURES. — CROPPING. — CROPS.—ROTATIONS. — WEEDS. — FEEDING.—DAIRYING.—LIVE STOCK.—HORSES.—CATTLE. — SHEEP.—PIGS.—POULTRY,— FORESTRY.—HORTICULTURE.—MISCELLANEOUS.

"No farmer, and certainly no agricultural student, ought to be without this *multum-in-parvo* manual of all subjects connected with the farm."—*North British Agriculturist.*

"This little pocket-book contains a large amount of useful information upon all kinds of agricultural subjects. Something of the kind has long been wanted."—*Mark Lane Express.*

"The amount of information it contains is most surprising; the arrangement of the matter is so methodical—although so compressed—as to be intelligible to everyone who takes a glance through its pages. They teem with information."—*Farm and Home.*

THE ELEMENTS OF AGRICULTURAL GEOLOGY.

A Scientific Aid to Practical Farming. By PRIMROSE McCONNELL. Author of "Note-Book of Agricultural Facts and Figures," &c. Royal 8vo, cloth.
Net **21/O**

"On every page the work bears the impress of a masterly knowledge of the subject dealt with, and we have nothing but unstinted praise to offer."—*Field.*

BRITISH DAIRYING.

A Handy Volume on the Work of the Dairy-Farm. For the Use of Technical Instruction Classes, Students in Agricultural Colleges and the Working Dairy-Farmer. By Prof. J. P. SHELDON. With Illustrations. Second Edition, Revised. Crown 8vo, cloth **2/6**

"Confidently recommended as a useful text-book on dairy farming."—*Agricultural Gazette.*
"Probably the best half-crown manual on dairy work that has yet been produced."—*North British Agriculturist.*
"It is the soundest little work we have yet seen on the subject."—*The Times.*

MILK, CHEESE, AND BUTTER.

A Practical Handbook on their Properties and the Processes of their Production. Including a Chapter on Cream and the Methods of its Separation from Milk. By JOHN OLIVER, late Principal of the Western Dairy Institute, Berkeley. With Coloured Plates and 200 Illustrations. Crown 8vo, cloth.
7/6

"An exhaustive and masterly production. It may be cordially recommended to all students and practitioners of dairy science."—*North British Agriculturist.*
"We recommend this very comprehensive and carefully-written book to dairy-farmers and students of dairying. It is a distinct acquisition to the library of the agriculturist."—*Agricultural Gazette.*

SYSTEMATIC SMALL FARMING.

Or, The Lessons of My Farm. Being an Introduction to Modern Farm Practice for Small Farmers. By R. SCOTT BURN, Author of "Outlines of Modern Farming," &c. Crown 8vo, cloth. **6/0**
"This is the completest book of its class we have seen, and one which every amateur farmer will read with pleasure, and accept as a guide."—*Field.*

OUTLINES OF MODERN FARMING.

By R. SCOTT BURN. Soils, Manures, and Crops—Farming and Farming Economy—Cattle, Sheep, and Horses—Management of Dairy, Pigs, and Poultry—Utilisation of Town-Sewage, Irrigation, &c. Sixth Edition. In One Vol., 1,250 pp., half-bound, profusely Illustrated **12/0**

FARM ENGINEERING, The COMPLETE TEXT-BOOK of.

Comprising Draining and Embanking; Irrigation and Water Supply; Farm Roads, Fences and Gates; Farm Buildings; Barn Implements and Machines; Field Implements and Machines; Agricultural Surveying, &c. By Professor JOHN SCOTT. In One Vol., 1,150 pp., half-bound, with over 600 Illustrations.
12/0

"Written with great care, as well as with knowledge and ability. The author has done his work well; we have found him a very trustworthy guide wherever we have tested his statements. The volume will be of great value to agricultural students."—*Mark Lane Express.*

THE FIELDS OF GREAT BRITAIN.

A Text-Book of Agriculture. Adapted to the Syllabus of the Science and Art Department. For Elementary and Advanced Students. By HUGH CLEMENTS (Board of Trade). Second Edition, Revised, with Additions. 18mo, cloth **2/6**
"It is a long time since we have seen a book which has pleased us more, or which contains such a vast and useful fund of knowledge."—*Educational Times.*

TABLES and MEMORANDA for FARMERS, GRAZIERS, AGRICULTURAL STUDENTS, SURVEYORS, LAND AGENTS, AUCTIONEERS, &c.

With a New System of Farm Book-keeping. By SIDNEY FRANCIS. Fifth Edition. 272 pp., waistcoat-pocket size, limp leather **1/6**
"Weighing less than 1 oz., and occupying no more space than a match-box, it contains a mass of facts and calculations which has never before, in such handy form, been obtainable. Every operation on the farm is dealt with. The work may be taken as thoroughly accurate, the whole of the tables having been revised by Dr. Fream. We cordially recommend it."—*Bell's Weekly Messenger.*

THE ROTHAMSTED EXPERIMENTS AND THEIR PRACTICAL LESSONS FOR FARMERS.

Part I. STOCK. Part II. CROPS. By C. J. R. TIPPER. Crown 8vo, cloth.
3/6

"We have no doubt that the book will be welcomed by a large class of farmers and others interested in agriculture."—*Standard.*

FERTILISERS AND FEEDING STUFFS.

Their Properties and Uses. A Handbook for the Practical Farmer. By BERNARD DYER, D.Sc. (Lond.). With the Text of the Fertilisers and Feeding Stuffs Act of 1893, The Regulations and Forms of the Board of Agriculture, and Notes on the Act by A. J. DAVID, B.A., LL.M. Fourth Edition, Revised. Crown 8vo, cloth. *[Just Published.* **1/0**

"This little book is precisely what it professes to be—'A Handbook for the Practical Farmer.' Dr. Dyer has done farmers good service in placing at their disposal so much useful information in so intelligible a form."—*The Times.*

BEES FOR PLEASURE AND PROFIT.

A Guide to the Manipulation of Bees, the Production of Honey, and the General Management of the Apiary. By G. GORDON SAMSON. With numerous Illustrations. Crown 8vo, wrapper **1/0**

BOOK-KEEPING for FARMERS and ESTATE OWNERS.

A Practical Treatise, presenting, in Three Plans, a System adapted for all Classes of Farms. By JOHNSON M. WOODMAN, Chartered Accountant. Fourth Edition. Crown 8vo, cloth. *[Just Published.* **2/6**

"The volume is a capital study of a most important subject."—*Agricultural Gazette.*

WOODMAN'S YEARLY FARM ACCOUNT BOOK.

Giving Weekly Labour Account and Diary, and showing the Income and Expenditure under each Department of Crops, Live Stock, Dairy, &c., &c. With Valuation, Profit and Loss Account, and Balance Sheet at the End of the Year. By JOHNSON M. WOODMAN, Chartered Accountant. Second Edition. Folio, half-bound . . . . *Net* **7/6**

"Contains every requisite for keeping farm accounts readily and accurately."—*Agriculture.*

THE FORCING GARDEN.

Or, How to Grow Early Fruits, Flowers and Vegetables. With Plans and Estimates for Building Glasshouses, Pits and Frames. With Illustrations. By SAMUEL WOOD. Crown 8vo, cloth **3/6**

"A good book, containing a great deal of valuable teaching."—*Gardeners' Magazine.*

A PLAIN GUIDE TO GOOD GARDENING.

Or, How to Grow Vegetables, Fruits, and Flowers. By S. WOOD. Fourth Edition, with considerable Additions, and numerous Illustrations. Crown 8vo, cloth **3/6**

"A very good book, and one to be highly recommended as a practical guide. The practical directions are excellent."—*Athenaeum.*

MULTUM-IN-PARVO GARDENING.

Or, How to Make One Acre of Land produce £620 a year, by the Cultivation of Fruits and Vegetables ; also, How to Grow Flowers in Three Glass Houses, so as to realise £176 per annum clear Profit. By SAMUEL WOOD, Author of "Good Gardening," &c. Sixth Edition, Crown 8vo, sewed . . . **1/0**

THE LADIES' MULTUM-IN-PARVO FLOWER GARDEN.

And Amateur's Complete Guide. By S. WOOD. Crown 8vo, cloth . **3/6**

POTATOES: HOW TO GROW AND SHOW THEM.

A Practical Guide to the Cultivation and General Treatment of the Potato. By J. PINK. Crown 8vo **2/0**

MARKET AND KITCHEN GARDENING.

By C. W. SHAW, late Editor of "Gardening Illustrated." Crown 8vo, cloth. **3/6**

AUCTIONEERING, VALUING, LAND SURVEYING, ESTATE AGENCY, ETC.

INWOOD'S TABLES FOR PURCHASING ESTATES
AND FOR THE VALUATION OF PROPERTIES,

Including Advowsons, Assurance Policies, Copyholds, Deferred Annuities, Freeholds, Ground Rents, Immediate Annuities, Leaseholds, Life Interests, Mortgages, Perpetuities, Renewals of Leases, Reversions, Sinking Funds, &c., &c. 27th Edition, Revised and Extended by WILLIAM SCHOOLING, F.R.A.S., with Logarithms of Natural Numbers and THOMAN'S Logarithmic Interest and Annuity Tables. 360 pp., Demy 8vo, cloth.

[Just Published. Net **8/0**

" Those interested in the purchase and sale of estates, and in the adjustment of compensation cases, as well as in transactions in annuities, life insurances, &c., will find the present edition of eminent service."—*Engineering.*

" This valuable book has been considerably enlarged and improved by the labours of Mr. Schooling, and is now very complete indeed."—*Economist.*

" Altogether this edition will prove of extreme value to many classes of professional men in saving them many long and tedious calculations."—*Investors' Review.*

THE APPRAISER, AUCTIONEER, BROKER, HOUSE
AND ESTATE AGENT AND VALUER'S POCKET ASSISTANT.

For the Valuation for Purchase, Sale, or Renewal of Leases, Annuities, and Reversions, and of Property generally ; with Prices for Inventories, &c. By JOHN WHEELER, Valuer, &c. Sixth Edition, Re-written and greatly Extended by C. NORRIS. Royal 32mo, cloth **5/0**

" A neat and concise book of reference, containing an admirable and clearly-arranged list of prices for inventories, and a very practical guide to determine the value of furniture, &c."—*Standard.*

" Contains a large quantity of varied and useful information as to the valuation for purchase, sale, or renewal of leases, annuities and reversions, and of property generally, with prices for inventories, and a guide to determine the value of interior fittings and other effects."—*Builder.*

AUCTIONEERS: THEIR DUTIES AND LIABILITIES.

A Manual of Instruction and Counsel for the Young Auctioneer. By ROBERT SQUIBBS, Auctioneer. Second Edition, Revised. Demy 8vo, cloth . **12/6**

" The work is one of general excellent character, and gives much information in a compendious and satisfactory form."—*Builder.*

" May be recommended as giving a great deal of information on the law relating to auctioneers, in a very readable form."—*Law Journal.*

THE AGRICULTURAL VALUER'S ASSISTANT.

A Practical Handbook on the Valuation of Landed Estates ; including Example of a Detailed Report on Management and Realisation ; Forms of Valuations of Tenant Right ; Lists of Local Agricultural Customs ; Scales of Compensation under the Agricultural Holdings Act, and a Brief Treatise on Compensation under the Lands Clauses Acts, &c. By TOM BRIGHT, Agricultural Valuer. Author of "The Agricultural Surveyor and Estate Agent's Handbook." Fourth Edition, Revised, with Appendix containing a Digest of the Agricultural Holdings Acts, 1883 and 1900. Crown 8vo, cloth . *Net* **6/0**

" Full of tables and examples in connection with the valuation of tenant-right, estates, labour, contents and weights of timber, and farm produce of all kinds."—*Agricultural Gazette.*

" An eminently practical handbook, full of practical tables and data of undoubted interest and value to surveyors and auctioneers in preparing valuations of all kinds."—*Farmer.*

POLE PLANTATIONS AND UNDERWOODS.

A Practical Handbook on Estimating the Cost of Forming, Renovating, Improving, and Grubbing Plantations and Underwoods, their Valuation for Purposes of Transfer, Rental, Sale or Assessment. By TOM BRIGHT. Crown 8vo, cloth **3/6**

" To valuers, foresters and agents it will be a welcome aid."—*North British Agriculturist.*

" Well calculated to assist the valuer in the discharge of his duties, and of undoubted interest and use both to surveyors and auctioneers in preparing valuations of all kinds."—*Kent Herald.*

AGRICULTURAL SURVEYOR AND ESTATE AGENT'S HANDBOOK.

Of Practical Rules, Formulæ, Tables, and Data. A Comprehensive Manual for the Use of Surveyors, Agents, Landowners, and others interested in the Equipment, the Management, or the Valuation of Landed Estates. By Tom Bright, Agricultural Surveyor and Valuer, Author of "The Agricultural Valuer's Assistant," &c. With Illustrations. Fcap. 8vo, Leather.

Net **7/6**

"An exceedingly useful book, the contents of which are admirably chosen. The classes for whom the work is intended will find it convenient to have this comprehensive handbook accessible for reference."—*Live Stock Journal.*

"It is a singularly compact and well informed compendium of the facts and figures likely to be required in estate work, and is certain to prove of much service to those to whom it is addressed."—*Scotsman.*

THE LAND VALUER'S BEST ASSISTANT.

Being Tables on a very much Improved Plan, for Calculating the Value of Estates. With Tables for reducing Scotch, Irish, and Provincial Customary Acres to Statute Measure, &c. By R. Hudson, C.E. New Edition. Royal 32mo, leather, elastic band **4/0**

"Of incalculable value to the country gentleman and professional man."—*Farmers' Journal.*

THE LAND IMPROVER'S POCKET-BOOK.

Comprising Formulæ, Tables, and Memoranda required in any Computation relating to the Permanent Improvement of Landed Property. By John Ewart, Surveyor. Second Edition, Revised. Royal 32mo, oblong, leather . **4/0**

"A compendious and handy little volume."—*Spectator.*

THE LAND VALUER'S COMPLETE POCKET-BOOK.

Being the above Two Works bound together. Leather **7/6**

HANDBOOK OF HOUSE PROPERTY.

A Popular and Practical Guide to the Purchase, Tenancy, and Compulsory Sale of Houses and Land, including Dilapidations and Fixtures: with Examples of all kinds of Valuations, Information on Building and on the right use of Decorative Art. By E. L. Tarbuck, Architect and Surveyor. Sixth Edition. 12mo, cloth **5/0**

"The advice is thoroughly practical."—*Law Journal.*

"For all who have dealings with house property, this is an indispensable guide."—*Decoration.*

"Carefully brought up to date, and much improved by the addition of a division on Fine Art. A well-written and thoughtful work."—*Land Agents' Record.*

LAW AND MISCELLANEOUS.

MODERN JOURNALISM.

A Handbook of Instruction and Counsel for the Young Journalist. By John B. Mackie, Fellow of the Institute of Journalists. Crown 8vo, cloth . **2/0**

"This invaluable guide to journalism is a work which all aspirants to a journalistic career will read with advantage."—*Journalist.*

HANDBOOK FOR SOLICITORS AND ENGINEERS

Engaged in Promoting Private Acts of Parliament and Provisional Orders for the Authorisation of Railways, Tramways, Gas and Water Works, &c. By L. L. Macassey, of the Middle Temple, Barrister-at-Law, M.I.C.E. 8vo, cloth . . . . **£1 5s.**

PATENTS for INVENTIONS, HOW to PROCURE THEM.

Compiled for the Use of Inventors, Patentees and others. By G. G. M. Hardingham, Assoc. Mem. Inst. C.E., &c. Demy 8vo, cloth . . **1/6**

CONCILIATION & ARBITRATION in LABOUR DISPUTES.

A Historical Sketch and Brief Statement of the Present Position of the Question at Home and Abroad. By J. S. Jeans. Crown 8vo, 200 pp., cloth **2/6**

EVERY MAN'S OWN LAWYER.

A Handy-Book of the Principles of Law and Equity. With a Concise Dictionary of Legal Terms. By A Barrister. Forty-first Edition, carefully Revised, and comprising New Acts of Parliament, including the *Motor Car Act*, 1903; *Employment of Children Act*, 1903; *Pistols Act*, 1903; *Poor Prisoners' Defence Act*, 1903; *Education Acts of* 1902 *and* 1903; *Housing of the Working Classes Act*, 1903, &c. *Judicial Decisions* pronounced during the year have also been duly noted. Crown 8vo, 800 pp., strongly bound in cloth.
[*Just Published.* **6/8**

** *This Standard Work of Reference forms* A COMPLETE EPITOME OF THE LAWS OF ENGLAND, *comprising* (*amongst other matter*):

THE RIGHTS AND WRONGS OF INDIVIDUALS

LANDLORD AND TENANT
VENDORS AND PURCHASERS
LEASES AND MORTGAGES
JOINT-STOCK COMPANIES
MASTERS, SERVANTS AND WORKMEN
CONTRACTS AND AGREEMENTS
MONEY LENDERS, SURETISHIP
PARTNERSHIP, SHIPPING LAW
SALE AND PURCHASE OF GOODS
CHEQUES. BILLS AND NOTES
BILLS OF SALE, BANKRUPTCY
LIFE, FIRE, AND MARINE INSURANCE
LIBEL AND SLANDER

CRIMINAL LAW
PARLIAMENTARY ELECTIONS
COUNTY COUNCILS
DISTRICT AND PARISH COUNCILS
BOROUGH CORPORATIONS
TRUSTEES AND EXECUTORS
CLERGY AND CHURCHWARDENS
COPYRIGHT, PATENTS, TRADE MARKS
HUSBAND AND WIFE, DIVORCE
INFANCY, CUSTODY OF CHILDREN
PUBLIC HEALTH AND NUISANCES
INNKEEPERS AND SPORTING
TAXES AND DEATH DUTIES

FORMS OF WILLS, AGREEMENTS, NOTICES, &C.

☞ *The object of this work is to enable those who consult it to help themselves to the law ; and thereby to dispense, as far as possible, with professional assistance and advice. There are many wrongs and grievances which persons submit to from time to time through not knowing how or where to apply for redress ; and many persons have as great a dread of a lawyer's office as of a lion's den. With this book at hand it is believed that many a SIX-AND-EIGHTPENCE may be saved ; many a wrong redressed ; many a right reclaimed ; many a law suit avoided ; and many an evil abated. The work has established itself as the standard legal adviser of all classes, and has also made a reputation for itself as a useful book of reference for lawyers residing at a distance from law libraries, who are glad to have at hand a work embodying recent decisions and enactments*

** OPINIONS OF THE PRESS.

"The amount of information given in the volume is simply wonderful. The continued popularity of the work shows that it fulfils a useful purpose."—*Law Journal.*

"As a book of reference this volume is without a rival."—*Pall Mall Gazette.*

"No Englishman ought to be without this book."—*Engineer.*

"Ought to be in every business establishment and in all libraries."—*Sheffield Post.*

"The 'Concise Dictionary' adds considerably to its value."—*Westminster Gazette.*

"It is a complete code of English Law written in plain language, which all can understand. . . . Should be in the hands of every business man, and all who wish to abolish lawyers' bills."—*Weekly Times.*

"A useful and concise epitome of the law, compiled with considerable care."—*Law Magazine.*

"A complete digest of the most useful facts which constitute English law."—*Globe.*

"Admirably done, admirably arranged, and admirably cheap."—*Leeds Mercury.*

"A concise, cheap, and complete epitome of the English law. So plainly written that he who runs may read, and he who reads may understand."—*Figaro.*

"A dictionary of legal facts well put together. The book is a very useful one."—*Spectator*

LABOUR CONTRACTS.

A Popular Handbook on the Law of Contracts for Works and Services. By DAVID GIBBONS. Fourth Edition, with Appendix of Statutes by T. F. UTTLEY, Solicitor. Fcap. 8vo, cloth **3/6**

BRADBURY, AGNEW, & CO. LD., PRINTERS, LONDON AND TONBRIDGE.

www.ingramcontent.com/pod-product-compliance
Lightning Source LLC
Chambersburg PA
CBHW021404210326
41599CB00011B/1004